Parallel Computations

This is Volume 1 in

COMPUTATIONAL TECHNIQUES

Edited by BERNI J. ALDER and SIDNEY FERNBACH

Parallel Computations

Edited by

GARRY RODRIGUE
Computation Department
Lawrence Livermore National Laboratory
University of California
Livermore, California

 1982

ACADEMIC PRESS
A Subsidiary of Harcourt Brace Jovanovich, Publishers

New York London
Paris San Diego San Francisco São Paulo Sydney Tokyo Toronto

COPYRIGHT © 1982, BY ACADEMIC PRESS, INC.
ALL RIGHTS RESERVED.
NO PART OF THIS PUBLICATION MAY BE REPRODUCED OR
TRANSMITTED IN ANY FORM OR BY ANY MEANS, ELECTRONIC
OR MECHANICAL, INCLUDING PHOTOCOPY, RECORDING, OR ANY
INFORMATION STORAGE AND RETRIEVAL SYSTEM, WITHOUT
PERMISSION IN WRITING FROM THE PUBLISHER.

ACADEMIC PRESS, INC.
111 Fifth Avenue, New York, New York 10003

United Kingdom Edition published by
ACADEMIC PRESS, INC. (LONDON) LTD.
24/28 Oval Road, London NW1 7DX


```
Library of Congress Cataloging in Publication Data
Main entry under title:

   Parallel computations

   (Computational techniques; v. 1)
   Bibliography: p.
   Includes index.
   1. Parallel processing (Electronic computers)
2. Science--Data processing.  I. Rodrigue, G.
QA76.6.P348    1982       001.64'2       82-8805
ISBN  0-12-592101-2                      AACR2
```


PRINTED IN THE UNITED STATES OF AMERICA

82 83 84 85   9 8 7 6 5 4 3 2 1

Contents

List of Contributors	ix
Preface	xi

A Guide to Parallel Computation and Some Cray-1 Experiences

T. L. Jordan

I.	Introduction	1
II.	Hardware	3
III.	Theoretical Considerations	14
IV.	Applications	23
	Appendix A. A Register Assignment for Sparse-Banded Matrix Multiply	47
	Appendix B. Factor and Forward Substitution	47
	Appendix C. Backward Substitution	48
	Appendix D. Factorization Only	49
	References	49

Vectorizing the FFTs

Paul N. Swarztrauber

I.	Introduction	51
II.	Preliminaries	53
III.	The Complex FFT Algorithms	57
IV.	Vectorizing Multiple Transforms	67
V.	Transforming Real Sequences	68
VI.	The Symmetric Transforms	71
VII.	Software and Summary	78
	References	83

Solution of Single Tridiagonal Linear Systems and Vectorization of the ICCG Algorithm on the Cray-1

David Kershaw

I.	A Vector Algorithm for Tridiagonal Linear Systems	85
II.	An Incomplete Cholesky Conjugate Gradient (ICCG) Algorithm for the Cray-1 Computer	90
III.	Cyclic Reduction on Future Machines	98
	References	99

An Implicit Numerical Solution of the Two-Dimensional Diffusion Equation and Vectorization Experiments

Garry Rodrigue, Chris Hendrickson, and Mike Pratt

I.	Introduction	101
II.	Spatial Differencing	105
III.	Matrix Formulation	110
IV.	Properties of the Matrix A	113
V.	Method of Lines	114
VI.	The Generalized Conjugate Gradient Algorithm	117
VII.	Computational Example	120
VIII.	Comments and Conclusions	126
	References	127

Swimming Upstream: Calculating Table Lookups and Piecewise Functions

Paul F. Dubois

I.	Introduction to Table Lookup	129
II.	Evaluating Algorithms on Vector Processors	131
III.	Basic Processes on Vector Processors	134
IV.	One-Dimensional Problems	135
V.	Two-Dimensional Problems: Equations of State	144
	References	151

Trade-Offs in Designing Explicit Hydrodynamical Schemes for Vector Computers

Paul R. Woodward

I.	Introduction	153
II.	Why Vectorization of Explicit Hydrodynamical Schemes Should Be Easy	154
III.	Why Vectorization of Explicit Hydrodynamical Schemes Can Be Difficult	155

IV.	Alternative Approaches and Their Costs on Vector Computers	160
V.	The Example of the Interaction of Two Blast Waves	162
VI.	Conclusions	170
	References	171

Vectorized Computation of Reactive Flow

Jay P. Boris and Niels K. Winsor

I.	Introduction and Statement of the Problem	173
II.	Vectorization and Optimization	180
III.	Techniques for Modeling Fast Time Scales	187
IV.	Techniques for Modeling Short Space Scales	195
V.	Techniques for Dealing with Physical and Geometric Complexity	202
VI.	Programming Guidelines and Summary of Parallelism Principles	208
	References	214

A Fully Implicit, Factored Code for Computing Three-Dimensional Flows on the ILLIAC IV

Harvard Lomax and Thomas H. Pulliam

I.	Introduction	217
II.	Basic Equations	219
III.	ILLIAC Architecture	230
IV.	Data-Base Considerations	231
V.	The ILLIAC Code ARC3	239
VI.	Results	243
VII.	Concluding Remarks	248
	References	249

A Time-Split Difference Scheme for the Compressible Navier–Stokes Equations with Applications to Flows in Slotted Nozzles

John C. Strikwerda

I.	Introduction	251
II.	The Difference Scheme	252
III.	The Application	257
IV.	The Implementation	260
V.	Results	263
	Appendix. Numerical Grid Generation	265
	References	267

Geophysical Fluid Simulation on a Parallel Computer

James G. Welsh

I.	Introduction	269
II.	The Salient Characteristics of the ASC	270
III.	The FORTRAN Compiler on the ASC	270
IV.	The Physical Processes of a Model	271
V.	Estimating Parallelism in Models	275
VI.	Conclusion	277

Experiences with a Floating Point Systems Array Processor

Kenneth G. Wilson

I.	Introduction	279
II.	Scientific Computing beyond the CDC 7600	282
III.	The AP-190L Installation at Cornell	286
IV.	FPS Array Processors and Parallel Computing	292
V.	Examples of Optimal Programming for the AP	297
VI.	The Two-Machine Environment	305
VII.	Practical Problems of AP Ownership	309
VIII.	Conclusions	312
	References	313

A Case Study in the Application of a Tightly Coupled Multiprocessor to Scientific Computations

Neil S. Ostlund, Peter G. Hibbard, and Robert A. Whiteside

I.	Introduction	315
II.	Tightly Coupled Multiprocessors	316
III.	Case Studies	329
IV.	Conclusions	346
	Appendix. Implementing Parallel Algorithms	348
	References	363

Computer Modeling in Plasma Physics on the Parallel-Architecture CHI Computer

Robert W. Huff, John M. Dawson, and G. J. Culler

I.	Introduction	365
II.	Formulation of the Simulation Problems	367
III.	Design of the Computer System	375
IV.	Programming for Efficiency	380
V.	Observations and Speculations	390
	References	396

Index   397

List of Contributors

Numbers in parentheses indicate the pages on which the authors' contributions begin.

JAY P. BORIS (173), Laboratory for Computational Physics, Naval Research Laboratory, Washington, D.C. 20375

G. J. CULLER (365), CHI Systems, Goleta, California 93117

JOHN M. DAWSON (365), Department of Physics, University of California, Los Angeles, California 90024

PAUL F. DUBOIS (129), Mathematics and Statistics Division, Lawrence Livermore National Laboratory, University of California, Livermore, California 94550

CHRIS HENDRICKSON (101), Computation Department, Lawrence Livermore National Laboratory, University of California, Livermore, California 94550

PETER G. HIBBARD (315), Department of Computer Science, Carnegie-Mellon University, Pittsburgh, Pennsylvania 15213

ROBERT W. HUFF (365), Department of Physics, University of California, Los Angeles, California 90024

T. L. JORDAN (1), Los Alamos National Laboratory, Los Alamos, New Mexico 87545

DAVID KERSHAW (85), Lawrence Livermore National Laboratory, University of California, Livermore, California 94550

HARVARD LOMAX (217), Ames Research Center, National Aeronautics and Space Administration, Moffet Field, California 94035

NEIL S. OSTLUND (315), Department of Computer Science, Carnegie-Mellon University, Pittsburgh, Pennsylvania 15213

MIKE PRATT (101), Computation Department, Lawrence Livermore National Laboratory, University of California, Livermore, California 94550

THOMAS H. PULLIAM (217), Ames Research Center, National Aeronautics and Space Administration, Moffet Field, California 94035

GARRY RODRIGUE (101), Computation Department, Lawrence Livermore National Laboratory, University of California, Livermore, California 94550

JOHN C. STRIKWERDA (251), Computer Sciences Department, University of Wisconsin-Madison, Madison, Wisconsin 53706

PAUL N. SWARZTRAUBER (51), National Center for Atmospheric Research, Boulder, Colorado 80307

JAMES G. WELSH (269), The Geophysical Fluid Dynamics Laboratory, The National Oceanic and Atmospheric Administration, Princeton University, Princeton, New Jersey 08540

ROBERT A. WHITESIDE (315), Department of Computer Science, Carnegie-Mellon University, Pittsburgh, Pennsylvania 15213

KENNETH G. WILSON (279), Newman Laboratory of Nuclear Studies, Cornell University, Ithaca, New York 14853

NIELS K. WINSOR* (173), Laboratory for Computational Physics, Naval Research Laboratory, Washington, D.C. 20375

PAUL R. WOODWARD (153), Lawrence Livermore National Laboratory, University of California, Livermore, California 94550

---

* Present address: GT-Devices, Alexandria, Virginia 22312.

Preface

The introduction of vector, array, and multiprocessors into scientific computing in the past few years is the beginning of a new era. The invention of new languages and algorithms will be required to ensure realization of the potential of these and future architectural improvements in computers. Already the use of parallel computers has given rise to studies in concurrency factors, vectorization, and asynchronous procedures. These have led to severalfold increases in speed over conventional serial machines after the calculations have been rearranged to take advantage of the specific hardware. The intent of this book is to collect these experiences of molding algorithms in a variety of numerical and physical applications and for as many different types of parallel computers as were available.

The first set of five articles considers vectorizing frequently used mathematical tasks. In the first of these articles the author classifies parallel computers and then concentrates on techniques for performing matrix operations on them. In the next article fast Fourier transforms are considered. Then the tridiagonal linear system and the incomplete Cholesky conjugate gradient (ICCG) method are described. Different versions of the conjugate gradient method for solving the time dependent diffusion equation are presented in the following article. Lastly, the table lookup problem as well as the evaluation of piecewise functions is described.

The next group of five articles describes two- and three-dimensional fluid flow calculations. In the first of these the author outlines the principal issues in designing efficient numerical methods for hydrodynamic calculations. In the next article the decisions that a numerical modeler must make to optimize chemically reactive flow simulations are discussed in detail. Next a specific discussion on how to handle disk-to-core data transfer and storage allocation is given for the solution of the implicit equations for three-dimensional flows. Then the time-split finite difference scheme is described for solving the two-dimensional Navier–Stokes equation for flows through slotted nozzles. In the last of these five articles, experiences in converting programs for atmospheric and oceanographic models to one of the first parallel computers are recounted.

The last set of three articles is primarily concerned with adapting particle simulation codes to the new breed of computers. In the first, the author explains how to carry out Monte Carlo calculations economically on a small computer with an array processor. In the following article the authors report on how to adapt such Monte Carlo and molecular dynamics calculations to a multiprocessor. Finally, the large-scale stimulation of plasmas, as carried out on a small computer with an array processor, is discussed.

A Guide to Parallel Computation and Some Cray-1 Experiences

T. L. Jordan

Los Alamos National Laboratory
Los Alamos, New Mexico

I. INTRODUCTION

A slowdown in the rate of growth of computing power available from a single processor and a dramatic decrease in the hardware cost of executing an arithmetic operation have stimulated both users and developers of large-scale computers to investigate the feasibility of highly parallel computation. It has been estimated that the rate of growth of computing power from a single processor has declined from a factor of 100 in the 1960s to 10 in the 1970s. At present, we are unable to deal adequately with three-dimensional problems. At the same time, problems have grown in difficulty and complexity to require this kind of detailed treatment. Thus, unless something dramatic and unforeseen happens with the single-processor capability, the necessary computing power must be supplied by parallel processing. It is the purpose of this paper to examine some basic principles of parallel processing and discuss experiences with the Cray-1 vector computer.

If we look at the total dependency graph of a sizable computation, it would certainly offer many possibilities for selecting independent tasks that could be processed in parallel. In the past we have looked for independence at the most elementary level, say, statement level, to keep independent functional units and pipelines busy. A current trend in keeping asynchronous multiprocessors busy is to decompose existing codes into independent tasks and transfer data in sequence from one process to another in some round-robin fashion. This global approach has an immediate advantage of not having to alter codes in detail. In contrast, we are more concerned with basic parallel algorithms than with dissecting problems into large independent tasks.

We have included in Section II a simple taxonomy of parallel computers that contains existing parallel computers as well as some under development. The Cray-1 vector features are discussed in sufficient detail to allow the reader to understand algorithms provided in Section IV that were designed for the Cray-1. To understand better the performance prospects and limitations of various architectures, a subsection is devoted to hardware modeling.

It is the purpose of Section III to study the parallel complexity of some fundamental classes of operation. Concerning complexity, a distinction is required between algorithms best suited for sequential processing and those best suited for parallel processing. In sequential processing, the time required to process an algorithm is well correlated with its complexity, whereas for parallel processing the time required is more likely to be determined by the number of parallel steps, that is, parallel complexity, needed to implement a given algorithm. Despite our attention to this subject in only the ideal computer setting, we believe that it helps to identify the expected improvement of various operations in terms of speedup and efficiency. We also identify troublesome operations specific to the Cray-1 and other pipeline computers.

Specific Cray-1 experiences with some difficult kernels from applications important to the Los Alamos National Laboratory are studied in Section IV. We focus on problems that cannot be coded optimally in FORTRAN. We address the following subjects:

(a) Vectorized table lookup procedure.

(b) The complexity necessary for vectorized interpolation to be profitable.

(c) An algorithm for achieving super vector performance with sparse-banded (9-point) matrix multiplication.

(d) Solution of bidiagonal and tridiagonal systems of equations by odd–even reduction and modified Gaussian elimination. Times are given for two codes.

(e) r-cyclic reduction as applied to tridiagonal systems of equations; asymptotic times are estimated. This algorithm appears to be asymptotically the fastest.

(f) Algorithms for the vectorization of the 9-point incomplete Cholesky conjugate gradient algorithm. Its performance is estimated for a Cray-1 assembly language implementation (CAL).

(g) Parallel applications that do not adapt well to synchronized parallel processing.

II. HARDWARE

We are interested in computers that can exploit the parallelism of large-scale scientific calculations as well as those research prototypes that are designed specifically to investigate parallel computation. It is important that these systems have the potential of having all their power applied to a single application as opposed to processing many completely independent jobs. With no pretense of supplying an exhaustive list of such systems, we shall identify and classify them using the simplest of taxonomies. This allows us to provide some simple performance models. It is hoped that we can then extrapolate to more complex models since the directions hardware development is taking are myriad.

A. *Classification of Parallel Computers*

We classify parallel computers as either SIMD (single instruction–multiple data) or MIMD (multiple instruction–multiple data). In the class of SIMD machines, we distinguish between vector and array computers. This distinction is based primarily on the way data are communicated to elements of the system. Processors in array computers typically access data from their own memories and those of their nearest neighbors (see Section II.C.1). Vector computation has been supported through pipelining or streaming and by synchronous multiprocessing. A rigorous taxonomy of parallel computers would surely exclude pipeline computers. However, effective use of them depends heavily on the design of efficient parallel algorithms, particularly the regular parallelism that dominates most large-scale scientific applications. Figure 1 summarizes our classification. Hardware characteristics of some advanced architectures appear in Jordan (1979).

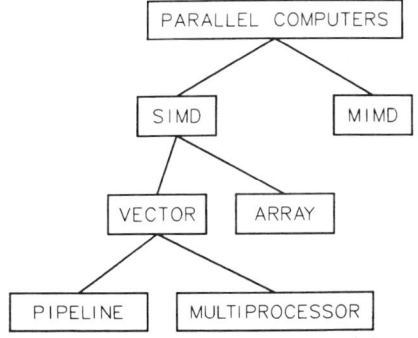

FIG. 1. A simple taxonomy of parallel computers.

TABLE I

PARTIAL LIST OF PARALLEL PROCESSORS

SIMD	MIMD
Vector pipeline	Burroughs "Wind Tunnel Proposal"
CDC STAR-100, Cyber 203, 205	Carnegie-Mellon CM*
CRI Cray-1	CDC "Wind Tunnel Proposal"
TI ASC	DYNELCOR HEP
Vector multiprocessor	Goodyear MPP
Burroughs BSP	Lawrence Livermore Laboratory S-1
Burroughs PEPE	NASA Langley Finite Element Machine
Array	NOSC/UNIVAC Multiple AP-120s
Burroughs ILLIAC IV	University of Texas TRAC
ICL DAP	

If we were to form subclasses of MIMD computers based upon types of data communication, we would find the architectures to be very disparate as well as hybrid within an architecture. Further refinement of an already inadequate classification scheme seems unprofitable. For example, there are already designs for asynchronous multiprocessing systems in which each processor contains vector capabilities. Table I contains a partial list of parallel computers in both the SIMD and MIMD class.

In addition to the computers listed in Table I, there is a class of computers that fits our definition of vector computers that are commonly called array computers, but do not fit our classification for array computers. These are processors that are usually attached to host computers to perform effectively vector-oriented tasks with emphasis on signal analysis. Computers of this type include the following: Floating Point Systems AP-120B; CSPI Inc., CSPI-MAP; Control Data Corporation CDC-MAP III; International Business Machines IBM-3838; and Data West MATP. These computers are fast and cost effective, but generally not well suited to complete large-scale computation. Work is being done to connect several of these computers into a single system (see NOSC/UNIVAC in Table I for an example).

New and exotic MIMD architectures, called data flow computers, that depart dramatically from the classical von Neumann architecture are in the design stage. The concept is based upon the dependency graph of a computation. Operands flow to the functional units and fire (operate) whenever all the operands are present and the result can be accommodated. The output with new instruction packets is then forwarded to the next functional unit. Hence, data need not have any permanent residence. Since the dependency graph depicts all the parallelism of a computation whether or not the parallelism is regular, the data flow architecture offers a high potential for capturing parallelism. Texas Instruments has a four-processor laboratory model of a data flow machine.

B. *Cray*-1

To understand the algorithms described in Section IV, we present some salient features of the Cray-1 architecture. Figure 2 shows a block diagram of the Cray-1 Computer. Of all the registers shown, only the address registers (eight 24-bit registers) and scalar registers (eight 64-bit registers) are considered conventional. The unique features of the Cray-1 include (1) backup

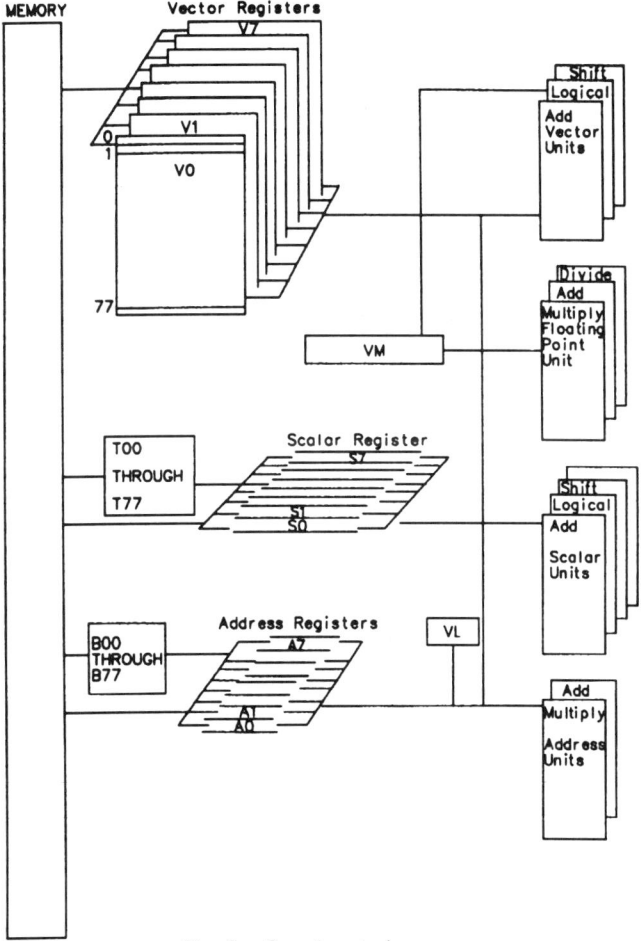

Fig. 2. Cray-1 central processor.

or scratch-pad memories for both the address and scalar registers, (2) a set of eight vector registers, each containing 64 words of 64 bits, and (3) the ability to perform, under certain conditions, several vector operations simultaneously. In particular, if two separate functional units are used and certain other conditions are met, the result of one vector operation may be used as an input of a subsequent vector operation even before the first operation is complete. This concept, called "chaining," is discussed in the following paragraphs. Other Cray-1 features are that it has a 12.5-nsec clock period (cycle time) and its memory is fully interleaved in 16 banks. Each bank has up to 65,536 64-bit words with a bank cycle time of four clock periods.

We shall discuss only those capabilities that relate to vector processing. All vector data must pass through the vector registers; that is, the data cannot move from memory to a functional unit or vice versa without using a vector register. To transfer data between memory and one of the vector registers V0, V1, ..., V7, instructions are issued to load a vector length (<65) into the vector length (VL) register, the starting address in memory in address register A0, and the memory increment (positive or negative) in another address register before issuing the data transfer instruction. Words are transferred to or from one of the vector registers. The transfer always starts with the first word of the vector register and continues to the length specified in VL when the transfer instruction is issued. In performing vector operations, we normally must use as a result register one which is distinct from the operand registers.

Consider the following sequence of two vector instructions:

$$V0 = V1 * V2 \quad \text{and} \quad V3 = V4 + V5.$$

The first instruction requests a multiply of operands in V1 and V2 with the result going to V0. Once this instruction is issued, Cray-1 will attempt to process the second instruction even while the first multiply is in progress. If we assume that V3, V4, and V5 are available and the add unit is free, the vector add instruction will begin almost immediately, so that the two operations are being done simultaneously. Because VL was not reset, the add operation uses the same VL as the multiply. The functional units that can perform the vector operations are fully segmented; that is, they can accept a new pair of 64-bit operands each clock period. Each functional unit has a "functional unit time," which is the number of clock periods required to produce a particular result, that is, the number of segments in the pipeline. Each pair of operands requires one cycle to move from its vector registers to the functional unit, plus the functional unit time to traverse the unit, and another cycle to return to the result vector register. This time is called the chain slot time. Each functional unit has a chain slot time precisely equal to its functional unit time plus two clock periods.

Now we consider the two vector instructions

$$V0 = V1 * V2 \quad \text{and} \quad V4 = V0 + V3.$$

These instructions cannot be issued in immediate succession because register V0 will not be ready for the second instruction. Although it is not necessary for the second instruction to wait until the first is complete, instruction issue is delayed until the chain slot time of the multiply operation has arrived. Thus at the same clock period that a result is returned to V0, a copy of it can be sent back to the add unit. For a floating point multiply, the functional unit time is seven clock periods; for a floating point addition, it is six clock periods. To be precise, the following conditions must be met before a vector instruction can be issued.

(a) The functional unit must be free.
(b) The result register must be free.
(c) The operand registers must be free or at chain slot time.
(d) The memory must be quiet if the instruction references memory.

The chain slot time is a window that is one clock period wide. If a contingent vector instruction cannot be issued at chain slot time, it must wait until both operand registers are free (the previous vector operation is complete). However, if the contingent vector instruction is ready before chain slot time and all other conditions are satisfied, it is delayed and then issued exactly at chain slot time. To illustrate the power of chaining, we present the following sequence of vector instructions, assuming VL is set to 64.

$$V0 = \text{load from memory,}$$

$$V2 = S1 * V0,$$

$$V3 = V1 + V2.$$

The first instruction loads V0 from memory and issues at clock period zero. The chain slot time for a memory read is 9 clock periods, so that the multiply of the scalar in register S1 times V0 issues at clock period 9. The chain slot time of the multiply is 9 clock periods, so the add starts at clock period 18. The add completes in a time of VL (which is 64) plus the functional unit time of the add (which is 6) plus 2 for a total of 72 clock periods. Because the add started at clock period 18, the entire sequence of instructions is completed in 90 clock periods. Incidentally, no operation can be chained to the memory read when the increment is a multiple of 8 because memory bank conflicts will prevent the vector register from loading at the rate of one word per clock period. Also, a store to memory can never be chained because it cannot always be known at chain slot time that there will not be any memory bank conflicts.

Three performance levels on the Cray-1 are characterized and estimated below.

(a) Scalar (S) is a performance level that does not use vector operations; it uses only A, B, S, and T registers. We estimate codes using only these registers will perform about two times faster than the CDC-7600.

(b) Vector (V) is a performance level that uses vector operations in which memory references are the limiting factor. Based on current timing data, an approximate 4:1 performance ratio between the Cray-1 and the CDC-7600 seems reasonable for V-level computations.

(c) Super Vector (SV) is a performance level that uses vector operations in which performance is limited by availability of the functional units and/or vector registers. SV performance is available in algorithms in which a considerable amount of arithmetic is done on a few operands, so that the vector registers can serve as the primary memory. We estimate the SV performance level to be about three times faster than that of the V performance. Matrix multiplication, polynomial evaluation of a vector, and the solution of linear equations are some fundamental problems that can be performed at this level.

An illustration of these performance levels for LU factorization is presented in Fig. 3. A representation of S performance is given by a FORTRAN code compiled in scalar mode. Two illustrations of V performance are given by (1) the same FORTRAN code vectorized and (2) a FORTRAN code calling assembly-coded basic linear algebra subroutines (BLAS). Details are given by Lawson *et al.* (1979). Finally, SV performance is demonstrated for a code written entirely in Cray assembly language (CAL).

Floating point SV performance on Cray-1, which is achieved by a balanced and chained use of the memory and the floating point multiply and add functional units, occurs when multiply and add (or subtract) operations alternate and can be chained to a vector load. Under these circumstances, the optimal performance for each multiply and add pair is approximately $1 + 2f$ cycles per result, where $6/n \leq f \leq 10/n$, $1 \leq n \leq 64$, and n is the VL. Thus f may be regarded as the startup time for a functional unit prorated over the number of elements the unit processes per startup. We shall condense this form of timing approximation by defining a *chime* to be sets of concurrent vector operations. This may occur as chained or independent operations. Thus a chime takes a cycle per element plus a prorated estimate of the total startup time.

C. Hardware Modeling

For some time we have had the opportunity to perform calculations mixed with at least two unequal processing rates. Scalar versus vector or array rates

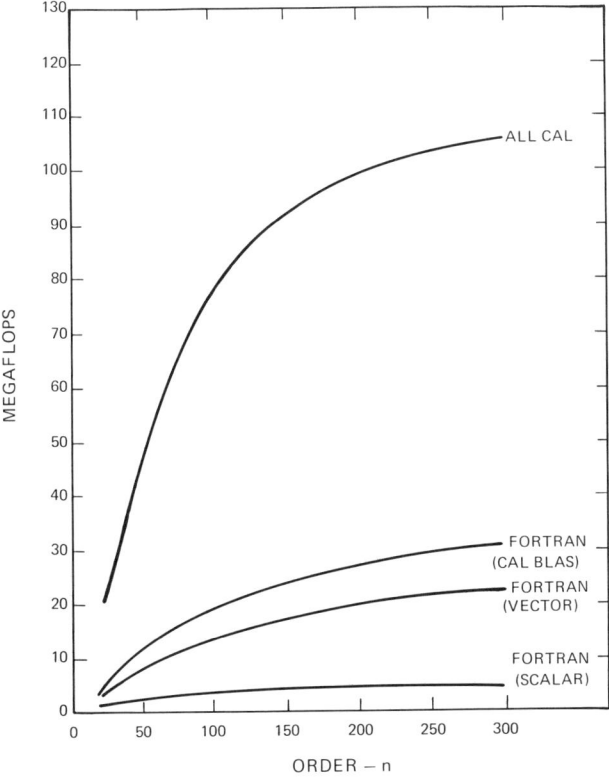

FIG. 3. Cray-1 performance on LU factorization.

are perhaps the most relevant. Amdahl (1967) modeled the result rate r from two processors given the fractions f_h and f_l of results from each processor and times t_h and t_l to generate the results from high- and low-speed processors, respectively. If $f_h + f_l = 1$, then

$$r = \frac{1}{f_h t_h + f_l t_l} \tag{1}$$

$$= \frac{r_l}{f_h t_h/t_l + f_l} \tag{2}$$

$$= \frac{r_h}{f_h + f_l t_l/t_h}, \tag{3}$$

where $r_l = 1/t_l$ and $r_h = 1/t_h$ are the results rates from the low- and high-speed processors, respectively. From Eq. (2) we observe that at best a factor of 2 speedup is available if only 50% of the work is done in the high-speed processor. From Eq. (3) we obtain the graphs in Fig. 4 and observe that the more disparate the t_l and t_h, the greater the f_h must be to achieve, say, 50% of the result rate of the high-speed processor.

If we time a simple vectorizable DO loop in a scalar mode, the time within the loop is fixed and may be considered the operation time of the loop. Hence we take as our scalar model $t = nr_s$, where r_s is the operation time of the loop and n is the number of times through the loop. Let us use as our pipeline model $t = ms + nr_v$ where r_v is the result time from the pipe, s is the startup time, and $m = \lceil n/p \rceil$, where p is the number of elements processed as a subvector. We represent the synchronous p-processor system by $t = mr_v$, where m is $\lceil n/p \rceil$ and r_v is the operation time to process p elements concurrently. The characteristics of these three models are shown in Fig. 5. We make no attempt to represent the asynchronous multiprocessor system.

1. Synchronous Computers

Pipeline computers to date have introduced an additional parameter s, known as startup time. Efficiency and effectiveness of a vector computation are affected by startup time and the average vector length. We now study these effects.

The *efficiency e* of vectorization in a pipelined computer is defined by

$$e = \frac{\text{computation time}}{\text{total time}}$$

$$= \frac{nr_v}{ms + nr_v}$$

$$= \left(1 + \left(\frac{s}{r_v}\right)\left(\frac{1}{n/m}\right)\right)^{-1},$$

where n is the total number elements to be processed; $m = n/p$, where p is the number of elements processed as a vector; s is the startup time; and r_v is the time per element. The factor (s/r_v) shows the dependence upon startup time, and the factor n/m describes the dependence of average vector length upon the computation. Then, if $x = (r_v/s)(n/m)$, the model $e = 1/(1 + 1/x)$ describes the combined effects of startup and average vector lengths. Thus it is desirable for x to be as large as possible.

With pipeline computers we are faced with the question of whether to compute in the scalar or vector mode. Vectorization is profitable when $n > ms/(r_s - r_v)$. The *break-even* point on the Cray-1 is perhaps as small as a 3-element vector for most vectorizable computations.

Before vector processors, the most common measure of computer speed was MIPS (millions of instructions per second). Since vector instructions

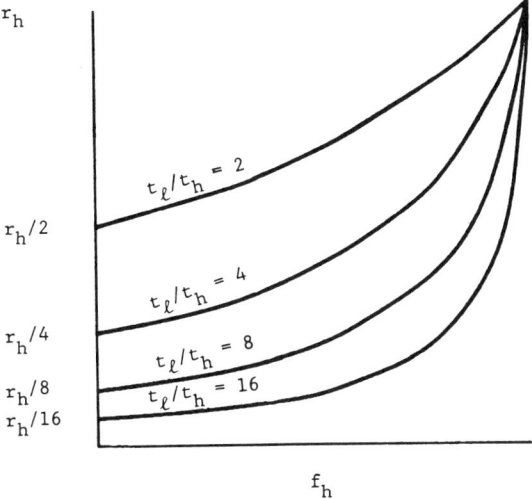

FIG. 4. Multiprocessor performance versus fraction of work executed at high speed.

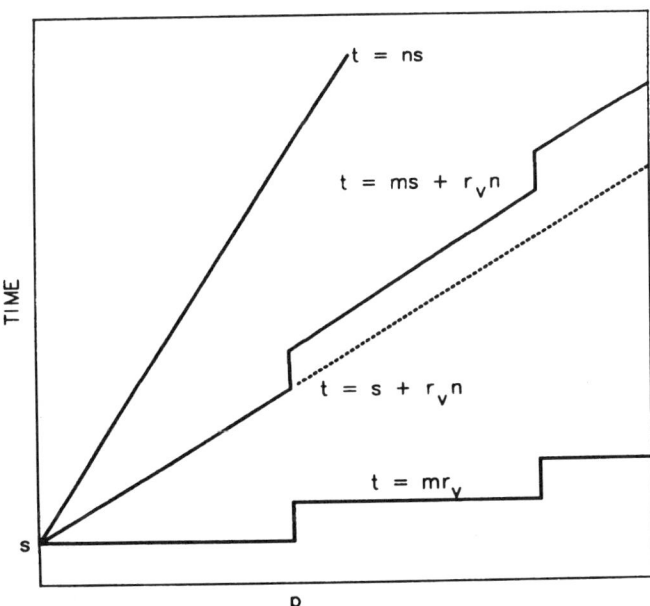

FIG. 5. Timing model of scalar, pipeline, and synchronous multiprocessor systems.

eliminate most of the bookkeeping instructions, MFLOPS (millions of floating point operations per second) is the accepted measure and is closely related to the pipeline efficiency of vectorization. If $t = ms + nr_v$ and n is the total number of floating point operations, then

$$\text{MFLOPS} = \frac{n}{t} = \frac{n}{ms + nr_v}$$

$$= \frac{1}{r_v}\left(1 + \left(\frac{s}{r_v}\right)\left(\frac{1}{n/m}\right)\right)^{-1} = \frac{1}{r_v}e,$$

where t is given in microseconds.

An analysis of a vector computer supported by synchronous multiprocessors differs only slightly from that of the vector pipeline computers. A p-processor computation may be timed by $t = ms + \lceil n/p \rceil r_p$, where s is the startup time, m is the number of startups, n is the number of elements processed, and r_p is the time to process p elements. The time to process from one to p elements is the same.

Vector computers could be thought of as one-dimensional systems operating primarily from a common memory, although caches or vector registers may be used to relieve data traffic to and from main memory. There is another type of synchronous computer that has a two-dimensional aspect to its parallelism. We have classified computers of this type as array computers; they are distinguished by restricted connectivity to the total memory. In contrast to the vector computer, each processor has immediate access to its own memory and to a limited amount of memory associated with neighboring processors. Although the timing model of the multiprocessor vector computer is adequate for the arithmetic performed in an array processor, an additional and important timing consideration, called data transfer time, must be entered into the model. The data transfer time is very dependent upon the application solved on such a computer.

The Burroughs ILLIAC IV uses the nearest neighbor connection of processors arranged in an 8×8 array as illustrated in Fig. 6, where each processor can concurrently access data from its N, S, E, or W neighbors. Boundary connections might well differ depending upon the computer design and/or the applications for which it is intended. This connection is useful for finite difference approximations to various PDE problems.

A more recently favored interconnection network that appears to be more useful for a larger class of algorithms is one based on the perfect shuffle and nearest neighbor. It appears well suited to applications that require fan-in or FFT-like data transformation. The perfect shuffle illustrated in Fig. 7 for an 8-processor system and all connections are defined by

$$P_n \leftrightarrow P_{n+1} \ (\text{mod } n) \quad \text{and} \quad P_n \leftrightarrow P_{\sigma(n)},$$

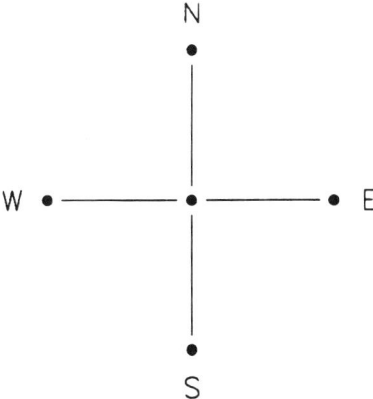

FIG. 6. Nearest neighbor connection.

where

$$\sigma(n) = \begin{cases} 2n & \text{if } 0 \leq n < N/2 \\ 2n + 1 - N & \text{if } N/2 \leq n < N. \end{cases}$$

It is generally recognized that data communication is a serious problem for multiprocessors.

2. Asynchronous Computers

In practice to date, we have addressed parallel constructs that are in some sense regular. This is perhaps the most frequent and important parallelism to capture. Nevertheless, there are important codes with parallelism that is either impossible to regularize or is unprofitable even if captured (see Section IV.E). Hence architectures that can exploit asynchronous parallelism are attractive. The asynchronous multiprocessors have this potential. However, the problems faced by a purely asynchronous system appear substantial. Not only is the data communication problem present, but we face new problems just because the system is asynchronous. We shall describe these problems.

In an environment of completely asynchronous multiprocessors, concurrent tasks may be required to synchronize before proceeding. The size of the tasks executed between synchronization steps will be referred to as the

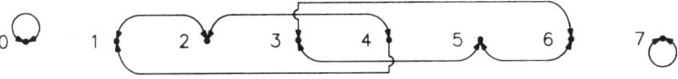

FIG. 7. A perfect shuffle connection.

granularity of the calculation. This is an important consideration in asynchronous multiprocessors for two reasons:

1. Communication is required at synchronization points, perhaps requiring many cycles.
2. When a task is completed, it must await the completion of cooperating tasks. This adds wait time to the overhead.

These costs are inevitable in MIMD architectures and are potentially very costly even for identical tasks and identical processors. Baudet (1978) illustrates that even the highly parallel Jacobi iteration

$$x^{i+1} = Ax^i + b$$

has a wait time t_w that behaves for large p like

$$t_w \sim 1 - 1/\ln p,$$

where p is the number of processors. This assumes that the time to compute each x^i is distributed according to the same exponential distribution and that x^i is computed in the ith processor. In the limit all processors are almost always idle. As a consequence of these kinds of arguments, there has been considerable attention paid to chaotic relaxation techniques.

The speedup s of a multiprocessor system is defined to be $s = t_1/t_p$ where t_i is the time required to solve a given problem with i processors. The efficiency e is given by $e = s/p$. It is commonly assumed that speedup is impossible for performing a sequential set of nonparallel tasks. However, Baudet (1978) has shown that speedup greater than one is possible by taking advantage of the fluctuations in process time that occur in asynchronous multiprocessors. In fact, under ideal assumptions, he shows the speedup is $O(\sqrt{p})$, where p is the number of processors and both p and the number of sequential tasks are large.

III. THEORETICAL CONSIDERATIONS

Switching from sequential processing (scalar, pipeline) to truly parallel computation motivates the search for new algorithms that take less time than before if executed in parallel. To measure the expected performance of an algorithm, we have in the past used rough measures of operation count, broadly termed complexity, to compare the expected performance of algorithms. For example, we currently perform LU factorization of a dense matrix rather than compute the inverse because the operation count is three times greater for the inverse when both are computed using Gaussian elimination. However, when performed in parallel, ignoring pivoting, both algorithms can be done in $O(n)$ parallel steps, and the code within a step is simpler for inversion than for LU factorization. Furthermore, the forward and back substitution is replaced by the simpler matrix multiplication; both have parallel complexity $O(\log_2 n)$. Also observe that if pivoting is required, time is now dominated by pivoting, not arithmetic, and requires $O(n \log_2 n)$ parallel steps. Hence the total parallel complexity of general LU is $O(n \log_2 n)$. Incidentally, because the $O(n \log_2 n)$ parallel complexity is far from the established lower bound of $O(\log_2 n)$ for inversion, there is an active search for a faster, stable, parallel algorithm to solve this problem. Generally, we need to look at primitive operations and basic algorithms to anticipate the kind of speedup we get when they are performed in parallel.

A. *Complexity*

To investigate the parallel complexity of fundamental operations, we place the operations of interest to us into three levels of parallel complexity: (1) constant-time operations, (2) $\log_2$-time operations, and (3) large-time operations. We assume an ideal computer (sometimes called a paracomputer) in which there are unlimited, immediately coupled processors with no memory access problems.

1. *Constant-Time Operations*

There are a large number of fundamental operations or calculations that can be done concurrently whenever there are many to be done, for example, $z_i = x_i + y_i$ for $i = 1, 2, \ldots, n$. Thus in our ideal computer the time required to process an arbitrary number of such calculations is the same independent of the number n. We shall refer to these as *constant-time* operations. Constant-time operations provide the greatest impetus for developing parallel hardware since in principle large problems made up of these kinds of operations could be done in the same time as small problems; that is, the speed could be as large as the number of processors p available. This does not mean that all nontrivial constant time operations will produce a speedup of p. For example, the table lookup of a value y in an ordered table has a speedup of only $O(\log_2 n)$ because the sequential complexity of a binary search is $O(\log_2 n)$ and the parallel complexity is $O(1)$.

Table II shows some primitive, constant-time operations that occur frequently and their performance on a Cray-1. The startup times are valid only for vectors less than 65, and the times for the FORTRAN intrinsics do not

TABLE II
Some Constant Time Operations with Cray-1 Times per Element

Operation	Execution time (cycles)	Startup time (cycles)
Register to register		
Add	1	8
Multiply	1	9
Reciprocation		
half precision	1	16
full precision	1	34
ABS	1	4
AINT	2	16
AMOD	4	68
FLOAT	1	14
FIX	1	14
LOGICAL (.AND., .OR., .NOT.)	1	4
Merge vectors on mask	1	4
SHIFT (end off-zero fill)	1	6
SIGN	2	12
Memory to memory		
Copy a vector	2	14
Swap two vectors	4	18
Difference ($y_i = x_i - x_{i+1}$)	3	21
Average ($y_i = 0.5(x_i + x_{i-1})$)	3	30
Sum of squares ($z_i = x_i^2 + y_i^2$)	3	35
mth degree polynomial evaluation of a vector $P_m(x)$ in		
nested form	1	$26m$

necessarily reflect those of a particular compiler. These operations all lead to a speedup of p on the ideal computer.

Certain constant-time operations have not performed well on synchronous parallel computers; often there is no advantage over scalar computation. Because of the frequency of their occurrence, notable ones are the GATHER and IF ... THEN ... ELSE operations.

If we have a calculation that involves, say, x_{j_i}, where j is a vector of indices and j_i is the ith one, then the elements of the vector x are generally unordered in memory. To operate on this vector in a parallel mode, it may be necessary to GATHER the vector into an ordered one before proceeding with the calculation. The implementation problems in array computers like the ILLIAC IV are most obvious. Furthermore, pipeline computers like the CDC STAR-100 and Cray-1 do not perform this operation well. Consequently, the complexity of the parallel computation dependent upon x_{j_i} must be considered to determine whether it is profitable to perform the

A Guide to Parallel Computation and Some Cray-1 Experiences

GATHER operation to do the calculation in a parallel mode. For an application in which this question arises on the Cray-1, see the table lookup and interpolation problem in Section IV.A. Applications in which this type of addressing appears extensively include sparse matrix solvers, particle-in-cell (PIC) codes, equation-of-state calculations, and Monte Carlo neutronics.

Although IF statements provide the most significant capability in automated computing, they are a nemesis of synchronous parallel computation. The complexity of IF statements is arbitrary; however, consider the common statement

$$\text{IF } l0_i \text{ THEN } y_i = l1_i \text{ ELSE } y_i = l2_i,$$

where $l0_i$, $l1_i$, and $l2_i$ are independent expressions computable in parallel. Because $l0_i$, $l1_i$, and $l2_i$ require different instruction streams, they must be evaluated sequentially in SIMD environments. Since the vector $l1$ ($l2$) is evaluated only where $l0_i$ is true (false), either $l1$ and $l2$ are both computed for all i and the results merged (as on the Cray-1), or alternatively new compressed vectors are formed, computed separately, and then merged to form y. In either case, considerable extra work is performed that may make a vector calculation unprofitable in a pipeline computer. These problems do not arise in an asynchronous multiprocessor because each processor can perform different instruction streams. However, processor utilization will depend upon how well $l1$ and $l2$ balance, for we assume the granularity of this computation would not ordinarily justify releasing and then reacquiring a processor to increase processor utilization.

2. $\log_2$-Time Operations

Perhaps the most frequently occurring operations that may reduce the anticipated speedup in a multiprocessor system are those for which the parallel complexity is $O(\log_2 t)$, where t is the time required if performed sequentially. These include the following operations:

(a) Reduction operations on vectors, sums (SSUM), norms (ISAMAX, SASUM, SNRM2), maximum and minimum (ISMAX, ISMIN), and dot products (SDOT). For descriptions of the BLAS routines, see Lawson et al. (1979).

(b) Applications requiring low-order linear recursion, such as first-order linear recursion or, equivalently, solving a bidiagonal system of equations. An example is the forward-and-back substitution steps in solving a tridiagonal system of equations. Second-order linear recursion can be used to factor tridiagonal systems of equations (see Section IV.C.1.c).

(c) Polynomial evaluation of single variable.

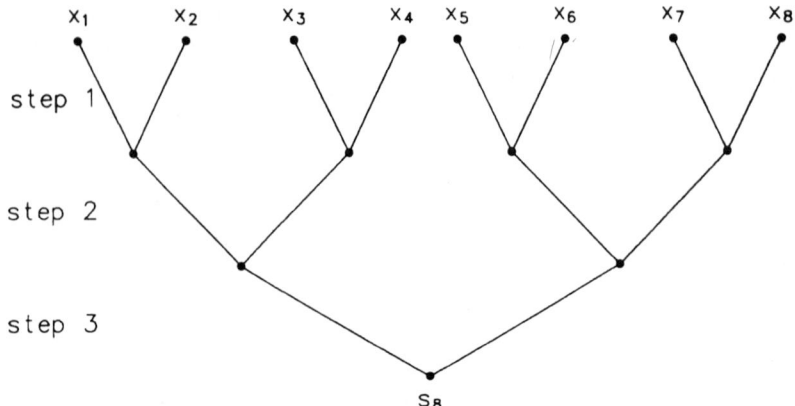

FIG. 8. Fan-in graph to sum eight numbers and with the same number of processors.

The $\log_2$-time operations are optimally performed through divide-and-conquer strategies best represented by an associative fan-in algorithm (Heller, 1978). Consider summing $n = 2^m$ elements $x_1, x_2, \ldots, x_n$. The sum is commonly expressed recursively by

$$s_i = s_{i-1} + x_i, \quad i = 1, 2, \ldots, n,$$

where $s_0 = 0$. The fan-in graph illustrating a parallel solution to this example of linear recursion is given in Fig. 8 for $m = 3$. Note that in the same time and with the same number of processors, we can also obtain all eight partial sums s_1–s_8 as shown in Fig. 9. Although the work (number of operations) is the same to compute the single sum sequentially or in parallel, this is not the case when computing all sums. In the sequential case only n additions are required, whereas $O(n \log_2 n)/2$ are adequate in our ideal computer. Although redundancy and reduced accuracy are often well correlated, note that this is not true here. The individual partial sums are more accurate when conducted with the parallel algorithm than with the sequential algorithm.

The sum example is a special case of the general linear recurrence relation

$$x_i = a_i x_{i-1} + b_i, \quad i = 2, 3, \ldots, n,$$

where $x_1 = b_1$. Also observe that this is equivalent to solving the simplest nontrivial linear (bidiagonal) equation $Ax = b$, where

$$A = \begin{bmatrix} 1 & & & & \\ -a_2 & 1 & & & \\ & -a_3 & 1 & & \\ & & \ddots & \ddots & \\ & & & -a_n & 1 \end{bmatrix} \quad \text{and} \quad b = \begin{bmatrix} b_1 \\ b_2 \\ b_3 \\ \vdots \\ b_n \end{bmatrix}.$$

A Guide to Parallel Computation and Some Cray-1 Experiences

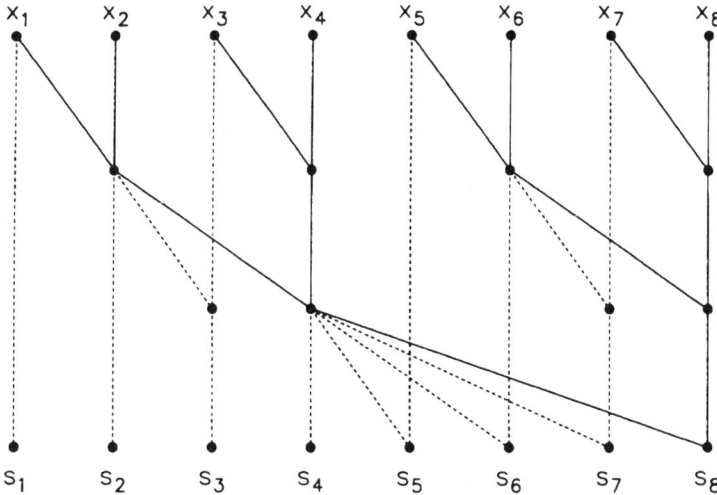

FIG. 9. Graph for all partial sums.

Although the minimum time required for the various fan-in operations mentioned above is $O(\log_2 n)$, this does not imply that the speed of these operations in a finite p-processor system is necessarily degraded by a function dependent upon n. To see that this is true for the fan-in operations mentioned, we shall illustrate a general approach with the evaluation of a single polynomial. Let

$$Q(x) = a_0 + a_1 x^1 + a_2 x^2 + \cdots + a_n x^n.$$

Consider $n = p \log_2 p$, where p is the number of processors available and is a power of 2. Rewrite

$$Q(x) = a_0 + Q_1(x) + x^{\log_2 p} Q_2(x) + \cdots + x^{(p-1)(\log_2 p)} Q_p(x),$$

where

$$Q_i(x) = a_k x + a_{k+1} x^2 + \cdots + a_{k+\log_2 p} x^{\log_2 p}$$

and

$$k = (i-1)p + 1.$$

Algorithm:

1. Distribute the coefficients $a_1, a_2, \ldots, a_{\log_2 p}$ to processor P_1; $a_{\log_2 p + 1}, a_{\log_2 p + 2}, \ldots, a_{2 \log_2 p}$ to P_2, etc.
2. Compute Q_i sequentially in P_i. This can be done in time proportional to the degree, or $O(\log_2 n)$.

3. Distribute x to all the processors and form $x, x^2, \ldots, x^p$ by the fan-in process. This requires $O(\log_2 p)$ steps.

4. Multiply $x^i Q_i$ in P_i in one step.

5. Our last parallel step sums the p terms $x^i Q_i$ by the fan-in process. This requires $O(\log_2 n)$ steps.

6. Finally, add in a_0 in one step.

Since the speedup s can be written

$$s = \frac{t_{\text{seq}}}{t_{\text{para}}} = \frac{K_1 p \log_2 p + c_1}{K_2 \log_2 p + c_2},$$

we conclude that asymptotically the speedup is $(K_1/K_2)p$ and is therefore $O(1)$ for a p-processor system. In this sense, polynomial evaluation is completely parallelizable.

Note that $K_1 \leq K_2$. Hence the speedup will depend upon not only p but also K_1 and K_2 as well. The bigger K_2/K_1, the more processors are wanted to compensate for redundant computation and/or more complex algorithms. On the other hand, as $p \log_2 p$ gets larger, the number of values of $n < p$ grows and speedup achieved using all resources is reduced. Hence optimal speedup for a process of this kind requires balancing these competing parameters.

If the SCATTER problem is defined by

$$y_{j_i} = f(x_i, y_{j_i}),$$

the complexity is arbitrary depending upon f. When solving a system of linear equations by Gaussian elimination with partial pivoting, performing the factorization with implicitly rather than explicitly interchanged rows leads to a scatter problem in which the j_i are just a permutation of the integers $1, 2, \ldots, n$. This SCATTER problem has complexity $O(1)$. The most common application takes the form $y_{j_i} = y_{j_i} + f_i x_i$ and repetition occurs in the set j_i. This can be thought of as bin summing and occurs in PIC codes, Monte Carlo neutron transport, etc. Since $j_i = $ constant leads to a problem of strict summation, the complexity of the bin summing problem is bounded above by $O(\log_2 n)$.

By replicating bin sum storage p times in a p-processor system, the parallel complexity may be reduced to $O(\log_2 p) + O(\log_2(n/p))$, where n is the total number of summands. The bin sum storage is generally proportional to the mesh size and may become too expensive for two- and three-dimensional problems. The SCATTER problem has been and probably will continue to be one difficult to speed up economically.

3. Large-Time Operations

For reasons based primarily on Cray-1 architecture, most of the functions cited with $\log_2$-time complexity perform better with algorithms that have the slightly higher square root parallel complexity. We expect these algorithms may also be useful in other synchronous systems. The approach is illustrated by the table lookup procedure (Section IV.A.1) and r-cyclic reduction (Section IV.C.3). Existing implementations of the various reduction operations use the same principle, but are also influenced by other Cray-1 hardware features.

A few operations with parallel complexity greater than $O(\log_2 n)$ are cited. Sorting has minimal scalar complexity $O(n \log_2 n)$, and Quick Sort is a well-known algorithm of this complexity. However, it is a contingent sort (that is, execution is data dependent) and generally is not best suited for synchronous multiprocessors. The best synchronous parallel sorts are noncontingent and of the Batcher type. The Batcher (1968) algorithm has sequential complexity $O(n \log_2^2 n)$ and parallel complexity $O(\log_2^2 n)$. We have not seen studies of this problem for MIMD computers. However, timing estimates have been derived for the CDC STAR-100 (Stone, 1975) and the CRI Cray-1 (Sedgewick, 1978).

Another central problem is linear system solving by Gaussian elimination and partial pivoting. This algorithm has parallel complexity $O(n \log_2 n)$ and the dominant time is due to pivoting, which is not very time consuming in sequential processing. The potential speedup is $s = O(n^2/\log_2 n)$. However, this speedup would require $p = (n - 1)^2$ computers and the efficiency $e = s/p$ is $O(1/\log_2 n)$. Finally, an uncommon example of an operation for which there is no speedup is the computation of x^{2^n}. There is a minimum of n multiplies, and each must be done in sequence.

B. *Miscellaneous Topics*

So far our discussion of parallelism has been organized around particular algorithms and how well they may be handled individually. This is likely to be one of many choices of parallelism. For example, we may have the choice of solving one tridiagonal system of equations (one FFT) many times or solving concurrently many systems (many FFTs). On the Cray-1, it is about three times faster to solve tridiagonal systems concurrently than it is to solve them one at a time despite implementations that are both fully vectorizable. For further discussion of the tridiagonal problem, see Section IV.C.1. Concurrent FFTs are also faster than many one at a time. However, the improvement is not so dramatic. These examples illustrate the new parameter associated with choosing the most appropriate of the *many directions of parallelism.*

Another decision parameter that appears much more frequently in the parallel environment is the need for additional *hybrid algorithms*. To form an optimal code, we are often forced into the implementation of more than one algorithm. Many times we must have codes with mixed scalar and vector algorithms. Examples requiring hybrid algorithms on the Cray-1 include the reduction operators (sums, maxima and minima, and norms), linear and bilinear recursion, single tridiagonal systems, sorting, and table lookup. The latter requires three algorithms on Cray-1 to be optimal for all table lengths. See Section IV.A for more detail.

As we switch from sequential processing, the question of *accuracy* will reappear. Both good and bad things can happen. Rarely do we compute the dot product $x^T y$, from a fan-in algorithm in a scalar computer. Yet in a multiprocessor that is likely to be the way to compute it. At any rate, a floating point analysis says that if done in parallel,

$$|\text{fl}(x^T y) - x^T y| \leq (1 + \log_2 n)\varepsilon |x|^T |y|,$$

whereas, if accumulated sequentially,

$$|\text{fl}(x^T y) - x^T y| \leq n\varepsilon |x|^T |y|,$$

where fl signifies floating point computation and ε is the unit roundoff of a floating point operation. However, it will also be common that parallel algorithms may require more arithmetic operations per result (*redundancy*) and less accuracy is achieved. In the area of linear algebra, high-order redundancy appears to be well correlated with high loss of accuracy. For example, an algorithm for solving triangular systems in parallel in $O(\log_2^2 n)$ time and with speedup $O(n^2/\log_2^2 n)$ produces solutions with much larger error. In fact, if $\hat{x}$ is the computed solution of $Lx = b$, $\hat{x}$ may be the solution of a much more highly perturbed matrix than with the usual sequential method. The perturbations of L are

$$\|\delta L\| \leq \alpha \varepsilon K^2 \|L\| \quad \text{and} \quad \|\delta L\| \leq n\varepsilon \|L\|$$

for the parallel and sequential algorithms, respectively, where

$$\alpha = O(n^2 \log_2 n),$$

K is the condition number of L, and ε is unit machine roundoff (Sameh and Brent, 1977).

It is usually important in pipeline computers to ensure that the complexity of a parallel algorithm is the same as that of a good sequential one since, asymptotically, the time required will be proportional to the complexity. Such algorithms have been termed consistent algorithms; that is, according to Lambiotte and Voigt (1974), a vector implementation of an algorithm for solving a problem of size n is said to be *consistent* if the number of mathematical operations required by this implementation is the same order of magnitude as required by the usual implementation on a serial computer.

A Guide to Parallel Computation and Some Cray-1 Experiences

IV. APPLICATIONS

In this section, we discuss a variety of experiences on the Cray-1. We describe the more difficult aspects of vectorization and, as a minimum, also cite work where exceptional performance has been achieved.

A. Table Lookup and Interpolation

1. Table Lookup

A simple yet fundamental problem is to evaluate a function that is defined by tabulating values at a finite number of points. If the data are not uniformly spaced, then a search, commonly referred to as table lookup, is required. We consider this problem first.

Problem: Given an ordered table $x = \{x_i | x_1 \leq x_2 \leq \cdots \leq x_n\}$, find a vector of integers $j = \{j_1, j_2, \ldots, j_n\}$ such that $x_{j_i} \leq y_i < x_{j_i+1}$, where y is an arbitrary vector, $y = \{y_1, y_2, \ldots, y_m\}$.

For simplicity, let $n = pq$, where p and q are approximately equal integers. Then consider x arranged into the $p \times q$ array

$$
\begin{array}{cccc}
x_{11} & x_{12} & \cdots & x_{1q} \\
x_{21} & x_{22} & \cdots & x_{2q} \\
\vdots & \vdots & & \vdots \\
x_{p1} & x_{p2} & \cdots & x_{pq}
\end{array}
$$

where $x_1 = x_{11}, x_2 = x_{21}, \ldots, x_n = x_{pq}$.

The algorithm locates first a column index l_i for each y_i by successively comparing elements of the first row with the vector y. The sign bits of the differences are accumulated to obtain the column index l_i. Also we use l_i to compute the base address of the column that brackets each y_i. Next compute a row index k_i for each y_i by successively comparing y_i with the column starting at $x(1, l_i)$. This is done by forming a vector mask of the signs of the differences and then doing a population count. Finally, with (k_i, l_i) we are able to compute j_i. All these operations can be performed in a vector mode.

Also of considerable frequency is the search problem that requires table lookup of only one y at a time. A vectorizable code similar to the one outlined above is possible by determining both k and l in the same fashion that k_i was computed above. However, since y is compared to the first row of the x array, care must be taken not to choose p to be a multiple of 8 because of the bank conflict problem on the Cray-1.

These algorithms have complexity $O(\sqrt{n})$ compared to $O(\log_2 n)$ for binary search. However, the vectorization of binary search for a vector y requires a GATHER operation. The GATHER time so dominates the total time that the improvement over a scalar code is disappointing. A "best" algorithm will require linear search (one chime on the Cray-1) for small n, square root search for intermediate n, and binary search for very large n. The linear search can be performed in 1 chime, but has complexity $O(n)$.

2. Interpolation

Interpolation may be broken down into three basic steps:

1. Locate the interval that contains the point at which interpolation is desired. Techniques range from table search to direct computation (for example, equal interval case).
2. Collect the data (coefficients, function values, etc.) needed for interpolation associated with the interval located in step 1.
3. Evaluate the interpolating function using the data collected in step 2.

Prior to computing with parallel architectures, the three steps outlined above were performed sequentially for each interpolant. Furthermore, steps 2 and 3 were combined, and indirect addressing was used to reference the coefficient data in function evaluation. Indirect addressing makes the calculation nonvectorizable even if there are many points to be interpolated concurrently. To perform step 3 in a vector mode, we separately gather the elements into a directly addressable vector.

Step 2 is inherently a slow process on the Cray-1. Only if the function evaluation has considerable arithmetic in it (that is, computational complexity) does it pay to separate steps 2 and 3 to operate with vectors in step 3. For example, if step 3 requires only a small amount of arithmetic, memory references so dominate steps 2 and 3 that the arithmetic is usually overlapped with the memory accesses. Consequently, scalar performance can be as good as the vectorized computation when the computational complexity is small.

To help determine whether an interpolatory problem has enough complexity to make vectorization worthwhile, we describe our experience with bilinear interpolation.

3. *Bilinear Interpolation*

In this problem, vectorization wins modestly at the expense of coding in assembly language and gathering all coefficients simultaneously from a common index vector that locates the intervals of interest. Bilinear interpolation appears in major calculations, such as PIC and hydrodynamics codes, among many other applications.

Let F be defined at the four corners of a rectangle on $\{(0, 1) \times (0, 1)\}$ by F_j, F_{j+1}, F_{j+k}, and F_{j+k+1}, where j is a pointer to the cell to be interpolated and k is constant for all j. Also let (x, y) be the point in the cell at which interpolation is desired. Then

$$F(x_i, y_i) = u_i + x_i(v_i - u_i), \tag{4}$$

where

$$u_i = F_{j_i} + y_i(F_{j_i+1} - F_{j_i})$$

and

$$v_i = F_{j_i+k} + y_i(F_{j_i+k+1} - F_{j_i+k}).$$

An assembly language code was written to load four vector registers with vectors of F, F_{j+1}, F_{j+k}, and F_{j+k+1}, and then to compute the interpolant from Eq. (4) in a vector mode. This was compared with a straightforward scalar FORTRAN implementation, and measured less than a factor of 2 improvement. Note all four function vectors are gathered concurrently on one index vector j.

This formulation appears somewhat differently in the PIC code. The equations that require GATHERs are written in the form

$$F(x_i, y_i) = w1_i F_{j_i} + w2_i F_{j_i+1} + w3_i F_{j_i+k} + w4_i F_{j_i+k+1}, \tag{5}$$

where $w1, w2, w3, w4$ are the appropriate multipliers of the corresponding Fs derived from Eq. (4). An implementation of Eq. (5), which called a CAL subroutine to collect one vector at a time, followed by a vectorized FORTRAN calculation of Eq. (5) did not perform as well as a completely scalar FORTRAN code.

4. Equation-of-State Study

Considerable computer time is required for equation-of-state (EOS) calculations in a variety of one- and two-dimensional codes. Functions describing the equation of state are kept in a tabular form, and the function values are approximated by numerical interpolation. Given material density and temperature, other related physical quantities (pressure, energy, and opacity) and their partial derivatives with respect to density and temperature may be estimated by two-dimensional interpolation in the tables of density and temperature.

However, the computational complexity of the EOS interpolation is sufficiently large to outweigh the extra cost of gathering the data required in step 2. Step 3 requires two logarithms (argument transformations) and one exponentiation (result transformations) as well as bilinear interpolation per

TABLE III

TIME IN MILLISECONDS FOR VARIOUS EOS RUNS

	Scalar performance				Vector performance restructured code
	Original code		Restructured code		
Number of cells	7600	Cray-1	7600	Cray-1	Cray-1
2	0.118	0.040	0.105	0.066	0.048
4	0.188	0.079	0.202	0.130	0.055
8	0.321	0.157	0.394	0.259	0.076
16	0.545	0.314	0.824	0.517	0.121
32	1.087	0.629	1.557	1.033	0.213
64	2.225	1.257	3.157	2.066	0.396
128	4.402	2.515	6.311	4.130	0.787
256	8.855	5.027	12.541	8.259	1.568
512	17.641	10.055	25.163	16.516	3.130
1024	35.290	20.110	50.495	33.031	6.255

point. In addition, we were able to reorganize the input tables in a way that allowed several of the scattered data items to be collected simultaneously in a vector mode rather than in scalar mode. There were 15 separate tables of input operands that could be organized as a two-dimensional table in which each column contains an input operand. Thus, these 15 operands could be collected concurrently by fetching a row from the new table. The code was completely vectorizable except for two scalar loops that fetched indirectly addressed data. One loop collected three items and the other four.

An original code was restructured to achieve vectorization. Each code was run in scalar mode on both the CDC 7600 and the Cray-1. The restructured code was also run on the Cray-1 in a vector mode. In each case, the compiler used was the optimal one for the mode selected on each computer. Table III contains the timing data for these runs, and Table IV gives the ratios of these times for evaluating EOS at 64 points simultaneously. These ratios are

TABLE IV

TIME RATIOS FOR EOS RUNS WITH 64 CELLS

	1	2	3	4	5
7600 original	1.0	—	—	—	—
7600 restructured	0.70	1.0	—	—	—
Cray-1 restructured scalar	1.07	1.52	1.0	—	—
Cray-1 original scalar	1.75	2.51	1.64	1.0	—
Cray-1 restructured CFT	5.64	8.07	5.28	3.21	1.0

essentially the same as the asymptotic ratios. The ratios are only modestly degraded when 32 cells are computed.

B. *Matrix Multiplication*

Matrix multiplication is probably the most important of basic operations with nontrivial parallel complexity $O(\log_2 n)$. It is an essential ingredient to solving linear systems of equations by both direct and indirect methods. We shall outline a procedure for the Cray-1 that produces SV performance for performing matrix multiplication with the symmetric, positive definite matrix that derives from a 9-point difference operator. Other work on this problem for both sparse and dense systems relevant to parallel processing will be cited.

1. *Sparse-Banded Multiplication*

The usual approaches to matrix multiplication that work with rows and or columns are not satisfactory for sparse-banded matrices that arise from 5- and 9-point difference operators when implemented on vector processors. This is due to the short and/or sparse vectors in a row or column. This motivated Karush *et al.* (1975) to derive a procedure for matrix multiplication by diagonals suitable to the CDC STAR-100. Since the diagonals for two-dimensional problems are long vectors, optimal performance on the STAR-100 can be easily achieved with straightforward coding. This same coding achieves V performance on the Cray-1. SV performance can be achieved for symmetric matrices by exploiting the symmetry and the Cray-1 hardware. In fact, the time is essentially determined by the multiply time or 9 chimes for 9 diagonals.

Let A be symmetric with nine nonzero diagonals and, temporarily, in block tridiagonal form as follows:

$$A = \begin{bmatrix} A_1 & B_2^T & & & & \\ B_2 & A_2 & B_3^T & & & \\ & \ddots & \ddots & \ddots & & \\ & & B_{n-1} & A_{n-1} & B_n^T \\ & & & B_n & A_n \end{bmatrix}, \qquad (6)$$

where

$$A_i = \begin{bmatrix} a_{1i} & b_{2i} & & & & \\ b_{2i} & a_{2i} & b_{3i} & & & \\ & \ddots & \ddots & \ddots & & \\ & & b_{(n-1)i} & a_{(n-1)i} & b_{ni} \\ & & & b_{ni} & a_{ni} \end{bmatrix}$$

and

$$B_i = \begin{bmatrix} d_{1i} & c_{1i} & & & & \\ e_{2i} & d_{2i} & c_{2i} & & & \\ & \ddots & \ddots & \ddots & & \\ & & e_{(n-1)i} & d_{(n-1)i} & c_{(n-1)i} \\ & & & e_{ni} & d_{ni} \end{bmatrix}.$$

By processing in sequence the set of blocks

$$\begin{bmatrix} A_{i-1} & B_i^T \\ B_i & \end{bmatrix} \begin{bmatrix} x_{i-1} \\ x_i \end{bmatrix} = \begin{bmatrix} y_{i-1} \\ y_i \end{bmatrix}, \quad i = 1, 2, \ldots, n-1,$$

it is possible to achieve SV performance for banded matrix multiplication on the Cray-1. By exploiting the symmetry in the positive definite case, the time is determined by the nine multiply times plus startups.

Let $f^-, f^<$, and $f^>$ designate the vector f at $i-1$, shifted left one element in a vector register, and shifted right one element in a vector register, respectively. The expressions computed in the loop are

$$y^- = y^- + ax^- + (bx^-)^> + b(x^-)^< + (cx)^> + dx + ex^<$$

and

$$y = c(x^-)^< + dx^- + (ex^-)^>.$$

An implementation of the 9 chimes is given in the accompanying tabulation. It is assumed that x^-, x, and y^- are left in the vector registers after the last execution of the loop. A register allocation is given in Appendix A. Let $*, +, <,$ and $>$ designate the operators multiply, add, shift left one, and shift right one, respectively.

Chime	Operations
1	Load b; $bx^- = b*x^-$; $(bx^-)^> = >bx^-$; $y^- = y^- + (bx^-)^>$
2	$(x^-)^< = <x^-$; Load d; $dx = d*x$; $y^- = y^- + dx$
3	Load c; $b(x^-)^< = b*(x^-)^<$; $y^- = y^- + b(x^-)^<$
4	Load a; $cx = c*x$; $(cx)^> = >cx$; $y^- = y^- + (cx)^>$
5	$ax^- = a*x^-$; $y^- = y^- + ax^-$; $x^< = <x$
6	Load e; $ex^< = e*x^<$; $y^- = y^- + ex^<$
7	$(x^-)^< = <x^-$; $y = c(x^-)^<$; Store y^-
8	Load d; $ex^- = e*x^-$; $(ex^-)^> = >ex^-$; $y = y + (ex^-)^>$
9	$dx^- = d*x^-$; $y = y + dx^-$

We now observe that the missing terms in the result sum can be obtained by loops of the form

DO $1\ j = 2, n$

$y(1, j) = y(1, j) + b(1, j) * x(n, j - 1)$

1 Continue

When the 9-diagonal matrix is nonsymmetric, the performance is degraded from 9 to 11 chimes. A similar approach permits an implementation whose chimes are determined by 11 memory references (the 9 loads of the diagonals, a load of x, and a store of y^-). The worst chime in both instances has a startup time no greater than 30 for vectors less than 65. This portion of the code (the dominant part) should then perform at up to 120 megaflops in the symmetric case and about 100 megaflops in the general case. The work not accounted for will perform at V speed and comprises about 2% of the total work. Hence we expect a total performance above 100 megaflops.

2. General Matrix Multiplication

Not surprisingly, all the advanced computers that we are aware of can perform the multiplication of general dense matrices at speeds that exploit the architecture. Somewhat more surprisingly, these computers require algorithmic revision and special attention to the hardware capabilities to achieve optimal performance. The Cray-1 is no exception. To achieve SV speeds (well in excess of 100 megaflops), it is necessary to code this function in CAL. An algorithm for achieving SV speed is described by Fong and Jordan (1979) and will not be given here. Further work that provides techniques for maintaining high processing rates with matrices of small order appears in Calahan et al. (1979).

C. Linear Equations

Contrary to the results just obtained for matrix multiplication, results for the inverse problem, solving systems of linear equations by direct methods, are mixed not only for the Cray-1 but for parallel computers in general. As noted earlier, we are far from achieving the lower bound of complexity $O(\log_2 n)$ with stable algorithms, and some sparse and tridiagonal systems offer little or no improvement over scalar performance on the Cray-1. We shall analyze in some detail the performance of tridiagonal systems on the Cray-1 and cite work done on more dense systems. Dense systems can be solved nearly as effectively as matrix multiplication on the Cray-1.

1. Bidiagonal and Tridiagonal Systems

The solution of a single tridiagonal system of linear equations, $Ax = y$, has been a problem on pipeline computers whenever speedup of this function is important. The recursion in classical Gaussian elimination does not permit vectorization. Hence, on computers like the CDC STAR-100, TI ASC, or Cray-1, for which the time required to solve such a system is proportional to the order, we must measure the lesser work of Gaussian elimination in scalar mode against the vector performance of cyclic reduction in each architecture. This requirement led to the development of a hybrid code on the STAR-100 for which Gaussian elimination is used for $n \leq 175$ and cyclic reduction is used for larger orders (Madsen and Rodrigue, 1976). The crossover point on the TI ASC for codes written entirely in FORTRAN is about 25 (Boris, 1976).

Since the Cray-1 has a very short startup time for vector operations, the "odds-on" favorite might appear to be cyclic reduction. Calahan *et al.* (1979) have written such an algorithm in CAL, and its times are given in Table V. These routines are written using the faster 9-digit divide. We would expect this to be a proper decision when speed is of concern because these codes appear most frequently in iterative procedures for which this precision would provide adequate accuracy in most problems.

In our analysis of the most productive things to be done in the early days of the Cray-1, this problem was set aside for two reasons. First, many important applications needing the solution to systems of tridiagonal equations could be easily converted to use a readily vectorizable routine that concurrently solves many simultaneous tridiagonal systems. This was done, and the rate obtained is

$$\text{cycles per equation} \cong \begin{cases} 100 + 6.5n \text{(factor)} \\ 100 + 8.7n \text{(solve)} \end{cases}$$

when 64 equations are solved concurrently.

TABLE V

MACHINE CYCLES PER EQUATION FOR CYCLIC REDUCTION

Order	Factorization	Forward and back substitution	Total
15	86.7	134.6	221.3
31	57.7	88.7	146.4
63	38.9	58.8	97.8
127	28.0	40.8	68.8
255	22.3	30.4	52.7
511	19.2	25.0	44.2
1023	17.7	22.0	39.7
2047	16.8	20.2	37.0
4095	16.3	19.1	35.5

A second reason for not writing a routine to handle a single tridiagonal system is that it probably would require some detailed coding of both methods to select the best method. After observing the number of clock periods to solve a system of tridiagonal equations with cyclic reduction, we were motivated to analyze classical Gaussian elimination written reasonably well in CAL. We believe the following work confirms our suspicions about the difficulty in selecting the "best" method on a pipeline computer.

a. Algorithm I. The first routine considered is one that might be used as a substitute, if pivoting is not required, for the LINPACK (Dongarra et al., 1979) tridiagonal solver in which a single call both factors the matrix and solves the system. To exploit the segmented functional units in a scalar mode, we divide the matrix into two halves and factor the upper half into an $L_1 U_1$ factorization and the lower half into $U_2 L_2$, where the diagonal elements of L_1 and L_2 are 1; that is, for n even

$$A = \begin{bmatrix} a_1 & b_1 & & & & & & \\ c_2 & a_2 & b_2 & & & & & \\ & \ddots & \ddots & \ddots & & & & \\ & & c_{n/2} & a_{n/2} & b_{n/2} & & & \\ \hdashline & & & & c_{n/2+1} & a_{n/2+1} & b_{n/2+1} & \\ & & & & & \ddots & \ddots & \ddots \\ & & & & & c_{n-1} & a_{n-1} & b_{n-1} \\ & & & & & & c_n & a_n \end{bmatrix}$$

and can be factored into the following form:

$$\begin{bmatrix} d_1 & u_1 & & & & & & \\ c_2 & d_2 & u_2 & & & & & \\ & \ddots & \ddots & \ddots & & & & \\ & & c_{n/2} & p & q & & & \\ \hdashline & & & r & s & b_{n/2+1} & & \\ & & & & \ddots & \ddots & \ddots & \\ & & & & & u_{n-1} & d_{n-1} & b_n \\ & & & & & & u_n & d_n \end{bmatrix}$$

We now solve the middle 2×2 system (p, q, r, s) before doing the "back substitutions" from the middle out. If n is odd, we have a 3×3 system in the middle to solve. We prefer to mock up the system to always be even and handle the mockup within the vector registers.

The FORTRAN factor and forward-substitution loop to the middle part that most closely models the CAL inner loop outlined in Appendix A is given by

$$a_1 = 1/a_1$$
$$b_1 = b_1 a_1$$
$$y_1 = y_1 a_1$$

DO $\quad i = 2, n/2$

Loop 1 $\quad b_i = b_i/(a_i - c_i b_{i-1} a_{i-1})$

$\quad y_i = y_i - c_i y_{i-1}/(a_i - c_i b_{i-1} a_{i-1}).$

The back-substitution loop for the upper half is given by

DO $\quad i = n/2 - 1, 1$

Loop 2 $\quad y_i = y_i - b_i y_{i+1}.$

A CAL implementation is outlined in Appendix B. We describe only the code for working with the upper half of the matrix. Code for the lower half follows from symmetry. This half of the loop uses S0–S3 as scalar registers, leaving S4–S7 for the bottom half in an almost symmetric way.

Procedure:

1. Load the vector registers V0–V3 with c, a, b, y from the top half and V4–V7 with corresponding elements from the bottom half. Load them backward so that the loop control index can also serve to address the V-registers. This requires 4.25 cycles per equation for vectors of length 64.

2. Execute the scalar loop outlined in Appendix B that performs Gaussian elimination on both halves of the matrix. Note that because S0 and S4 behave differently, a code that stores S0 into a T register and later retrieves this datum is not necessary in the other half of the code. We estimate that 40 cycles are required to process an equation from each half of the matrix. Hence 20 cycles are required for the factorization and forward substitution (Loop 1).

3. Store both halves of b and y in main memory. This requires 2.13 cycles per equation.

4. Load both halves of b and y into V0–V1 and V4–V5. This requires 2.13 cycles per equation. Steps 3 and 4 could be eliminated if $n \leq 128$.

5. Next perform the back substitution (Loop 2) with a code similar to that outlined in Appendix C. Some unfolding of the basic loop should be done

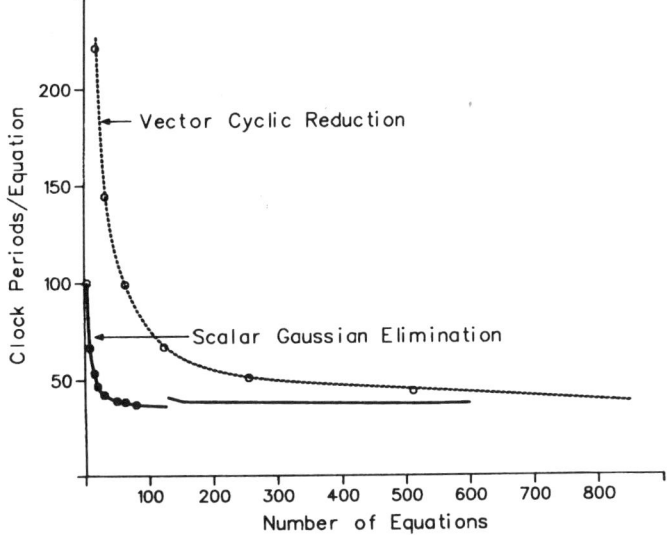

FIG. 10. Timing comparison of single-system tridiagonal equation solvers.

to hide the jump back to the beginning of the loop and to provide time for loop indexing. We would unfold the loop four times, and hence estimate the required time to be 8 cycles per equation.

6. Store Y. This requires 1.06 cycles per equation.

This totals to an estimate of 38 cycles per equation as an asymptotic rate.

A slightly improved implementation of this algorithm has been written by Brown (1980). Its performance compared to cyclic reduction is given in Figure 10, and verifies the estimated times.

b. Algorithm II. Can Gaussian elimination be competitive in the situation in which we wish to first factor the equation and then solve using these factors? We shall factor the matrix into the form $A = LDU$, where DU is just the U of the usual LU factorization. The FORTRAN code might be written in the following form:

Factorization: $\quad a_1 = 1/a_1$

$\qquad\qquad\qquad b_1 = b_1 a_1$

$\qquad\qquad$ DO $\quad i = 2, n/2$

Loop 3 $\qquad\qquad a_i = 1/(a_i - c_i b_{i-1} a_{i-1})$

$\qquad\qquad\qquad b_i = b_i a_i$

$\qquad\qquad\qquad c_i = c_i a_{i-1}$

Forward substitution: $y_1 = a_1 y_1$

DO $i = 2, n/2$

Loop 4 $y_i = y_i - c_i y_{i-1}$

Diagonal solution: DO $i = 1, n/2$

Loop 5 $y_i = a_i y_i$

Back substitution: DO $i = n/2 - 1, 1$

Loop 6 $y_i = y_i - b_i y_{i+1}$.

Procedure:

1. Load c, b, and a into V0–V2 for the top half and V4–V6 for the lower half. Also form the product $b_{i-1} c_i$ for both halves in V3 and V7. This can be done in 3.2 cycles per equation.

2. Perform the factorization with the scalar code outlined in Appendix D. This corresponds to Loop 3 working to and from the vector registers and can be done in 27 cycles to process an equation in each half of the matrix. Hence 13.5 cycles per equation are required for this function.

3. Compute $b_i = b_i a_i$ and $c_i = c_i a_{i-1}$ in vector mode and store vectors b, c, and a. This requires 3.2 cycles.

4. Load c_i and y_i into V0–V1 and V4–V5 for each half of the matrix. This requires 2.1 cycles per equation.

5. Use Appendix C to do the linear recursion of Loop 4. Perform Loop 5 in vector mode. This costs 9 cycles. Loading a_i comes free.

6. Store y_i. This costs 1.07 cycles.

7. Load y_i and b_i into V0–V1 and V4–V5 for each half. This costs 2.1 cycles.

8. Use Appendix B code to do the linear recursion of Loop 6. This codes costs 8 cycles.

9. Store y_i. This costs 1.07 cycles.

This totals to about 20 cycles for the factorization and 23 cycles for the solution.

c. *r-Cyclic Reduction.* We believe the best asymptotic performance of back-and-forward substitution on the Cray-1 comes from an algorithm that uses *r*-cyclic reduction. For simplicity, let the order $n = rs$, where r and s are nearly equal and r is not a multiple of 8. Let us consider a bidiagonal system of equations with unit diagonal (first-order linear recursion), for example:

A Guide to Parallel Computation and Some Cray-1 Experiences

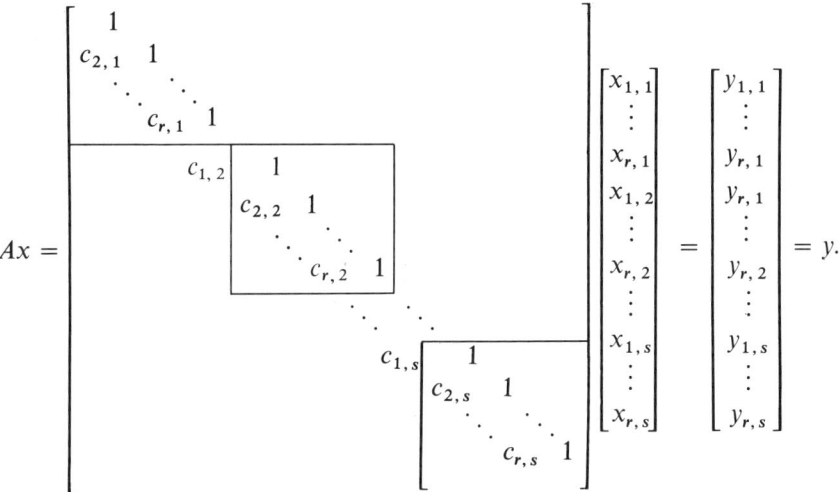

The following FORTRAN-like code:

Loop 7
$$\begin{aligned}
&\text{DO} \quad i = 2, r \\
&\quad \text{DO} \quad j = 1, s \\
&\qquad y'_{i,j} = y_{i,j} - c_{i,j} y_{i-1,j} \\
&\qquad b_{i,j} = -c_{i-1,j} c_{i,j}
\end{aligned}$$

transforms this system of equations into the following form:

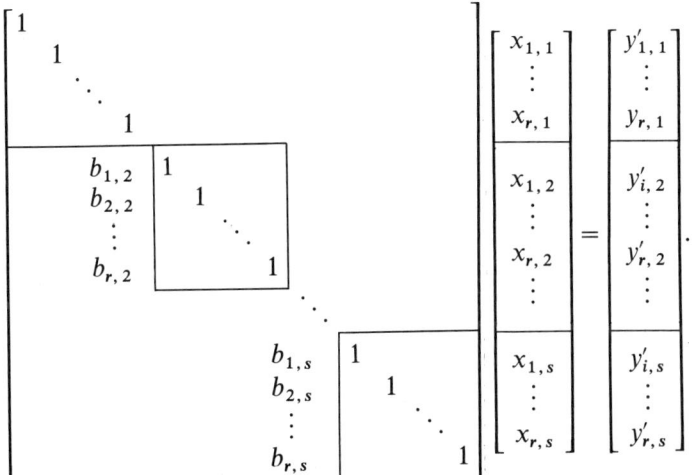

The inner loop is fully vectorizable. The following FORTRAN-like double loop can be applied to this system to obtain the final solution.

$$\text{DO} \quad j = 2, s$$
$$\text{DO} \quad i = 1, r$$

Loop 8 $\qquad x_{i,j} = y'_{i,j} - y'_{r,j-1} b_{i,j}$

The asymptotic time required to execute Loop 7 and Loop 8 is 7 chimes or, say, 8 machine cycles, if implemented in a straightforward way. For a factorization described in Algorithm II, the two recursions and the diagonal multiply could be done in about 18 cycles per equation for r and s both large. However, r and s behave like $\sqrt{n}$. Hence rs must be quite large before we reach the asymptotic rate.

It appears to us that a best code would be polyalgorithmic, using the scalar kernels for small and intermediate n- and r-cyclic reduction for large n. If only one algorithm were to be implemented, we believe the scalar algorithm best covers the most common orders since the asymptotic rate of the scalar code is reached much more quickly.

The factorization $A = LDU$ can be expressed as a second-order linear recurrence to compute D. Furthermore, the usual scalar divides required to form D can be replaced by vector divides. Let $d_i = p_i/p_{i-1}$. Then the equation for the diagonal

$$d_i = a_i - b_{i-1} c_i / d_{i-1}, \qquad i = 2, n,$$

becomes

$$\frac{p_i}{p_{i-1}} = \frac{a_i p_{i-1} - b_{i-1} c_i p_{i-2}}{p_{i-1}},$$

where $p_0 = p_1 = 1$. Hence

$$p_i = a_i p_{i-1} - b_{i-1} c_i p_{i-2}, \qquad i = 2, n.$$

To avoid overflow, let $a_i = 1$ for $1 \leq i \leq n$. We obtain

$$p_i = p_{i-1} - q_i p_{i-2}, \qquad i = 2, n,$$

where $q_i = b_{i-1} c_i$. The q_i are computable in a vector mode. Take coefficients of the last $n - 1$ equations in $n + 1$ unknowns, and block as follows for r-cyclic reduction.

$$\begin{array}{c}
\begin{array}{|cccccc|}\hline
q_{11} & -1 & 1 & & & \\
q_{21} & -1 & 1 & & & \\
q_{31} & & -1 & 1 & & \\
& & & \ddots & & \\
q_{r1} & & & -1 & 1 & \\
\hline
\end{array}
\end{array}$$

(block diagonal structure with subsequent blocks for q_{i2} and q_{is})

Now solve simultaneously by Gaussian elimination the matrix equation

$$A_j[e_j\,f_j] = [g_j\,h_j],$$

where

$$g_j^{\mathrm{T}} = [-q_{1j}, 0, 0, \ldots, 0],$$
$$h_j^{\mathrm{T}} = [1, -q_{2j}, 0, \ldots, 0],$$

and

$$A_j = \begin{bmatrix} 1 & & & & \\ -1 & 1 & & & \\ q_{3j} & -1 & 1 & & \\ \vdots & & & & \\ q_{rj} & & & -1 & 1 \end{bmatrix}$$

to form

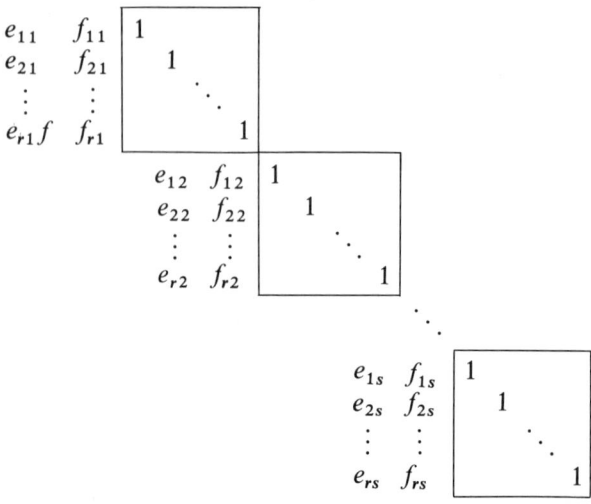

Finally with vector recursion, solve for p with the following loops:

$$\text{DO} \quad j = 1, s$$
$$\quad \text{DO} \quad i = 1, r$$
$$p_{ij} = e_{ij} p_{(r-1)(j-1)} + f_{ij} p_{r(j-1)}$$

The inner loop is now a linear combination of two vectors. Once the ps are obtained, we can form the reciprocal ds in a vector mode and subsequently scale the cs and bs to obtain the LDU factorization.

Pipeline computers appear to offer little advantage over scalar algorithms for this problem. This stems from the extra work performed in a vector algorithm and the shorter average vector length. Also note that the algorithms discussed assume no pivoting is required. Hence they are limited to a more restricted class of problems. For a discussion of stable parallel algorithms, see Sameh and Kuck (1978).

2. *Banded and General Linear Systems*

Performance in solving linear systems of equations on the Cray-1 improves as we increase bandwidths and/or order. Bandwidths or orders equal to 64 give SV performance at almost 100 megaflops (Fong and Jordan, 1977). With special attention paid to the programming of small systems, excellent performance may also be achieved (see Calahan et al., 1979). To achieve this kind of performance, it is currently necessary to program the Cray-1 in CAL.

A Guide to Parallel Computation and Some Cray-1 Experiences

Drastically different performances are achieved depending upon the language and whether the algorithm is implemented in vector mode (see Fig. 3). As with matrix multiplication, the ability to achieve SV speed derives from the formation of linear combinations of vectors in the vector registers. High speeds for small systems are achieved by interleaving vector operations in such a way that the startup time associated with chaining is avoided (Orbits and Calahan, 1978). In essence this allows startup times to be overlapped. We should also point out that the usual dot product implementation does not allow SV speed since it requires two memory references and has only two floating point operations.

D. Iterative Methods

Some of the better iterative methods for solving systems of linear equations are not attractive for parallel computation. As a consequence, it may be better to use a more slowly convergent algorithm that is well suited to parallel computation to expoit the parallel speed of advanced computers. The better algorithms usually are implicit and require some form of recursion. However, in some cases it may be possible to reformulate the good algorithms so that they can be implemented mostly, if not entirely, with parallel hardware. We do this for the incomplete Cholesky conjugate gradient (ICCG) algorithm and a kernel from a hydrodynamics code that is recursive to demonstrate techniques for achieving parallelism.

1. ICCG—Incomplete Cholesky Conjugate Gradient

Consider the system of equations

$$Ax = b,$$

where A is symmetric, positive definite, large, and sparse. The ICCG is a popular method used to solve this system and is but one of several that are classified as preconditioned conjugate gradient methods. The method combines the conjugate gradient method of Hestenes and Stiefel (1952) with an incomplete Cholesky factorization of A. Successive approximations x_i to the solution x are generated by iterating the following steps until convergence:

$$\alpha = (r_i, M^{-1}r_i)/(p_i, Ap_i),$$

$$x_{i+1} = x_i + \alpha p_i,$$

$$r_{i+1} = r_i - \alpha Ap_i,$$

$$\beta = (r_{i+1}, M^{-1}r_{i+1})/(r_i, M^{-1}r_i),$$

$$p_{i-1} = p_i + \beta p_i.$$

The matrix M for ICCG is an incomplete factorization of $A = LDL^T$ such that element $l_{ij} = 0$ if $a_{ij} = 0$. The factorization is defined recursively by

$$l_{ij} = a_{ij} - \sum_{k=1}^{i-1} l_{ik} l_{jk} d_k;$$

$$d_i = 1/l_{ii}.$$

The matrix A is a 9-diagonal matrix of the form given in Eq. (6).

A detailed examination of the algorithm shows that the operation count is given by two dot products (SDOT), three expressions of the form $y = y + ax$(SAXPY), a 9-diagonal matrix multiply, and a solution to $Mx = LDL^T x = b$, where L is now the incomplete factorization. We have described in Section IV.B.1 how the matrix multiplication can be done at SV speed in 9 chimes. Two chimes are required for SDOT and 3 for SAXPY for a total of 23 chimes for all functions exclusive of solving the linear system.

A vector algorithm for the solution of $LDL^T x = b$, where A is 5-diagonal, is given by Greenbaum and Rodrigue (1977). The solution of this system is found as usual by solving $Lz = b$, $Dw = z$, and $L^T x = w$ in sequence. We give a similar algorithm for a 9-diagonal A and estimate the time required to solve $Lz = y$, where

$$Lz = \begin{bmatrix} D_1 & & & \\ L_2 D_2 & & & \\ & & \ddots & \\ & & & L_m D_m \end{bmatrix} \begin{bmatrix} Z_1 \\ Z_2 \\ \vdots \\ Z_m \end{bmatrix} = \begin{bmatrix} Y_1 \\ Y_2 \\ \vdots \\ Y_m \end{bmatrix} = y,$$

$$D_j = \begin{bmatrix} 1 & & & \\ b_{2j} & 1 & & \\ & & \ddots & \\ & & b_{mj} & 1 \end{bmatrix}, \quad Y_j = \begin{bmatrix} y_{1j} \\ y_{2j} \\ \vdots \\ y_{mj} \end{bmatrix},$$

$$L_j = \begin{bmatrix} d_{1j} c_{1j} & & & \\ e_{2j} d_{2j} c_{2j} & & & \\ & & \ddots & \\ & & e_{mj} d_{mj} & \end{bmatrix}, \quad \text{and} \quad Z_j = \begin{bmatrix} z_{1j} \\ z_{2j} \\ \vdots \\ z_{mj} \end{bmatrix}.$$

A Guide to Parallel Computation and Some Cray-1 Experiences

The following DO loops describe the vector algorithm, where we assume z_1 has been computed:

DO $j = 2, m$

    DO $i = 1, m$

Loop 1 $z_{ij} = y_{ij} - d_{ij} * z_{i(j-1)}$

    DO $i = 2, m$

        $z_{(i-1)j} = z_{(i-1)y} - c_{(i-1)j} * z_{i(j-1)}$

Loop 2 $z_{ij} = z_{ij} - e_{ij} * z_{(i-1)(j-1)}$

    DO $i = 2, m$

Loop 3 $z_{ij} = z_{ij} - b_{ij} * z_{(i-1)j}$

Observe that Loop 1 and Loop 2 are trivially vectorizable. Loop 3 is a first-order linear recurrence and can be vectorized with r-cyclic reduction. Hence the ICCG algorithm is fully vectorizable. Loop 3 performance on the Cray-1 has been discussed in Section IV.C.1 and has little payoff when compared to good scalar code.

Loop 1 and Loop 2, if coded together in assembly language, can be executed in 4 chimes, as illustrated by the following vector operations, where we have assumed that z^- has been stored but need not be reloaded in a vector register and that z will be left in a vector register for processing by Loop 3.

Chime	Operations
1	Load $y; (z^-)^< = <z^-$
2	Load $c; c(z^-)^< = c * (z^-)^<; z = y - c(z^-)^<$
3	Load $d; dz^- = d * z^-; z = z - dz^-$
4	Load $e; ez^- = e * z; (ez^-)^> = >ez^-; z = z - (ez^-)^>$

In Section IV.C.1 we estimated the linear recursion of Loop 3 to take 8 cycles (conservatively 8 chimes) for a total of 12 chimes for the solution of $Lz = y$. A similar time is required for the back substitution. One chime is required for the diagonal multiply since it can be done as part of the forward solve before the forward solution is stored. This amounts to 25 chimes for computing $M^{-1}r$ in the ICCG algorithm, or a total of 48 chimes (about 60 cycles per equation) for the complete algorithm exclusive of convergence testing. Observe that one third of the total time (8 chimes) is required for

Loop 3 and its counterpart in the back substitution. Despite this slow aspect of calculation, there are 23 multiply–add combinations or 46 floating point operations being performed in 60 cycles, or approximately at a 60-megaflop rate.

To our knowledge, the incomplete Cholesky factorization has not previously been profitably vectorized. Also, since it is not involved in the iterative procedure, it is not a time-consuming aspect of the ICCG algorithm. Nevertheless, we believe it is instructive to illustrate the approach to vectorization. The code is completely vectorizable with less than 20% of the work only marginally effective on the Cray-1 compared to good scalar code.

Let A be given by Eq. (6) and assume that A_1 is already factored. Then the following loops of the algorithm define completely vectorizable code. Let α, β, γ, δ, and ε be the diagonal factors corresponding to diagonals a, b, c, d, and e of A:

DO $\quad j = 2, n$

Form factors in row block j

$$\varepsilon_{ij} = e_{ij}$$
$$\delta_{ij} = d_{ij} - \alpha_{(i-1)(j-1)}\beta_{i(j-1)}\varepsilon_{ij}$$
$$\gamma_{ij} = c_{ij} - \gamma_{(i-1)(j-1)}\beta_{i(j-1)}\delta_{ij}$$
$$\beta_{ij} = b_{ij} - \alpha_{i(j-1)}\gamma_{(i-1)j}\delta_{ij} - \alpha_{(i-1)(j-1)}\delta_{(i-1)j}\varepsilon_{ij}$$
$$\alpha_{ij} = a_{ij} - \alpha_{(i+1)(j-1)}\gamma_{ij}^2 - \alpha_{i(j-1)}\delta_{ij}^2 - \alpha_{(i-1)(j-1)}\varepsilon_{ij}^2$$
$$\alpha_{ij} = 1/(\alpha_{ij} - \alpha_{(i-1)j}\beta_{ij}^2)$$

Scale the off-diagonal elements

$$\varepsilon_{ij} = \alpha_{i(j-1)}\varepsilon_{ij}$$
$$\delta_{ij} = \alpha_{i(j-1)}\varepsilon_{ij}$$
$$\gamma_{ij} = \alpha_{i(j-1)}\varepsilon_{ij}$$
$$\beta_{ij} = \alpha_{ij}\beta_{ij}$$

Each equation corresponds to a DO loop on i, where the appropriate range of i is assumed. Elements in the corners of the blocks present no real problem. Unfortunately, as in the case of the vectorized forward and back substitutions, the vector lengths are determined by the square root of the problem size for two-dimensional problems. Although these lengths are satisfactory for

A Guide to Parallel Computation and Some Cray-1 Experiences

the Cray-1, they may not be for other vector computers. Note that the only recursion is in the sixth equation of the algorithm, and it can be converted to a bilinear recursion as was done for the factorization of a tridiagonal system.

Two recent improvements have been made to the ICCG algorithm. First, Eisenstadt (1980) has shown how to transform the problem so that the 9-diagonal matrix multiply may be replaced by a strictly diagonal matrix multiply. This removes 8 chimes from the Cray-1 algorithm given, or about one sixth of the estimated time. In other computing environments where there is less disparity between the matrix multiply time and solution time, the time saved could be much larger. Second, the problem can be converted from an ICCG to MICCG (Gustavson, 1978) at the expense of four additional adds with no extra storage cost and at an improved convergence rate.

2. A Frontal Approach to Vectorization

Consider the following double DO loop excerpted from a two-dimensional hydrodynamics calculation for updating pressure and velocity in two-space. It is an implicit relaxation scheme that contains recursion.

DO $j = 2, J - 1$

    DO $i = 2, I - 1$

$$dp_{ij} = a_i u_{ij} + b_i u_{(i-1)j} + cv_{ij} + dv_{i(j-1)}$$

$$p_{ij} = p_{ij} + \alpha dp_{ij}$$

$$u_{ij} = u_{ij} + \beta dp_{ij}$$

$$u_{(i-1)j} = u_{(i-1)j} + \beta dp_{ij}$$

$$v_{ij} = v_{ij} + \gamma dp_{ij}$$

$$v_{i(j-1)} = v_{i(j-1)} + \gamma dp_{ij}$$

Observe that recursion appears in both the i and j directions and consequently might appear not to be vectorizable.

One pass through the loop performs the following transformation:

$(u^{1/2}, v^{1/2})$ (u^0, v^0) (u^1, v^1) $(u^{1/2}, v^{1/2})$
•$(i-1)j$ •ij •$(i-1)j$ •ij

$\longrightarrow$

$(u^1, v^{1/2})$ (u^1, v^1)
•$i(j-1)$ •$i(j-1)$

Consequently, it is possible to process in parallel all of the points that lie along diagonals $k, k + 1$ of the grid, since they have the following time values:

$$
\begin{array}{cc}
k & k+1 \\
(\tfrac{1}{2}, \tfrac{1}{2}) & (0, 0) \\
(\tfrac{1}{2}, \tfrac{1}{2}) & (0, 0) \\
\ddots & \ddots \\
(\tfrac{1}{2}, \tfrac{1}{2}) & (0, 0) \\
& (1, \tfrac{1}{2})
\end{array}
$$

Vector performance is degraded by (1) average vector lengths of $n/2$ and (2) $2n - 1$ startups.

We illustrate how the algorithm can be rewritten to obtain another parallel solution to the problem. If we substitute the expression for dp_{ij} into the equation for u_{ij} and make explicit the time dependencies, we have

$$
\begin{aligned}
u_{ij}^{1/2} &= u_{ij}^0 + \beta(a_i u_{ij}^0 + b_i u_{(i-1)j}^{1/2} + cv_{ij}^0 + dv_{i(j-1)}^{1/2}) \\
&= e_i u_{(i-1)j}^{1/2} + f_i,
\end{aligned}
$$

where

$$
e_i = \beta b_i \quad \text{and} \quad f_i = u_{ij}^0 + \beta(a_i u_{ij}^0 + cv_{ij}^0 + dv_{i(j-1)}^{1/2}).
$$

Note that $e_i, f_i,$ and $v_{ij}^{1/2}$ are simple vector computations. The expression for $u_{ij}^{1/2}$ is a first-order linear recurrence. As shown in Section IV.C.1, this process is vectorizable but is surely less effective on Cray-1 than the previous calculations. We may now compute $u_{(i-1)j}^1$ and $v_{i(j-1)}^1$ in a vector mode. For this formulation, the average vector length, excluding the linear recursion, is n and the startups are half that of the previous algorithm.

E. Asynchronous Applications

There are common applications for which there has been little or no successful vectorization. Most notable are adaptive quadrature procedures and stochastic processes, such as Monte Carlo simulation of neutron transport. Yet these are problems for which there is abundant parallelism. However, there has been little, if any, success in capturing this parallelism for use on synchronous devices.

1. Numerical Quadrature

The fundamental problem of approximating $\int_R f(x)\,dx$ is one of computing $\sum_{i=1}^n w_i f(x)$ for a judicious selection of points x_i in R. If the points x_i are known in advance, then the complexity is $O(\log_2 n)$ due to the problem of summing. The complexity of evaluating f is problem dependent, but it is

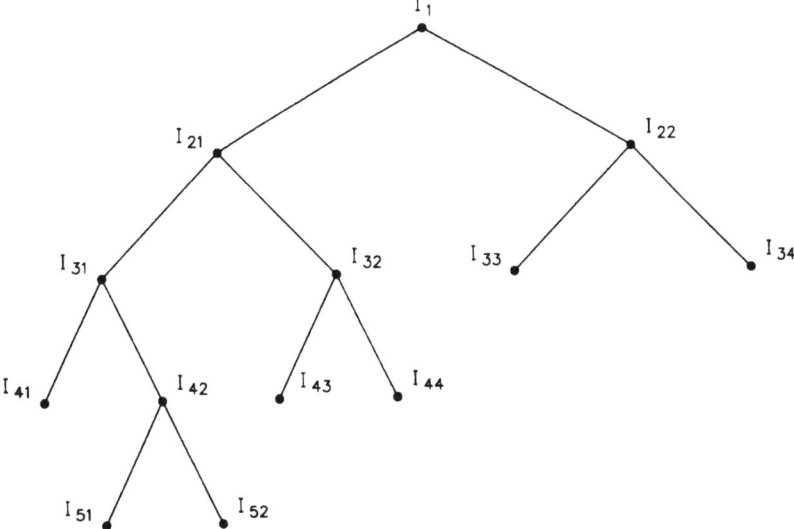

FIG. 11. Example tree of an adaptive quadrature.

likely the most important aspect of the computation and will usually be the part of the computation for which exploiting parallelism will provide the greatest speedup.

Due to the importance of minimizing function evaluations, much attention has been given to "judicious point selection." This has led to very sophisticated adaptive techniques that compute integrals over subregions to an accuracy consistent with the overall accuracy requirement. Thus, some subregions are terminated before others and the continuing subregions may become very disjoint; consequently, there is additional parallel complexity. For example, there are many more steps in the adaptive processes than $O(\log_2 n)$, where n is the final number of points. The process is a "divide-and-conquer" strategy as illustrated in Fig. 11. Thus $I = I_{41} + I_{51} + I_{52} + I_{43} + I_{44} + I_{33} + I_{34}$, or the sum of the bottom leaves of the tree.

There would appear to be two fundamental problems of exploiting the available parallelism of adaptive techniques with SIMD hardware. Presumably, the parallelism that we would attempt to exploit is to compute concurrently all of the new function evaluations required at a given level, Fig. 11, to benefit from synchronous parallelism that may exist in the function evaluation. Thus vector performance in the quadrature routine itself would be degraded by

1. point collection (GATHER) for function evaluation, and
2. shorter vectors for the function evaluation and even shorter vectors for the subsums needed.

Little work is found on the parallel quadrature problem in either software for synchronous parallel architecture or the literature.

For asynchronous hardware there may be a more fruitful approach to adaptive quadrature by assigning new processors to each node of the tree as the nodes are encountered. When a processor becomes the bottom leaf, it sends the result up the tree and the leaf is pruned (for example, the processor is freed). The burden of work is then placed on a language to accommodate the task description as well as an operating system for spawning tasks and pruning the tree.

2. Monte Carlo

The Monte Carlo neutron transport problem does not lack for parallelism; in fact, the total computation is almost completely parallel. The tracking of each particle and its progeny is independent of all other histories. Only the accumulation of statistics and the use of statistics in sample biasing couple one particle to another. This is not to say that the computation performed on each history is the same. The randomness and variety of possible events make it very difficult to develop and process queues of similar computations. Today's codes are replete with conditional and case statements that define the many possible reactions in complex and varied geometries. Given the physical input parameters, the whole calculation is a function of a single random variable.

We must be careful not to extrapolate this computational independence to all codes that might use the Monte Carlo method. As particle dependence or coupling between particles increase, the lengths of the chains of independent computation may decrease. This will in turn decrease the efficiency of multiprocessing. At another extreme some problems in which there is tight coupling (for example, a many-body problem in which one particle's behavior affects every other one similarly) have been highly vectorized.

The problem is an ideal example of asynchronous parallelism. In an environment of n asynchronous multiprocessors, the code could be easily modified so that each processor could independently process large batches of particles before rejoining to accumulate statistics. When the processors have finished their tasks, one or more processors could be used to accumulate total statistics.

Given only a glimpse of future asynchronous architectures and a Monte Carlo application that has ideal asynchronous parallelism, we are yet unable to estimate the effort that will be required to fit the problem to the machine. We know that memory requirements may be exorbitant in those cases where the processor count is large. In addition, it is unlikely that the more profitable global parallelism will be discovered and selected automatically by the compiler. We believe that new syntax will be needed to assist the programmer by (1) describing the parallelism available to the compiler and (2) making it easier to increase the dimensionality of the problem without rewriting code.

APPENDIX A. A REGISTER ASSIGNMENT FOR SPARSE-BANDED MATRIX MULTIPLY

TABLE A

Register	0	1	2	3	4	5	6	7
Chime								
0	x^-	x	y^-	—	—	—	—	—
1	—	—	—	$y^- = y^- + (bx^-)^>$	—	$(bx^-)^>$	bx^-	$\uparrow b$
2	—	—	$y^- = y^- + dx$	—	$\uparrow d$	dx	$(x^-)^<$	—
3	—	—	—	$y^- = y^- + b(x^-)^<$	$\uparrow c$	$b(x^-)^<$	—	—
4	—	—	$y^- = y^- + (cx)^>$	—	—	$(cx)^>$	cx	$\uparrow a$
5	—	—	—	$y^- = y^- + ax^-$	—	ax^-	$x^<$	—
6	—	—	$y^- = y^- + ex^<$	—	—	$ex^<$	—	$\uparrow e$
7	—	—	$\downarrow y^-$	$y = c(x^-)^<$	—	—	$(x^-)^<$	—
8	—	—	$y = y + (ex^-)^<$	—	$\uparrow d$	ex^-	$(ex^-)^>$	—
9	$\uparrow$ new x	next x^-	—	$y = y + dx^-$	—	—	dx^-	—

See Table A. The symbols $\uparrow, \downarrow, >$, and $<$ correspond to the vector operations load, store, shift right one element, and shift left one element, respectively. x^- and y^- are subvectors preceding x and y.

APPENDIX B. FACTOR AND FORWARD SUBSTITUTION

Vector registers V0–V3 are assumed to contain c, a, b, and y loaded backward from memory. A check mark indicates a comparable instruction for the lower half of the matrix. Left column is cycle number.

1 Loop 1 S2   S2 * FS3 y_{i-1}/a_{i-1}
              ✓
        S3   S1 * FS3 b_{i-1}/a_{i-1}
              ✓
        S1   V0, A1   Load c_i from V0
              ✓
        A1   A1−1 $i = i − 1$

48   T. L. Jordan

8		V3, A2	S2 ✓	Store y_{i-1}/a_{i-1} in V3
		V2, A2	S3 ✓	Store b_{i-1}/a_{i-1} in V2
12		S3	S1 * FS3 ✓	$c_i b_{i-1}/a_{i-1}$
		S2	S1 * FS2 ✓	$c_i y_{i-1}/a_{i-1}$
		S1	T.A ✓	Get a_i from T
		A2	A2–1	$(i-1) = (i-1) - 1$
19		S3	S1–FS3 ✓	$a_i - c_i b_{i-1}/a_{i-1}$
		T.Y	S0	Get Y_i from S0 (upper half only).
		S1	T.Y	
		S2	S1–FS2 ✓	$y_i - c_i y_{i-1}/a_{i-1}$
25		S3	/H3 ✓	$1/a_i$
		S_1	V1, A1 ✓	Load a_i from V1
		S0	V3, A1	Load y_i from V3
		A0	A1	i to A0 for exit test
		T.A.	S1 ✓	Save y_i
		S1	V2, A1 ✓	Load b_i from V2
39		JAN ⋮	LOOP1	

APPENDIX C. BACKWARD SUBSTITUTION

Vector registers V0–V1 are assumed to contain b and y loaded forward from memory. A check mark indicates a comparable instruction from the lower half of the matrix.

1	Loop 2	S3	S2 * FS1 ✓	$b_i y_{i+1}$
		S2	V1, A1 ✓	Load y_i
		V0, A2	S1 ✓	Store y_{i+1}
		A1	A1–1	$i = i - i$

8	S1	S2–FS3 ✓	$y_i = y_i - b_i y_{i+1}$
	S2	V0, A1 ✓	Load b_{i-1}
	A2	A2–1	
	unused		
	unused		
14	S3	S2 ∗ F1	Start next cycle(s)

for as many times as one chooses to unfold.

APPENDIX D. FACTORIZATION ONLY

Vector registers V0–V3 are assumed to contain c_i, a_i, and b_i and $c_i b_{i-1}$ loaded backward from memory. A check mark indicates there is a comparable instruction from the lower half of the matrix.

1	Loop 3	S2	S2 ∗ FS3 ✓	$c_i b_{i-1}/a_{i-1}$
		V1, A2	S3	Store $1/a_{i-1}$
		A1	A1–1	$i = i + 1$
		A2	A2–1	$(i - 1) = (i - 1) + 1$
		A0	A1	
8		S2	S1–FS2 ✓	$a_i - c_i b_{i-1}/a_{i-1}$
		0, A3	S3 ✓	Save $1/a_{i-1}$ in memory—a 2-cycle instruction
14		S3	/HS2 ✓	$1/a_{i-1}$
		S2	V3, A1 ✓	Load $c_i b_{i-1}$
		S1	V1, A1 ✓	Load a_i
		A3	A3–1	Bump address of a_i in memory
		JAM	LOOP3	
28		⋮		

Acknowledgments

This work has been made possible by the support and helpful suggestions of B. L. Buzbee. We are indebted to W. J. Worlton for an expanded analysis of Amdahl's (1967) early paper.

References

Amdahl, G. (1967). Validity of the single processor approach to achieving large scale computing capabilities. *AFIPS Conf. Proc., Spring Joint Comput. Conf.* **30**, 483–485.

Batcher, K. E. (1968). Sorting networks and their applications. *AFIPS Conf. Proc., Spring Joint Comput. Conf.* **32**, 307–314.

Baudet, G. M. (1978). The design and analysis of algorithms for asynchronous computers. Rep. CMU-CS-78-116, Carnegie-Mellon University, Pittsburgh, Pennsylvania.

Boris, J. (1976). Vectorized tridiagonal solvers. Rep. NRL-3408, Naval Research Laboratory, Washington, D.C.

Brown, B. (1980). A high performance scalar tridiagonal equation solver for the CRAY-1. Personal communication, Dept. of Nuclear Engineering, University of Michigan, Ann Arbor.

Calahan, D. A., Ames, W. G., and Sesek, E. J. (1979). A collection of equation solving codes for the CRAY-1. Rep. SEL 133, University of Michigan, Ann Arbor.

Dongarra, J. J., Bunch, J. R., Moler, C. B., and Stewart, G. W. (1979). "LINPACK User's Guide." SIAM, Philadelphia, Pennsylvania.

Eisenstadt, S. G. (1980). Efficient implementation of a class of preconditioned conjugate methods. Res. Rep. No. 185, Yale University, New Haven, Connecticut.

Fong, K. W., and Jordan, T. L. (1977). Some linear algebraic algorithms and their performance on CRAY-1. Rep. LA-6774, Los Alamos National Laboratory, Los Alamos, New Mexico.

Greenbaum, A., and Rodrigue, G. (1977). The incomplete Cholesky conjugate gradient method for the star (5-point) operator. Rep. UCID 17574, Lawrence Livermore National Laboratory, Livermore, California.

Gustavson, I. (1978). A class of first-order factorization methods. *BIT* **18**, 142–156.

Heller, D. (1978). A survey of parallel algorithms in numerical linear algebra. *SIAM Rev.* **20**, No. 4, 740–777.

Hestenes, M. R., and Stiefel, E. (1952). Methods of conjugate gradients for solving linear systems. *J. Res. Nat. Bur. Standards* **49**, No. 6, 409–436.

Jordan, T. L. (1979). A performance evaluation of linear algebra software in parallel architectures. Rep. LA-8078-MS, Los Alamos National Laboratory, Los Alamos, New Mexico.

Karush, J. I., Madsen, N. K., and Rodrigue, G. H. (1975). Matrix multiplication by diagonals on vector/parallel processors. Rep. UCID 16899, Lawrence Livermore National Laboratory, Livermore, California.

Lambiotte, J. J., and Voigt, R. G. (1974). The solution of tridiagonal linear systems on the CDC STAR-100 computer. ICASE Report, Institute for Computer Applications in Systems Engineering, NASA, Langley Research Center, Hampton, Virginia.

Lawson, C., Hanson, R., Kincaid, D., and Krogh, F. (1979). Basic linear algebra subprograms for Fortran usage. *ACM Trans. Math. Software* **5**, No. 3,

Madsen, N. K., and Rodrigue, G. H. (1976). A comparison of direct methods for tridiagonal systems on the CDC-STAR 100. Rep. UCRL-76993, Lawrence Livermore National Laboratory, Livermore, California.

Orbits, D. A., and Calahan, D. A. (1978). A CRAY-1 simulator and its application to the development of high performance codes. *Proc. 1978 LASL Workshop Vector Parallel Processors*, Rep. LA-7491C, Los Alamos National Laboratory, Los Alamos, New Mexico.

Sameh, A., and Brent, R. (1977). Solving triangular systems of equation. *SIAM J. Numer. Anal.* **14**, No. 6, 1101–1113.

Sameh, A., and Kuck, D. (1978). On stable linear system solvers. *J. Assoc. Comput. Mach.* **25**, No. 1, 81–91.

Sedgewick, R. (1978). Sorting on CRAY-1: An overview. IDA-CRD Log. No. 80392, Institute for Defense Analyses, Arlington, Virginia.

Stone, H. S. (1975). Sorting on STAR. ICASE Rep. 75-13, NASA, Langley Research Center, Hampton, Virginia.

Vectorizing the FFTs[*]

Paul N. Swarztrauber

National Center for Atmospheric Research[†]
Boulder, Colorado

I. INTRODUCTION

For any individual who has an interest in scientific computing, the fast Fourier transform (FFT) is probably the most well known of all algorithms. It is dramatically superior to the slow transform and has applications in virtually all areas of scientific computing. Some interesting historical notes on the algorithm are given in Cooley *et al.* (1967). The term FFT was originally applied to a specific algorithm for the rapid computation of the discrete complex Fourier transform; however, it has become almost a generic term that is applied to any one of a large number of algorithms that compute the complex as well as other Fourier transforms.

Many algorithms exist for a given Fourier transform, and when they are applied to a particular sequence, the result is of course the same. However, the algorithms differ substantially in the ways that intermediate results are computed and stored. It is these important differences that provide the algorithms with unique properties that make one or the other more attractive for a particular application. Indeed, one of the main purposes of this chapter

---

[*] This chapter was written while the author was visiting the Scientific Computing Division at the National Bureau of Standards.

[†] Sponsored by the National Science Foundation.

is to match applications with algorithms. There are many factors that influence the selection of an algorithm, including:

(a) The Fourier transform itself. Is the sequence real or complex? Is the sequence even or odd? Does it have any other symmetries?

(b) The degree to which the algorithm can be vectorized. The FFT is an interesting algorithm from this standpoint since its vectorization is nontrivial.

(c) The number of transforms. Are there many sequences? How many? Do they have the same length? Can they be transformed simultaneously?

(d) The ordering properties of the algorithm. Ordering can take both computing time and storage. It will be shown that ordering can be avoided in most applications.

(e) The storage requirements of the algorithm. Is storage limited? This is a particular concern for multiple transforms or multidimensional transforms.

(f) The availability of software. Many algorithms have not been implemented in software. Unless the user can afford to develop custom software, this may be the most important consideration.

Each of these topics is discussed in this chapter. For some applications the choice of algorithm is clear; possibly several algorithms are equally appropriate for others; and for still others, no single algorithm can satisfy all the requirements. As an example of the latter case, no algorithm can be done in place with ordered coefficients for arbitrary N, although some come close (Singleton, 1969). Nevertheless, in Section VII an attempt is made to recommend algorithms for applications. They are presented in a "tree" format which the reader descends by answering questions related to the application.

In order to understand and motivate the algorithms that will be recommended, it is helpful to understand the concept of splitting which is at the heart of all FFT algorithms. The concept is quite straightforward and is presented in the next section together with the definition of the discrete complex transform and its aliases.

Six algorithms are presented in Section III for the complex transform of a single sequence. Although vectorization is of primary interest, we also examine in some detail the other properties of ordering and storage. The algorithms are presented by example for the case $N = 8$. All other algorithms are related, either implicitly or explicitly, to the complex transform. Hence, for the most part, the properties of any algorithm are the same as the properties of the underlying complex transform.

Most of the algorithms for a single sequence are "pseudo" vectorizable in the sense that the vectors change length as the computation proceeds.

The minimum vector length is about $\sqrt{N}$, which for most present-day applications is sufficiently small to cause some deterioration in the performance of the algorithms on a vector computer. For example, any sequence with less than 4096 points would have a minimum length vector less than 64, which is optimal for the Cray-1. Larger vectors are possible if a number of sequences (say M sequences of length N) are transformed at the same time. Two algorithms for multiple transforms are given in Section IV. All vectors have length M in the first algorithm, which is quite satisfactory for most cases; however, it can be inefficient if M is small relative to N. The vectors vary in length in the second algorithm, but they are always longer than M and attain a sizable maximum length that is proportional to MN.

The other Fourier transforms that are presented in this chapter are simply special forms that a complex transform assumes when the sequence is symmetric; i.e., if the sequence is real, then the transform is conjugate symmetric, and the complex transform can be written in real trigonometric form. If additional symmetries are present, then other forms are created; i.e., if the sequence is real and even, then the complex transform can be written as the cosine transform. Although the real transform is a symmetric transform, it is sufficiently important that it is considered separately in Section V.

In Section VI we consider four symmetric transforms, namely, the sine, cosine, and two quarter-wave transforms. The quarter-wave transforms correspond to sine and cosine transforms with only odd wave numbers. They have important applications to elliptic partial differential equations that are subject to mixed boundary conditions. Indeed, these transforms have been implemented, together with the other transforms, in a software package called FFTPACK that was originally developed for elliptic partial differential equations but which has now become a stand alone package that can be used for any application that requires these discrete Fourier transforms. FFTPACK is described in Section VII.

II. PRELIMINARIES

A. *Definitions*

1. *The Aliased Discrete Complex Transforms*

The discrete complex periodic transform and its aliases are defined in this section. In addition, it is shown that an N-point transform can be computed in terms of two $N/2$-point transforms. This procedure is called the splitting algorithm and is presented in order to simplify and motivate much of the material in the later sections.

Given a complex sequence x_n for $n = 1, \ldots, N$, the discrete complex periodic Fourier transform, in aliased form, is given by

$$X_k = \frac{1}{N} \sum_{n=1}^{N} x_n \exp\left(-ikn\frac{2\pi}{N}\right) \tag{1}$$

with inverse

$$x_n = \sum_{k=1}^{N} X_k \exp\left(ikn\frac{2\pi}{N}\right). \tag{2}$$

The identification of these forms as aliased is significant. Although they are commonly encountered in the literature, they may not be useful in many applications. For example, consider the simple problem of fitting a curve $x(\theta)$ to constant data $x_n = c$. By substituting this sequence into (1) we find that X_k is zero except for $X_N = c$. Hence we obtain $x(\theta) = ce^{iN\theta}$. This result is not intuitive since we would expect the result $x(\theta) = c$. The reason for this discrepancy is simply that any discrete set of data points has an infinite number of alternative characterizations, or *aliases*, in terms of continuous functions; i.e., for any integer m the data $x_n = c$ have the aliases $ce^{imN\theta}$, which pass identically through the data at $\theta_n = n(2\pi/N)$. This apparent ambiguity is easily eliminated by specifying that x_n is to be represented in terms of complex exponentials with the smallest possible wave numbers; i.e., instead of expanding x_n in terms of $e^{ik\theta}$ for $k = 1, \ldots, N$, we select $k = -(N/2) + 1, \ldots, N/2$ if N is even and $k = -(N-1)/2, \ldots, (N-1)/2$ if N is odd. This leads to unaliased forms of (1) and (2).

2. The Discrete Complex Transform

Given a complex sequence x_n for $n = 0, \ldots, N - 1$, the discrete complex periodic Fourier transform is given by

$$X_k = \frac{1}{N} \sum_{n=0}^{N-1} x_n \exp\left(-ikn\frac{2\pi}{N}\right). \tag{3}$$

If N is even, then the inverse is given by

$$x_n = \sum_{k=-(N/2)+1}^{N/2} X_k \exp\left(ikn\frac{2\pi}{N}\right), \tag{4}$$

and if N is odd, the inverse is given by

$$x_n = \sum_{k=-(N-1)/2}^{(N-1)/2} X_k \exp\left(ikn\frac{2\pi}{N}\right). \tag{5}$$

The coefficients X_k and data x_n, computed either from the aliased or unaliased forms, are equivalent modulo N; i.e., X_{-k} computed from (3) is equal to X_{N-k} computed from (1). Therefore, either form can be used to compute X_k. The difference becomes important only in the context of a particular application when the continuous form of the transform is used.

B. *The Splitting Algorithms*

1. *The Transform of an N-point Sequence in Terms of Two N/2-Point Sequences*

The concept of splitting is fundamental to all of the FFT algorithms. It is quite straightforward and our presentation will follow that given by Cochran *et al.* (1967) and Jenkins and Watts (1968). Although the aliased forms (1) and (2) are not suitable for most applications, they are nevertheless quite attractive for expository purposes since separate presentations for even and odd N are not required. This will be apparent in the following development of the splitting algorithm which provides a method for computing a transform of length N in terms of two transforms of length $N/2$. This approach is motivated by the fact that it will reduce the amount of computation by almost a factor of two.

Assume the sequence x_n is given and we wish to compute the forward transform X_k using (1). If for the moment we assume that N is even, then we can begin by splitting x_n into two $N/2$-point sequences consisting of the even and odd elements of x_n; i.e., we define

$$y_n = x_{2n}, \qquad n = 1, \ldots, N/2, \tag{6}$$

$$z_n = x_{2n-1}, \qquad n = 1, \ldots, N/2. \tag{7}$$

By replacing N by $N/2$ and $\exp(i2\pi/N)$ by w in the definition of the forward transform (1), we obtain the transforms Y_k and Z_k of y_n and z_n, respectively. Without loss of generality we can ignore the scale factor of $1/N$ in (1):

$$Y_k = \sum_{n=1}^{N/2} y_n w^{-2kn}, \qquad k = 1, \ldots, N/2, \tag{8}$$

$$Z_k = \sum_{n=1}^{N/2} z_n w^{-2kn}, \qquad k = 1, \ldots, N/2. \tag{9}$$

The approach is to express the desired transform X_k in terms of Y_k and Z_k. To this end we split (1) into its even and odd terms:

$$X_k = \sum_{n=1}^{N/2} [x_{2n} w^{-k(2n)} + x_{2n-1} w^{-k(2n-1)}] \tag{10}$$

or

$$X_k = \sum_{n=1}^{N/2} y_n w^{-2kn} + w^k \sum_{n=1}^{N/2} z_n w^{-2kn}. \tag{11}$$

From (8) and (9)

$$X_k = Y_k + w^k Z_k, \quad k = 1, \ldots, N/2. \tag{12}$$

For k greater than $N/2$ we have

$$X_{k+N/2} = Y_{k+N/2} + w^{k+N/2} Z_{k+N/2}. \tag{13}$$

Now

$$w^{k+N/2} = \exp\left[i\left(k + \frac{N}{2}\right)\frac{2\pi}{N}\right] = e^{i\pi} \exp\left(ik\frac{2\pi}{N}\right) = -w^k, \tag{14}$$

and Y_k and Z_k have period $N/2$, so that (13) can be written

$$X_{k+N/2} = Y_k - w^{k+N/2} Z_k. \tag{15}$$

The splitting algorithm can now be summarized; given the sequence x_n,

1. Define the sequences $y_n = x_{2n}$ and $z_n = x_{2n-1}$.
2. Compute the transforms Y_k and Z_k from (8) and (9).
3. Compute X_k from Y_k and Z_k using

$$X_k = Y_k + w^k Z_k, \quad k = 1, \ldots, N/2, \tag{16}$$

$$X_{k+N/2} = Y_k - w^k Z_k, \quad k = 1, \ldots, N/2. \tag{17}$$

This algorithm is preferred to computing X_k directly from (1) because it requires fewer operations. Step 2 above requires two $N/2$-point transforms or about $(N/2)^2 + (N/2)^2 = N^2/2$ multiplications and additions. Step 3 requires $N/2$ multiplications, $N/2$ additions, and $N/2$ subtractions; this becomes negligible compared to step 2 as N increases. Hence for large N the splitting algorithm is about twice as fast as computing X_k directly from (1).

The FFT results from the simple observation that the splitting algorithm can also be applied to the transforms Y_k and Z_k, i.e., if N is divisible by 4, then Y_k and Z_k can each be expressed in terms of two $N/4$-point transforms. If $N = 2^m$ for some integer m, then this process can be repeated m times until N transforms of length 1 are obtained. The algorithm of Cooley and Tukey (1965), which is based on this procedure, is described by example for $N = 2^3 = 8$ in the next section.

2. The Transform of an N-Point Sequence in Terms of Three N/3-Point Sequences

The generalization of the splitting algorithm to any composite N is straightforward. For example, assume that N is divisible by 3; then the splitting algorithm begins by defining the sequences

$$u_n = x_{3n-2}, \tag{18}$$

$$y_n = x_{3n-1}, \tag{19}$$

$$z_n = x_{3n}, \tag{20}$$

for $n = 1, \ldots, N/3$. These sequences have the transforms

$$U_k = \sum_{n=1}^{N/3} u_n w^{-3kn}, \tag{21}$$

$$Y_k = \sum_{n=1}^{N/3} y_n w^{-3kn}, \tag{22}$$

$$Z_k = \sum_{n=1}^{N/3} z_n w^{-3kn}. \tag{23}$$

Like the splitting algorithm for N divisible by 2, the X_k can be expressed in terms of U_k, Y_k, and Z_k:

$$X_k = w^{2k} U_k + w^k Y_k + Z_k; \tag{24}$$

$$X_{k+N/3} = w^{2(k+N/3)} U_k + w^{(k+N/3)} Y_k + Z_k; \tag{25}$$

$$X_{k+2N/3} = w^{2(k+2N/3)} U_k + w^{(k+2N/3)} Y_k + Z_k. \tag{26}$$

In general, if N is divisible by p, then X_k can be expressed in terms of p transforms of length N/p. The efficiency of the splitting decreases as p increases relative to N and is not possible if N is prime. Therefore it is important to choose a highly composite N whenever possible.

III. THE COMPLEX FFT ALGORITHMS

A. The Cooley–Tukey Algorithm

In this section, six forward FFT algorithms are presented. The results of each of these algorithms, applied to a particular sequence, is of course the same, but they differ substantially in the way that intermediate results are computed and stored. It is these important differences that provide the algorithms with unique properties that make one or the other more attractive for a particular application. The algorithms are presented by example for the case $N = 8$; however, they are valid for arbitrary composite N by using the generalized splitting algorithm given in the preceding section. Only the forward transforms are presented since the inverse transforms can be obtained simply by replacing w by w^{-1}.

The algorithm of Cooley and Tukey (1965) is a direct result of the repeated application of the splitting algorithm. Their algorithm is described by example in Table I. Column 1 contains the original sequence and column 7 contains the discrete complex forward transform of column 1, multiplied by N. Column 2 contains the split of column 1 into 2 sequences containing the even and odd elements of the original sequence. Column 3 contains four 2-point sequences that correspond to splits of the sequences in column 2. Finally, column 4 contains eight 1-point sequences that correspond to the splits of the sequences in column 3. From (1), the forward transform of a 1-point sequence is itself; therefore column 4 also corresponds to the transforms of eight 1-point sequences.

The combination phase begins in column 5, where the 1-point transforms in column 4 are combined into 2-point transforms in column 5. This combination continues, using step 3 of the splitting algorithm (Section II) until the transform of the original sequence is obtained in column 7. As an example, consider the calculation of the 4-point transform at the top of column 6

TABLE I

THE COOLEY-TUKEY FFT FOR $N = 8$

1	2	3	4	5	6	7
—	—	—	0.07	—	—	—
—	—	0.07	—	0.25	—	—
—	—	0.32	0.32	0.39	—	—
—	0.07	—	—	—	$-0.62 + 0.25i$	—
0.07	0.91	—	0.91	—	$0.81 + 0.00i$	$-0.05 - 0.10i$
0.40	0.32	0.91	—	-0.62	$-0.62 - 0.25i$	$-0.03 + 0.81i$
0.91	0.29	0.29	0.29	1.20	$1.59 + 0.00i$	$1.19 - 0.42i$
0.18	—	—	—	—	—	$0.30 + 0.00i$
0.32	—	—	—	—	—	$1.19 + 0.42i$
0.56	0.40	0.40	0.40	0.16	$0.57 + 0.16i$	$-0.03 - 0.81i$
0.29	0.18	0.56	—	0.96	$-0.03 + 0.00i$	$-0.05 - 0.10i$
0.75	0.56	—	0.56	—	$0.57 - 0.16i$	$3.48 + 0.00i$
—	0.75	—	—	—	$1.89 + 0.00i$	—
—	—	0.18	0.18	0.57	—	—
—	—	0.75	—	0.93	—	—
—	—	—	0.75	—	—	—

from the two transforms of length 2 at the top of column 5. Using the notation developed in Section II gives

$$Y_1 = -0.62, \quad Z_1 = 0.25, \tag{27}$$

$$Y_2 = 1.20, \quad Z_2 = 0.39. \tag{28}$$

The transform can be computed using (16) and (17) with $N = 4$:

$$X_1 = Y_1 + e^{i\pi/2}Z_1 = Y_1 + iZ_1 = -0.62 + 0.25i; \tag{29}$$

$$X_2 = Y_2 + e^{i\pi}Z_2 = Y_2 - Z_2 = 0.81 + 0.00i; \tag{30}$$

$$X_3 = Y_1 - e^{i\pi/2}Z_1 = Y_1 - iZ_1 = -0.62 - 0.25i; \tag{31}$$

$$X_4 = Y_2 - e^{i\pi}Z_2 = Y_2 + Z_2 = 1.59 + 0.00i. \tag{32}$$

All computation occurs in the combination phase that begins in column 5 and ends in column 7. For $N = 2^m$ there is a total of m such columns. Since each column requires $N/2$ multiplications and N additions, the total operation count is $(N/2)\log_2 N$ complex multiplications and $N \log_2 N$ complex additions. In practice, the ordering phase does contribute to the computational time required by the algorithm and cannot be ignored. Fortunately most problems can be formulated in such a way that ordering is not required. This topic is discussed further following the Gentleman–Sande FFT below. An important aspect of the Cooley–Tukey algorithm is that its combine phase can be implemented in place; i.e., no additional work storage is required. This is quite useful for applications in which storage is limited, including very large sequences, multiple sequences, and transforms on small computers.

As mentioned earlier, the Cooley–Tukey algorithm is "pseudo vectorizable" in the sense that the length of the vectors gets smaller as the computation proceeds. This can be seen at a glance from Table I, where the sequences, or vectors, have length $N/2 = 4$ in column 2 and proceed down to a length of 1 in column 4. This results in a significant increase in computing time when compared with algorithms such as matrix multiplication, which can be fully vectorized.

The combine phase is implemented in a program with two nested loops. When computing column 5 the outer loop is executed 4 times and the inner loop is executed once (two elements are computed in each pass of the inner loop). The situation is quite different when column 7 is computed and the outer loop is executed once and the inner loop is executed 4 times. The calculation of column 7 is more efficient than column 5 since the inner loop or vector lengths are longer. This calculation can be made more efficient by a simple but important modification to the program. The code that corresponds to the nested loops is duplicated but with the order of the loops inverted.

An IF test is made at run time and the code with the longest inner loop is selected to compute the column. The length of the inner loop still varies from column to column, but its minimum is about $\sqrt{N}$, which can be a substantial improvement over a minimum length of 1. The details are given by Temperton (1979a,b). Up to a 40% reduction in computing time on the Cray-1 has been obtained by using this technique.

B. *The Pease Algorithm*

The next two FFT algorithms differ from the Cooley–Tukey algorithm only in the way that intermediate results are stored. Consider the algorithm developed by Pease (1968) that is described by example in Table II. By comparison with Table I it can be seen that the ordering phase in columns 1–4 is the same; however, the results are stored elementwise in the combine phase, columns 5–7. The first sequence in column 5 of Table II contains the first element of each transform in column 5 of Table I and the second sequence contains the second element of each transform. Columns 6 and 7 are organized in the same manner.

The Pease algorithm is particularly suited to vectorization. Although four transforms are computed in column 6, they can be computed pairwise in a single loop. Similar results hold for the other columns, and hence the length of the vectors remain constant at $N/2 = 4$ throughout the calculation. In

TABLE II

THE PEASE FFT FOR $N = 8$

1	2	3	4	5	6	7
—	—	—	0.07	—	—	—
—	—	0.07	—	—	$-0.62 + 0.25i$	—
—	—	0.32	0.32	—	$0.57 + 0.16i$	—
—	0.07	—	—	0.25	—	—
0.07	0.91	—	0.91	-0.62	—	$-0.05 - 0.10i$
0.40	0.32	0.91	—	0.16	$0.81 + 0.00i$	$-0.03 + 0.81i$
0.91	0.29	0.29	0.29	0.57	$-0.03 + 0.00i$	$1.19 - 0.42i$
0.18	—	—	—	—	—	$0.30 + 0.00i$
0.32	—	—	—	—	—	$1.19 + 0.42i$
0.56	0.40	0.40	0.40	0.39	$-0.62 - 0.25i$	$-0.03 - 0.81i$
0.29	0.18	0.56	—	1.20	$0.57 - 0.16i$	$-0.05 - 0.10i$
0.75	0.56	—	0.56	0.96	—	$3.48 + 0.00i$
—	0.75	—	—	0.93	—	—
—	—	0.18	0.18	—	$1.59 + 0.00i$	—
—	—	0.75	—	—	$1.89 + 0.00i$	—
—	—	—	0.75	—	—	—

general, if N has the factorization $N = p_1 p_2 \cdots p_m$, then the length of the vector at the ith column is N/p_i. A disadvantage of the Pease algorithm is that it requires more storage than any of the other algorithms. The algorithm cannot be implemented in place and it requires a separate array of length N/p_i for each column. The storage requirement is proportional to $N \log N$. Nevertheless, it may be most appropriate algorithm for certain applications that are discussed in Section IV.

C. The Transposed Stockham Algorithm

The third algorithm, presented in Table III, was presented by Cochran et al. (1967). It has the decided advantage that the transform is automatically sorted, thereby eliminating a time-consuming part of the algorithm. Like the Pease algorithm, the transposed Stockham algorithm is also a reordering of the Cooley–Tukey algorithm. The columns in Table III are obtained from those in Table I by reordering the sequences. The grouping can be seen by comparing column 3 in Tables I and III. Note that the first two sequences in column 3 of Table III correspond to the first transforms obtained from the two larger sequences in column 2 and that the last two sequences in column 3 correspond to the last two sequences obtained from the two larger sequences in column 2.

TABLE III

The Transposed Stockham FFT for $N = 8$

1	2	3	4	5	6	7
—	—	—	0.07	—	—	—
—	—	0.07	—	0.25	—	—
—	—	0.32	0.40	0.39	—	—
—	0.07	—	—	—	$-0.62 + 0.25i$	—
0.07	0.91	—	0.91	—	$0.81 + 0.00i$	$-0.05 - 0.10i$
0.40	0.32	0.40	—	0.16	$-0.62 - 0.25i$	$-0.03 + 0.81i$
0.91	0.29	0.56	0.18	0.96	$1.59 + 0.00i$	$1.19 - 0.42i$
0.18	—	—	—	—	—	$0.30 + 0.00i$
0.32	—	—	—	—	—	$1.19 + 0.42i$
0.56	0.40	0.91	0.32	-0.62	$0.57 + 0.16i$	$-0.03 - 0.81i$
0.29	1.18	0.29	—	1.20	$-0.03 + 0.00i$	$-0.05 - 0.10i$
0.75	0.56	—	0.56	—	$0.57 - 0.16i$	$3.48 + 0.00i$
—	0.75	—	—	—	$1.89 + 0.00i$	—
—	—	0.18	0.29	0.57	—	—
—	—	0.75	—	0.93	—	—
—	—	—	0.75	—	—	—

The most noticeable aspect of this transform is that the order of the elements in column 4 is the same as the order of the elements in column 1 and that therefore the ordering phase of the algorithm does not have to be implemented and the algorithm begins with the combination phase in column 5. This is an important algorithm for applications that require the transform to be ordered or for use in general-purpose software when the ordering requirements cannot be anticipated. A disadvantage of the algorithm is that it requires an additional work array since it cannot be implemented in place. Its vectorization characteristics are the same as the Cooley–Tukey algorithm with a minimum vector length of about $\sqrt{N}$. Temperton (1977) has examined the autosort algorithm in great detail.

D. *The Gentleman–Sande Algorithm*

There are two ways to obtain an inverse transform. The first is simply to replace w by w^{-1} and the second is to invert the order of the operations in any of the algorithms given above. The latter process requires the inverse of the splitting algorithm. From Step 3 in the previous section we obtain

$$Y_k = (X_k + X_{k+N/2})/2, \tag{33}$$

$$Z_k = w^k(X_k - X_{k+N/2})/2. \tag{34}$$

An inverse can now be obtained by computing the columns in reverse order beginning with column 7. If we apply both of these methods for inverting the transform, then we obtain a forward transform that is quite distinct from those given above. The remaining three algorithms for the forward transform were obtained by using this double inversion procedure. The algorithm given by Gentleman and Sande (1966) is presented in Table IV and is obtained by applying the double inversion process to the Cooley–Tukey algorithm. The division by 2 in (33) and (34) is not performed so that the scaling will be the same for all six methods.

The vectorization and storage characteristics for this algorithm are the same as for the Cooley–Tukey algorithm. An important aspect of this algorithm is that the order and combine phase are reversed compared with the Cooley–Tukey algorithm. For many applications the Gentleman–Sande and inverse Cooley–Tukey algorithms can be used to eliminate ordering altogether. Since their ordering phases are just their mutual inverses, they can both be eliminated if the unordered transform is not important to the user.

For example, consider the following problem: suppose we wish to approximate the derivative of a periodic function $x(\theta)$ from its tabulation x_n.

TABLE IV

The Gentleman–Sande FFT for $N = 8$

1	2	3	4	5	6	7
—	—	—	$-0.05 - 0.10i$	—	—	—
—	—	$0.62 + 0.26i$	—	$-0.05 - 0.10i$	—	—
—	—	$0.57 + 0.16i$	$1.19 + 0.42i$	$1.19 + 0.42i$	—	—
—	—	—	—	—	$-0.05 + 0.10i$	—
0.07	$-0.18 - 0.18i$	—	—	—	$1.19 - 0.42i$	$-0.05 - 0.10i$
0.40	$0.00 - 0.16i$	—	$1.19 - 0.42i$	$1.19 - 0.42i$	$1.19 + 0.42i$	$-0.03 + 0.81i$
0.91	$-0.44 + 0.44i$	$-0.62 + 0.26i$	—	—	$-0.05 + 0.10i$	$1.19 - 0.42i$
0.18	$0.57 + 0.00i$	$0.57 - 0.16i$	$-0.05 + 0.10i$	$-0.05 + 0.10i$	—	$0.30 + 0.00i$
0.32	—	—	—	—	—	$1.19 + 0.42i$
0.56	$0.39 + 0.00i$	$0.00 - 0.81i$	$-0.03 + 0.81i$	$-0.03 + 0.81i$	$-0.03 + 0.81i$	$-0.03 - 0.81i$
0.29	$0.96 + 0.00i$	$-0.03 + 0.00i$	—	$-0.03 - 0.81i$	$0.30 + 0.00i$	$-0.05 - 0.10i$
0.75	$1.20 + 0.00i$	—	$-0.03 - 0.81i$	—	$-0.03 - 0.81i$	$3.48 + 0.00i$
—	$0.93 + 0.00i$	—	—	—	$3.48 + 0.00i$	—
—	—	$1.59 + 0.00i$	$0.30 + 0.00i$	$0.30 + 0.00i$	—	—
—	—	$1.89 + 0.00i$	—	$3.48 + 0.00i$	—	—
—	—	—	$3.48 + 0.00i$	—	—	—

Using the combine phase of the Gentleman–Sande algorithm we can compute the unordered Fourier coefficients $X_{p(k)}$, where $p(k)$ is a known permutation of the integers 1–N. The Fourier coefficients of the derivative can then be computed by multiplying each $X_{p(k)}$ by $ip(k)$. The derivative can then be tabulated using the inverse combine phase of the Cooley–Tukey algorithm, i.e., with w replaced by w^{-1}. This approach can be used for any problem in which the solution is desired in physical rather than spectral space and the Fourier coefficients are "invisible" to the user. Most problems, including the solution of partial differential equations, fall into this category. By operating on the unordered coefficients in spectral space, both ordering phases can be eliminated. This approach requires less storage than the Stockham autosort algorithm since the combine phase of both the Cooley–Tukey and Gentleman–Sande algorithms can be performed in place.

E. *The Transposed Pease Algorithm*

The next algorithm is obtained by applying the double inversion procedure to the Pease algorithm. It is presented in Table V, where, following the notation developed by Temperton (1977), it is called the "transposed" Pease algorithm. Like the Pease algorithm, the transposed Pease algorithm is highly vectorizable, but it requires more storage than the other algorithms. Like the Cooley–Tukey and Gentleman–Sande algorithms, the ordering phases in the Pease and transposed Pease algorithms can be eliminated when solving problems in which the Fourier coefficients are "invisible" to the user.

F. *The Stockham Algorithm*

The final algorithm of this section is shown in Table VI. It is presented by Cochran *et al.* (1967), who credit Stockham as the originator. It can be obtained from the transposed Stockham algorithm using the double inversion process. This autosort algorithm was originally presented for $N = 2^m$ and Glassman (1970) made the important generalization to arbitrary N. The properties of the Stockham and transposed Stockham are comparable and they seem equally appropriate where applicable.

This completes the presentation of the discrete transforms for a single complex sequence. However, there remain a number of other transforms for multiple and symmetric sequences that will be considered in the following sections.

TABLE V

THE TRANSPOSED PEASE FFT FOR $N = 8$

1	2	3	4	5	6	7
—	−0.18 − 0.18i	—	—	—	—	—
—	0.39 + 0.00i	—	—	—	—	—
—	—	—	−0.05 − 0.10i	−0.05 − 0.10i	—	—
—	—	0.62 + 0.26i	—	1.19 + 0.42i	—	—
0.07	—	−0.62 + 0.26i	1.19 + 0.42i	—	−0.05 − 0.10i	−0.05 − 0.10i
0.40	0.00 − 0.16i	0.00 − 0.81i	1.19 − 0.42i	—	1.19 − 0.42i	−0.03 + 0.81i
0.91	0.96 + 0.00i	1.59 + 0.00i	—	1.19 − 0.42i	1.19 + 0.42i	1.19 − 0.42i
0.18	—	—	−0.05 + 0.10i	−0.05 + 0.10i	−0.05 + 0.10i	0.30 + 0.00i
0.32	—	—	—	—	—	1.19 + 0.42i
0.56	−0.44 + 0.44i	0.57 + 0.16i	−0.03 + 0.81i	−0.03 + 0.81i	−0.03 + 0.81i	−0.03 − 0.81i
0.29	1.20 + 0.00i	0.57 − 0.16i	—	−0.03 − 0.81i	0.30 + 0.00i	−0.05 − 0.10i
0.75	—	−0.03 + 0.00i	−0.03 − 0.81i	—	−0.03 − 0.81i	3.48 + 0.00i
—	—	1.89 + 0.00i	—	—	3.48 + 0.00i	—
—	0.57 + 0.00i	—	0.30 + 0.00i	0.30 + 0.00i	—	—
—	0.93 + 0.00i	—	—	3.48 + 0.00i	—	—
—	—	—	3.48 + 0.00i	—	—	—

TABLE VI

The Stockham FFT for $N = 8$

1	2	3	4	5	6	7
—	—	—	—	—	—	—
—	—	—	—	—	—	—
—	—	$0.62 + 0.26i$	$-0.05 - 0.10i$	—	—	—
—	—	$0.57 + 0.16i$	$-0.03 + 0.81i$	$-0.05 - 0.10i$	—	—
—	$-0.18 - 0.18i$	—	—	$1.19 + 0.42i$	—	—
—	$0.00 - 0.16i$	—	$1.19 - 0.42i$	—	$-0.05 - 0.10i$	$-0.05 - 0.10i$
—	$-0.44 + 0.44i$	$0.00 - 0.81i$	$0.30 + 0.00i$	$-0.03 + 0.81i$	$1.19 - 0.42i$	$-0.03 + 0.81i$
—	$0.57 + 0.00i$	$-0.03 + 0.00i$	—	$-0.03 - 0.81i$	$1.19 + 0.42i$	$1.19 - 0.42i$
0.07	—	—	—	—	$-0.05 - 0.10i$	$0.30 + 0.00i$
0.40	—	—	$1.19 + 0.42i$	$1.19 - 0.42i$	—	$1.19 + 0.42i$
0.91	$0.39 + 0.00i$	$-0.62 + 0.26i$	—	$-0.05 + 0.10i$	$-0.03 + 0.81i$	$-0.03 - 0.81i$
0.18	$0.96 + 0.00i$	$0.57 - 0.16i$	$-0.03 - 0.81i$	—	$0.30 + 0.00i$	$-0.05 - 0.10i$
0.32	$1.20 + 0.00i$	—	—	$0.30 + 0.00i$	$-0.03 - 0.81i$	$3.48 + 0.00i$
0.56	$0.93 + 0.00i$	—	$-0.05 - 0.10i$	$3.48 + 0.00i$	$3.48 + 0.00i$	—
0.29	—	$1.59 + 0.00i$	—	—	—	—
0.75	—	$1.89 + 0.00i$	$3.48 + 0.00i$	—	—	—
—	—	—	—	—	—	—
—	—	—	—	—	—	—

IV. VECTORIZING MULTIPLE TRANSFORMS

A. *Vectors with Length Equal to the Number of Transforms*

As previously noted, most of the FFT algorithms are pseudovectorizable in the sense that the length of the vectors vary during the course of the calculation with a minimum length of approximately $\sqrt{N}$. The Pease algorithm provides some improvement from this point of view since all vectors have constant length $N/2$ when $N = 2^m$ and they are relatively constant when N is composite.

However, longer vectors are possible when multiple sequences are simultaneously available for transforming. Problems of this type are frequently encountered when solving elliptic partial differential equations (Hockney, 1969; Swarztrauber, 1977) or in any multidimensional spectral computation. Storage is also likely to be limited for these problems, and the in-place algorithms become more attractive. If M sequences are available, then the most straightforward approach is to modify any of the algorithms given in the previous section so that each operation is applied to all sequences with the result that all vectors have length M throughout the calculation. For M larger than $N/2$ some improvement can be expected over M separate transforms of length N using the Pease transform. Temperton (1979a,b) analyzes this approach in detail. He has implemented the method in both FORTRAN and Cray assembly language (CAL) programs. For $M = 64$ and $N = 1024$ the time per transform is less than half a millisecond. For many applications this approach is quite satisfactory; however, it can be quite inefficient if M is small relative to N.

B. *Vectors with Larger Lengths*

Another approach that is efficient even if M is small compared to N was presented by Korn and Lambiotte (1979). Additional results are presented by Fornberg (1981). If we examine Table I we note, for example, that column 3 consists of four sequences whose transforms are computed in column 5; i.e., columns 3-5 correspond to the solution of a multiple transform problem in which $N = 2$ and $M = 4$. Also columns 2-6 also correspond to a multiple transform problem in which $N = 4$ and $M = 2$. Hence the multiple transform can be implemented by a truncation of the FFT for a single sequence of length MN.

If $N = 2^m$, then the ith column will contain $M2^{i-1}$ sequences of length $N/2^{i-1}$. For M less than $N/4$ the minimum length is about $\sqrt{MN}$, and for M

greater than $N/4$ the minimum length is $2M$. In either case the maximum length is $MN/2$. For $M = 1$ the algorithm automatically reduces to the method for a single N-point sequence. Therefore this approach is always at least as efficient as the repeated use of the single transform to solve the multiple transform problem. On the Cray-1 there is not much gain for vector lengths greater than 64 and therefore the two methods are likely comparable if M and N are greater than or equal to 64. Long vectors can also be maintained for multidimensional transforms without an explicit transposition of the data (see Wang, 1980).

Any of the FFTs discussed in the previous section can be truncated for the multiple transform problem. Whenever possible, the problem should be formulated so that ordering is not required since the autosort algorithms require MN additional storage locations. In-place ordering is possible if $N = 2^m$. If N has a number of squared factors, then very little additional storage is required (Singleton, 1969). In either case, ordering will increase computing time, particularly on a vector computer.

For multidimensional partial differential equations, ordering is not required and storage is often limited. The truncated Gentleman–Sande and inverse Cooley–Tukey algorithms are attractive for these applications since they can be performed in place.

V. TRANSFORMING REAL SEQUENCES

A. *Preliminaries*

1. *The Real Periodic Transform*

In this section and the next we examine the discrete transform of symmetric sequences. Because of its importance, we devote this section to the transform of real sequences and defer discussion of the remaining symmetric transforms to the following section. If x_n is real, then the complex discrete Fourier transform is conjugate symmetric; i.e., $X_k = \bar{X}_{N-k}$. The complex transform can be modified to take advantage of this with the result that the computation can be reduced almost in half. Indeed, it is this reduction in computing time that has motivated the development of all the symmetric transforms.

The complex transform (1) can be used for any application that requires the discrete transform, including the transform of symmetric sequences. However, when the sequence x_n has one or more symmetries, the complex transform can be rewritten in a form that makes these symmetries quite evident. When the sequence is real, the complex transform can be rewritten in a real trigonometric form which is often preferred to the complex form.

The transform for real sequences is given below followed by a description of several efficient algorithms.

Given the sequence x_n for $n = 0, \ldots, N - 1$, if N is an even integer, then the discrete real periodic transform is given by

$$a_k = \frac{2}{N} \sum_{n=0}^{N-1} x_n \cos kn \frac{2\pi}{N}, \qquad k = 0, \ldots, N/2, \tag{35}$$

$$b_k = \frac{2}{N} \sum_{n=1}^{N-1} x_n \sin kn \frac{2\pi}{N}, \qquad k = 1, \ldots, N/2 - 1, \tag{36}$$

with inverse

$$x_n = \frac{1}{2} a_0 + \sum_{k=1}^{N/2-1} \left(a_k \cos kn \frac{2\pi}{N} + b_k \sin kn \frac{2\pi}{N} \right) + \frac{1}{2} a_{N/2}(-1)^n \tag{37}$$

If N is an odd integer, then the discrete real periodic transform is given by

$$a_k = \frac{2}{N} \sum_{n=0}^{N-1} x_n \cos kn \frac{2\pi}{N}, \qquad k = 0, \ldots, (N-1)/2, \tag{38}$$

$$b_k = \frac{2}{N} \sum_{n=1}^{N-1} x_n \sin kn \frac{2\pi}{N}, \qquad k = 1, \ldots, (N-1)/2, \tag{39}$$

with inverse

$$x_n = \frac{1}{2} a_0 + \sum_{k=1}^{(N-1)/2} \left(a_k \cos kn \frac{2\pi}{N} + b_k \sin kn \frac{2\pi}{N} \right). \tag{40}$$

2. The Real Transform in Terms of the Complex Transform

The FFT algorithms for the real transform are based on efficient variants of the complex transform. Therefore we proceed by first showing that the real transform can be determined from the complex transform. That is, given real x_n, we shall show that a_k and b_k can be computed from X_k, which is determined by (1). From (4) and the fact that $X_k = \bar{X}_{N-k}$,

$$x_n = X_0 + 2 \sum_{k=1}^{N/2-1} \mathrm{Re}\left(X_k \exp\left(ikn \frac{2\pi}{N}\right) \right) + (-1)^n X_{N/2}, \tag{41}$$

or

$$x_n = X_0 + 2 \sum_{k=1}^{N/2-1} \left(\mathrm{Re}\, X_k \cos kn \frac{2\pi}{N} - \mathrm{Im}\, X_k \sin kn \frac{2\pi}{N} \right) + X_{N/2}(-1)^n. \tag{42}$$

If N is odd, then

$$x_n = X_0 + 2 \sum_{k=1}^{(N-1)/2} \left(\operatorname{Re} X_k \cos kn \frac{2\pi}{N} - \operatorname{Im} X_k \sin kn \frac{2\pi}{N} \right). \quad (43)$$

By comparing (42) and (43) with (37) and (40) we see that

$$a_0 = 2X_0, \qquad a_k = 2 \operatorname{Re} X_k, \qquad b_k = -2 \operatorname{Im} X_k, \quad (44)$$

and if N is even, then $a_{N/2} = 2X_{N/2}$.

B. Vector Algorithms for the Real Transform

1. Transforming an Even Number of Sequences

Since the real trigonometric form can easily be determined from the complex transform, we can focus attention on the development of efficient algorithms for the complex transform of real sequences. Each of the complex algorithms given in Section III has a real counterpart. Theoretically the real algorithms should be twice as fast as the complex, but in practice this is not achieved on the Cray-1. This is due to the shorter vectors and some additional overhead that is required by the real transform.

There are currently three ways to compute real transforms; all of them are vectorizable, and each of them has its own unique properties. The first method combines two real sequences into a complex sequence. The complex transform of this sequence is then postprocessed to determine the transforms of the two real sequences. The details of this procedure are described in a number of places, including Brigham (1974). This method is quite satisfactory for applications that require an even number of transforms. The postprocessing requires on the order of N operations, and although it is asymptotically negligible, in practice it adds about 15% to the computing time.

2. Transforming a Sequence with an Even Number of Points

The second method can be used if N is even. We define a complex sequence c_n in terms of the real sequence x_n by setting $c_n = x_{2n} + ix_{2n-1}$. The complex transform of this sequence with $N/2$ elements is postprocessed to determine the transform of the original sequence x_n. The details of this procedure are also given in Brigham (1974).

Both these methods have the advantage that they can make use of existing software for the complex transform that minimizes the implementation effort. However, they both have nuisance requirements, namely, that there be an even number of sequences or that N be even. There is also the additional computation that is associated with the postprocessing.

3. Edson's Algorithm

The third method was published by Bergland (1968), who credits Edson as its originator. It is free from the nuisance requirements and the post-processing that is required by the previous methods. If we examine Table I, we note that the transform in column 7 has conjugate symmetry since the original sequence in column 1 is real. Indeed, all of the transforms in columns 5–7 have conjugate symmetry since they are the transforms of the corresponding real sequences in columns 1–3. For example, both transforms in column 6 satisfy $X_k = \overline{X}_{4-k}$. X_4 is real since $X_4 = \overline{X}_0 = \overline{X}_4$, where the last equality results from the fact that all sequences and transforms have period N, i.e., $X_k = X_{k+N}$. Therefore only half of each transform, or column, has to be computed, and the real transform can be derived from the complex transform by simply eliminating redundant calculations.

Six real algorithms can be developed by eliminating the redundant calculations in each of the six complex algorithms given in Section III. Just like its complex counterpart, each real algorithm has unique properties that make it appropriate for a particular application depending on whether it is a single or multiple transform problem and its vector, ordering, and storage requirements. Where storage is limited and ordering is not necessary, the real algorithm based on the Cooley–Tukey and inverse Gentleman–Sande algorithms would be appropriate. Where storage is not limited or ordering is necessary, Stockham's autosort algorithms would be appropriate. If storage is available and vectorization is important, then the Pease algorithm is appropriate. These recommendations would change for the problem of multiple transforms. The important point is that the real algorithm will have the same properties as its complex counterpart, and hence the appropriate real algorithm can be determined from the recommendations that have been given for the complex algorithms in the previous sections. More specific recommendations are presented in Section VII.

VI. THE SYMMETRIC TRANSFORMS

A. Preliminaries

In many applications the sequence x_n will not only be real, but will also have additional symmetries. For example, if x_n is an odd sequence, then it can be expanded in terms of a trigonometric series that only contains sine terms. Hence all the coefficients of the cosine terms in the trigonometric expansion are zero. We are therefore motivated to develop an algorithm for odd sequences that would not compute the coefficients of the cosine terms and be faster than the real periodic transform. Indeed, such algorithms do exist, as well as algorithms for other commonly encountered symmetries.

In this section we consider four symmetries, including both even and odd symmetries as well as the quarter-wave symmetries, which are not so common but are nevertheless important for a number of applications, including partial differential equations (Swarztrauber, 1977). Each of these algorithms contains a preprocessing phase that converts the symmetric sequence into a real periodic sequence. The results of the real transform are then postprocessed to obtain the symmetric transform. Hence there are six algorithms for each symmetric transform since any of the six real transforms can be used. However, the properties of the symmetric transform are not only determined by the properties of the corresponding real transform but also by the properties of their pre- and postprocessing phase.

In the previous section, it was shown that pre- and postprocessing for the real periodic transform could be eliminated if the transform was developed by modifying the complex transform. It seems possible that the real algorithm can also be modified to obtain the symmetric algorithms which would eliminate pre- and postprocessing. However, this approach appears to be more difficult that it was for the real transform, and such algorithms do not yet exist. Nevertheless, it is faster to use symmetric transforms than to transform the full real periodic sequence.

B. *The Sine Transform of Odd Sequences*

We begin with the transform for odd sequences. Given a real sequence x_n for $n = 1, \ldots, N - 1$, the discrete odd Fourier transform is given by

$$c_k = \frac{2}{N} \sum_{n=1}^{N-1} x_n \sin kn \frac{\pi}{N}, \qquad k = 1, \ldots, N - 1. \tag{45}$$

The inverse differs only by a multiplicative constant

$$x_n = \sum_{k=1}^{N-1} c_k \sin kn \frac{\pi}{N}, \qquad n = 1, \ldots, N - 1. \tag{46}$$

It is important to note that the sequence x_n is arbitrary. Just like the complex and real transforms, the symmetries are apparent only when the sequence is extended. If we compute x_n using (46) over an extended range of subscripts, then we can determine that the odd periodic sequence has the form $(0, x_1, x_2, \ldots, x_{N-1}, 0, -x_{N-1}, \ldots, -x_1)$. Therefore in order to determine the c_k using the real periodic transform, one would have to transform this $2N$-point sequence. Following Dollimore (1973), we shall show that the

c_k can be computed from a periodic transform of an N-point sequence. Given the x_n, the preprocessing consists of computing the sequence

$$d_n = \frac{1}{2}(x_n - x_{N-n}) + \sin n \frac{\pi}{N}(x_n + x_{N-n}), \qquad n = 1, \ldots, N-1. \quad (47)$$

If for the moment we assume that N is odd and substitute (46) into (47), we obtain

$$d_n = c_1 + \sum_{k=1}^{(N-1)/2} \left[(c_{2k+1} - c_{2k-1}) \cos kn \frac{2\pi}{N} + c_{2k} \sin kn \frac{2\pi}{N} \right]. \quad (48)$$

Therefore d_n is periodic and the real periodic transform in Section V can be used to compute a_k and b_k so that

$$d_n = \frac{1}{2} a_0 + \sum_{k=1}^{(N-1)/2} \left(a_k \cos kn \frac{2\pi}{N} + b_k \sin kn \frac{2\pi}{N} \right). \quad (49)$$

If we equate coefficients in (48) and (49) we obtain

$$c_{2k} = b_k, \quad (50)$$

$$c_1 = \tfrac{1}{2} a_0, \quad (51)$$

$$c_{2k+1} - c_{2k-1} = a_k. \quad (52)$$

A similar development is possible if N is even. The algorithm for computing the c_k can be summarized as follows:

1. Compute the sequence d_n from x_n using (47).
2. Compute a_k and b_k from the forward real transform of d_n.
3. Compute the c_{2k} from (50) and the c_{2k+1} from (52) but in recurrence form. Starting with $c_1 = \tfrac{1}{2} a_0$,

$$c_{2k+1} = c_{2k-1} + a_k. \quad (53)$$

Although this algorithm is satisfactory for most applications, the recurrence relation in step 3 is not efficiently vectorized. Although Stone (1973) has defined a parallel method for this recurrence that can be vectorized, like the FFT, it is not efficient since the vectors get small as the calculation proceeds. An alternative approach leads to a pre- and postprocessing algorithm that can be efficiently vectorized.

If we ignore the constant, then the sine transform (45) is its own inverse. Therefore, if we reverse the order of the calculations in the algorithm given above, then we also obtain a sine transform. The reverse of Dollimore's algorithm, which is therefore a sine transform, is due to Cooley et al. (1970). We assume that the c_k are given and that we wish to compute the x_n.

1. Compute a_k and b_k from $a_k = c_{2k}$ and $b_k = c_{2k+1} - c_{2k-1}$.
2. Compute the d_n from the inverse real periodic transform (37) or (40).
3. Compute x_n from the inverse of (47) that is given by

$$x_n = \frac{1}{2}(d_n - d_{N-n}) + \frac{1}{4 \sin n\pi/N}(d_n + d_{N-n}), \quad n = 1, \ldots, N-1. \quad (54)$$

Step 3 in Dollimore's algorithm is replaced by step 1 above which is fully vectorizable.

The division by $4 \sin n\pi/N$ in (54) is of interest. For large N and $n = 1$, the coefficient of the second term on the right of (54) will be about $N/4\pi$, which will result in a corresponding error growth. Although this division does not occur in Dollimore's algorithm, the solution of the recurrence relation can also result in an error that is proportional to N.

C. The Cosine Transform of Even Sequences

Consider now the transform of even sequences. Given a real sequence x_n for $n = 0, \ldots, N$, the discrete Fourier transform of an even sequence is given by

$$c_k = \frac{1}{N} x_0 + \frac{2}{N} \sum_{n=1}^{N-1} x_n \cos kn\frac{\pi}{N} + \frac{1}{N}(-1)^k x_N. \quad (55)$$

The inverse differs only by a multiplicative constant:

$$x_n = \frac{1}{2} c_0 + \sum_{k=1}^{N-1} c_k \cos kn\frac{\pi}{N} + \frac{1}{2}(-1)^n c_N. \quad (56)$$

It is important to note that the sequence x_n is arbitrary. Just like the complex and real transforms, the symmetries are apparent only when the sequence is extended. If we compute x_n using (56) over an extended range of subscripts, then we can determine that the even periodic sequence has the form $(x_0, x_1, \ldots, x_{N-1}, x_N, x_{N-1}, \ldots, x_1)$. Therefore, in order to determine the c_k using the real periodic transform, one would have to transform this $2N$-point sequence. Like the transform for odd sequences, the c_k can be computed using the real periodic transform with a pre- and postprocessing phase. The development is quite similar to that given for odd sequences and therefore we only state the algorithms. The first is due to Dollimore (1973).

1. Compute d_n from x_n using

$$d_n = \frac{1}{2}(x_n + x_{N-n}) - \sin n\frac{\pi}{N}(x_n - x_{N-n}), \quad n = 1, \ldots, N. \quad (57)$$

2. Compute a_k and b_k from the forward real transform of d_n using (35) and (37) or (38) and (39).

3. Compute c_i for some odd integer i using (55) in order to initialize the recurrence relation that is given below.

4. Compute c_k from $c_{2k} = a_k$ and the recurrence relation $c_{2k+1} = c_{2k-1} + b_k$ that was initialized in step 3.

Step 4 is not efficiently vectorized. Like the transform for odd sequences, the even transform is the inverse of itself to within a constant factor. The algorithm of Cooley et al. (1970), which is given below, is almost the inverse of Dollimore's algorithm. It provides an algorithm for the cosine transform that is efficiently vectorized.

1. Compute a_k and b_k from $a_k = c_{2k}$ and $b_k = c_{2k+1} - c_{2k-1}$.
2. Compute the d_n from the inverse real periodic transform (37) or (40).
3. For $n = 1, \ldots, N - 1$ compute x_n from the inverse of (57), which is given by

$$x_n = \frac{1}{2}(d_n + d_{N-n}) - \frac{1}{4 \sin(n\pi/N)}(d_n + d_{N-n}), \quad n = 1, \ldots, N - 1. \quad (58)$$

4. Compute x_0 and x_N from $x_0 = d_0 + \beta$ and $x_N = d_0 - \beta$, where

$$\beta = \sum_{k=0}^{N/2} c_{2k+1}. \quad (59)$$

Both the pre- and post processing are vectorizable on the Cray-1 and β can be computed at vector speed using the in-line FORTRAN function SSUM.

Next we consider the quarter-wave sine transform. It can be used to approximate functions that can be expanded in terms of a sine series with only odd wave numbers. These functions are zero at the left end of the interval and have a zero derivative at the right end. Each of the associated Legendre functions has either this form or one of the other symmetric forms that are given in this section. This transform can also be used to solve Poisson's equation when the solution is specified on the left boundary and the derivative is specified on the right boundary (Swarztrauber, 1977). Given the sequence x_n for $n = 1, \ldots, N$, the discrete odd quarter-wave Fourier transform is given by

$$c_k = \frac{2}{N} \sum_{n=1}^{N-1} x_n \sin(2k-1)n \frac{\pi}{2N} + \frac{1}{N}(-1)^{k+1} x_n, \quad k = 1, \ldots, N. \quad (60)$$

The inverse transform is given by

$$x_n = \sum_{k=1}^{N} c_k \sin(2k-1)n \frac{\pi}{2N}. \tag{61}$$

The sequence x_n for $n = 1, \ldots, N$ is arbitrary. Just like the complex and real transforms, the symmetries are apparent only when the sequence is extended. If we compute x_n using (61) over an extended range of subscripts, then we can determine that the real periodic sequence has the form $(0, x_1, \ldots, x_N, x_{N-1}, \ldots, x_1, 0, -x_1, \ldots, -x_N, -x_{N-1}, \ldots, -x_1)$. Therefore, in order to determine the c_k using the real periodic transform, one would have to transform this $4N$-point sequence.

As with the previous symmetric transforms the approach is to construct a periodic sequence d_n from the x_n. The coefficients a_k and b_k are computed by transforming d_n using the real periodic transform defined in Section V. The c_k are then determined from the a_k and b_k in the postprocessing phase. To this end we first compute the d_n from

$$d_n = (x_n + x_{N-n}) \cos n \frac{\pi}{2N} + (x_n - x_{N-n}) \sin n \frac{\pi}{2N}, \quad n = 1, \ldots, N. \tag{62}$$

If for the moment we assume that N is odd and substitute (61) into (62), then

$$d_n = c_1 + \sum_{k=1}^{(N-1)/2} \left[(c_{2k+1} - c_{2k}) \cos kn \frac{2\pi}{N} + (c_{2k+1} + c_{2k}) \sin kn \frac{2\pi}{N} \right]. \tag{63}$$

Therefore d_n is periodic and we can use the real periodic transform in Section V to compute a_k and b_k such that

$$d_n = \frac{1}{2} a_0 + \sum_{k=1}^{(N-1)/2} \left(a_k \cos kn \frac{2\pi}{N} + b_k \sin kn \frac{2\pi}{N} \right). \tag{64}$$

By setting the coefficients equal in (63) and (64)

$$c_1 = a_0, \tag{65}$$

$$c_{2k+1} - c_k = a_k, \tag{66}$$

$$c_{2k+1} + c_k = b_k. \tag{67}$$

These equations can be used to compute c_k; i.e., for $k = 1, \ldots, (N-1)/2$

$$c_{2k} = \tfrac{1}{2}(b_k - a_k), \tag{68}$$

$$c_{2k+1} = \tfrac{1}{2}(b_k + a_k). \tag{69}$$

A similar development is possible if N is even. The algorithm for computing the c_k can be summarized as follows:

1. Compute the sequences d_n from x_n using (62).
2. Compute a_k and b_k from the forward real transform of d_n using (35) and (36) or (38) and (39).
3. Compute c_k from a_k and b_k using (68) and (69).

The inverse transform (61) can be computed simply by reversing the order of the computations. To this end the inverse of (62) is

$$x_n = \frac{1}{2}\left[(d_n - d_{N-n})\cos n\frac{\pi}{2N} + (d_n + d_{N-n})\sin n\frac{\pi}{2N}\right]. \tag{70}$$

The fourth and last symmetric transform that we shall consider is the quarter-wave cosine transform. It can be used to approximate functions that can be expanded in terms of a cosine series with only odd wave numbers. These functions have a zero derivative at the left end of the interval and are zero at the right end. This transform can also be used to solve Poisson's equation when the derivative of the solution is specified on the left boundary and the solution is specified on the right boundary. Given the sequence x_n for $n = 0, \ldots, N - 1$, the discrete even quarter-wave Fourier transform is given by

$$c_k = \frac{1}{N}x_0 + \frac{2}{N}\sum_{n=1}^{N-1} x_n \cos(2k-1)n\frac{\pi}{2N}, \quad k = 1, \ldots, N. \tag{71}$$

The inverse transform is given by

$$x_n = \sum_{k=1}^{N} c_k \cos(2k-1)n\frac{\pi}{2N}, \quad n = 0, \ldots, N - 1. \tag{72}$$

The sequence x_n for $n = 0, \ldots, N - 1$ is arbitrary. Like the other transforms, the symmetries are apparent only when the sequence is extended. If we compute x_n using (71) over an extended range of subscripts, then we can determine that the real periodic has the form $(x_0, x_1, \ldots, x_{N-1}, 0, -x_{N-1}, \ldots, -x_0, -x_1, \ldots, -x_{N-1}, 0, x_{N-1}, \ldots, x_1)$. Therefore, in order to determine the c_k using the real periodic transform, one would have to transform this $4N$-point sequence.

The development of the algorithm for the forward transform is very similar to that given above for the discrete odd quarter-wave transform and hence we only summarize it. Given the sequence x_n,

1. Compute the sequence d_n from

$$d_n = (x_n + x_{N-n})\sin n\frac{\pi}{2N} + (x_n - x_{N-n})\cos n\frac{\pi}{2N}. \tag{73}$$

2. Compute a_k and b_k from the forward real transform of d_n using (35) and (36) or (38) and (39).

3. Compute the c from the a and b using

$$c_1 = \tfrac{1}{2}a_0, \qquad (74)$$

$$c_{2k} = \tfrac{1}{2}(a_k + b_k), \qquad (75)$$

$$c_{2k+1} = \tfrac{1}{2}(a_k - b_k), \qquad (76)$$

and if N is even

$$c_N = \tfrac{1}{2}a_N. \qquad (77)$$

The inverse transform (72) can be computed simply by reversing the order of the computations. To this end the inverse of (73) is

$$x_n = \frac{1}{2}\left[(d_n + d_{N-n})\cos n\frac{\pi}{2N} + (d_n - d_{N-n})\sin n\frac{\pi}{2N}\right]. \qquad (78)$$

The pre- and postprocessing algorithms for the quarter-wave transforms can be performed in place, and they are completely vectorizable with vector length $N/2$. Hence the properties of the complete algorithm are determined by the properties of the real periodic transform that is at the heart of the algorithm. This provides the same flexibility that is available for the complex and real transforms since any one of six choices can be made depending on the application.

VII. SOFTWARE AND SUMMARY

A. *FFTPACK: A Package of Subroutines for the Discrete Fourier Transforms*

In addition to Temperton's software for multiple transforms that was mentioned in Section IV, he has also developed programs for the complex and real periodic transforms that are vectorized for the Cray-1 and discussed in Temperton (1979a). Peterson (1978) has also developed programs in CAL for the complex and real transforms on the Cray-1.

From the preceding sections it can be seen that the implementation of all the Fourier transforms in a software package would be a very extensive project. The complex, real periodic and four symmetric transforms can each be implemented using any of the six algorithms given in Section III for a total of 36 transforms. If each of these is implemented for multiple sequences, then the total comes to 72. This number does not include the inverse transforms or the initialization programs that compute the trigonometric functions. The number of transforms greatly increases if multidimensional transforms are included since combinations of the various transforms are needed. For example, a three-dimensional problem may require the quarter-wave transform in one direction, the sine transform in another direction, and the real periodic in the third.

Although it may be some time before such an extensive package is available, a start has been made with the development of a package that includes a number of the Fourier transforms that have been discussed in this chapter. It was originally developed as a tool for solving elliptic partial differential equations (Swarztrauber, 1977); however, it is now an independent package called FFTPACK that can be used for any application that requires the discrete Fourier transform of complex, real periodic, and the symmetric transforms of a single sequence.

The package includes the following subroutines:

1. RFFTI   Initialize RFFTF and RFFTB
2. RFFTF   Forward transform of a real periodic sequence
3. RFFTB   Backward transform of a real coefficient array

4. EZFFTI   Initialize EZFFTF and EZFFTB
5. EZFFTF   A simplified real periodic forward transform
6. EZFFTB   A simplified real periodic backward transform

7. SINTI   Initialize SINT
8. SINT   Sine transform of a real odd sequence.

9. COSTI   Initialize COST
10. COST   Cosine transform of a real even sequence

11. SINQI   Initialize SINQF and SINQB (quarter-wave sine)
12. SINQF   Forward sine transform with odd wave numbers
13. SINQB   Inverse sine transform with odd wave numbers

14. COSQI   Initialize COSQF and COSQB (quarter-wave cosine)
15. COSQF   Forward cosine transform with odd wave numbers
16. COSQB   Inverse cosine transform with odd wave numbers

17. CFFTI   Initialize CFFTF and CFFTB
18. CFFTF   Forward transform of a complex periodic sequence
19. CFFTB   Inverse transform of a complex periodic sequence

FFTPACK has the following properties:

(a) The documentation is extensive. The package consists of 3400 lines of FORTRAN code including 865 lines of documentation.

(b) The programs are based on the autosort algorithms of Stockham as presented by Cochran *et al.*, (1967), and therefore the results are always ordered.

(c) There are no restrictions on N even for the symmetric transforms; however, the FFT algorithms are more efficient for composite N.

(d) The package has been PFORT verified as standard FORTRAN.

(e) The package is efficient on the scalar machines and vectorizes on the Cray-1. Computing times for the real and complex transforms are given in Table VII.

B. *Matching Applications with Algorithms: A Summary*

It is not as difficult as it may seem to summarize the results that have been presented. The properties of all of the transforms are for the most part determined by the properties of the corresponding complex transform, i.e., if the application calls for the sine transform, then the choice would be based on the properties of the corresponding complex transform. Hence we shall consider only the complex transform and recommend algorithms for a variety of applications. The recommendations will be presented using a "tree" format beginning with step 1.

1. If the application calls for the transform of many N-point sequences that are not in storage simultaneously go to step 4 (most problems fall into this category); otherwise go to step 2.

TABLE VII

COMPUTING TIME IN MILLISECONDS FOR THE REAL AND COMPLEX TRANSFORMS

	Real transform		Complex transform	
N	7600	Cray-1	7600	Cray-1
32	0.33	0.076	0.40	0.077
36	0.32	0.076	0.43	0.093
48	0.40	0.084	0.53	0.104
49	0.62	0.231	0.85	0.237
50	0.48	0.105	0.66	0.110
64	0.52	0.093	0.66	0.121
96	0.70	0.123	0.98	0.164
100	0.75	0.121	1.19	0.180
121	1.60	0.439	2.51	0.503
128	0.85	0.135	1.22	0.181
1024	7.30	0.648	12.10	1.177

2. If the application calls for multiple transforms of M sequences of length N that are available in storage simultaneously, go to step 11 (many problems fall into this category); otherwise go to step 3. For this application it is assumed that at least N extra locations are available for storing the trigonometric functions.

3. If the application calls for the transform of many sequences with different lengths, then the trigonometric functions must be recomputed for each transform. One approach is to compute the trigonometric functions using the algorithm based on the multiangle identities (see, e.g., Singleton, 1969). It is efficient, accurate and does not require any additional storage since the functions are computed only when they are needed. However, the algorithm is highly recursive and not efficiently implemented on a vector computer. A second approach is to compute the functions using the vectorized trigonometric library routines that are available on the Cray-1. This approach will minimize computing time; however, an additional array must be provided. In what follows, it is assumed that a decision has been made as to how the trigonometric functions will be computed. If storage is limited, then go to step 17; otherwise go to step 14.

4. If storage is limited, then go to step 8; otherwise proceed to step 5.

5. If the transform must be ordered, then go to step 7; otherwise proceed to step 6. Note that many problems can be reformulated so that ordering is not necessary. For more on this topic, see the discussion following the presentation of the Gentleman–Sande algorithm in Section III.

6. The Pease algorithm is quite satisfactory for this application, particularly on a vector computer. If $N = 2^m$, then the minimum vector length is $N/2$, compared with $\sqrt{N}$ for the other algorithms. However, mN extra storage locations are required to store the trigonometric functions.

7. Stockham's autosort algorithm and Glassman's generalization are appropriate for this application. The minimum vector length is about $\sqrt{N}$, which is less than the minimum length for the Pease algorithm.

8. There are two arrays of length N that can be of use to FFT programs. The first contains a tabulation of the trigonometric functions that are computed just once and then used for all subsequent transforms. This is the most common way that FFT programs are written. However, when storage is a severe problem, the trigonometric functions can be recomputed for each transform, but at some expense in computing time (see, e.g., Singleton, 1969). The algorithm for computing the trigonometric functions is based on the multiangle identities. It is quite straightforward and accurate; however, it is highly recursive and not efficient on a vector computer. Although there are vectorized versions of the trigonometric functions on the Cray-1, they require an N-point array.

A second N-point array is very useful if the transform must be ordered. If your application requires ordering, then proceed to step 10; otherwise go to step 9. Note that many problems can be reformulated so that ordering is not necessary. See the discussion following the presentation of the Gentleman–Sande algorithm in Section III.

9. The Cooley–Tukey and Gentleman–Sande algorithms can be performed in place without additional storage.

10. The Cooley–Tukey and Gentleman–Sande algorithms can be performed in place without additional storage. If $N = 2^m$, then the ordering can also be performed in place. If N has a number of factors that are squared, then a minimal amount of extra storage is required (see Singleton, 1969).

11. There are two methods for solving this problem that are discussed in Section IV. The first method extends the FFT for a single sequence to an algorithm for multiple sequences by applying each operation in the single transform to all M sequences. The vectors have length M throughout the calculation, and the algorithm is efficient when M is comparable or larger than N. In the second method all vectors have length at least M and increase in length to MN as the algorithm proceeds. Korn and Lambiotte (1979) discuss the relative merits of these methods in some detail. The recommendations that are given below apply to either method. If MN locations of work storage are available in addition to the MN locations used by the sequences, then proceed to step 12; otherwise go to step 13.

12. Either method in Section IV for multiple sequences can be used with Stockham's autosort algorithms described in Section III.

13. The Cooley–Tukey and Gentleman–Sande algorithms can be performed in place without additional storage. If $N = 2^m$, then the ordering can also be performed in place, or if N has a number of factors that are squared, then a minimal amount of extra storage is required (see Singleton, 1969).

14. If the transform must be ordered, then go to step 16; otherwise proceed to step 15. Note that many problems can be reformulated so that ordering is not necessary. See the discussion following the presentation of the Gentleman–Sande algorithm in Section III.

15. The Pease algorithm is quite satisfactory for this application, particularly on a vector computer. If $N = 2^m$, then the minimum vector length is $N/2$, compared with $\sqrt{N}$ for the other algorithms. However, it does require mN extra storage locations.

16. Stockham's autosort algorithms and Glassman's generalization are appropriate for this application. The minimum vector length is about $\sqrt{N}$, which is less than the minimum length for the Pease algorithm.

17. If your application requires ordering, then this point is discussed further in step 19; otherwise go to step 18. Note that many problems can be

reformulated so that ordering is not necessary. See the discussion following the presentation of the Gentleman–Sande algorithm in Section III.

18. The Cooley–Tukey and Gentleman–Sande algorithms can be performed in place without additional storage.

19. The Cooley-Tukey and Gentleman–Sande algorithms can be performed in place without additional storage. If $N = 2^m$, then the ordering can also be performed in place. If N has a number of factors that are squared, then a minimal amount of extra storage is required (see Singleton, 1969).

REFERENCES

Bergland, G. D. (1968). *Comm. ACM* **11**, 703–710.
Brigham, E. O. (1974). "The Fast Fourier Transform." Prentice-Hall, Englewood Cliffs, New Jersey.
Cochran, W. T., *et al.* (1967). *IEEE Trans. Audio Electroacoust.* **15**, 45–55.
Cooley, J. W., and Tukey, J. W. (1965). *Math. Comp.* **19**, 297–301.
Cooley, J. W., Lewis, P. A. W., and Welch, P. D. (1967). *Proc. IEEE* **55**, 1675–1677.
Cooley, J. W., Lewis, P. A. W., and Welch, P. D. (1970). *J. Sound Vib.* **12**, 315–337.
Dollimore, J. (1973). *J. Inst. Math. App.* **12**, 115–117.
Fornberg, B. (1981). *Math. Comp.* **36**, 189–191.
Gentleman, W. M., and Sande, G. (1966). *AFIPS Conf. Proc., Fall Joint Comput. Conf.* **29**, 563–578.
Glassman, J. A. (1970). *IEEE Trans. Comput.* **19**, 105–116.
Hockney, R. W. (1969). *In* " Methods of Computational Physics " (B. Adler, S. Fernbach, and M. Rotenberg, eds.), Vol. 9, pp. 136–211. Academic Press, New York.
Jenkins, G. M., and Watts, D. G. (1968). "Spectral Analysis and Its Applications." Holden-Day, San Francisco, California.
Korn, D. G., and Lambiotte, J. J., Jr. (1979). *Math. Comp.* **33**, 977–992.
Pease, M. C. (1968). *J. Assoc. Comput. Mach.* **15**, 252–264.
Peterson, W. P. (1978). Tech. Note 2240208, Cray Research, Minneapolis, Minnesota.
Singleton, R. C. (1969). *IEEE Trans. Audio Electroacoust.* **17**, 93–103.
Stone, H. S. (1973). *J. Assoc. Comput. Mach.* **20**, 27–38.
Swartztrauber, P. N. (1977). *SIAM Rev.* **19**, 490–501.
Temperton, C. (1977). Tech. Rep. 3. European Centre for Medium Range Weather Forecasts, Reading, Berkshire, England.
Temperton, C. (1979a). Internal Rep. No. 21. European Centre for Medium Range Weather Forecasts, Reading, Berkshire, England.
Temperton, C. (1979b). *In* "Super Computers," Vol. 2, pp. 360–379. Infotech International, Maidenhead, Berkshire, England.
Wang, H. H. (1980). *BIT* **20**, 233–243.

Solution of Single Tridiagonal Linear Systems and Vectorization of the ICCG Algorithm on the Cray-1

David Kershaw

Lawrence Livermore National Laboratory
University of California
Livermore, California

I. A VECTOR ALGORITHM FOR TRIDIAGONAL LINEAR SYSTEMS

When we examine the numerical algorithms used to solve the physics equations in codes which model laser fusion we find that a large number of subroutines require the solution of tridiagonal linear systems of equations. Radiation transport, thermal- and suprathermal-electron transport, ion thermal conduction, charged-particle and neutron transport all require the solution of tridiagonal systems of equations. The standard algorithm that has been used in the past on CDC 7600s will not vectorize and so cannot take advantage of the large speed increases possible on the Cray-1 through vectorization. There is, however, an alternative algorithm for solving tridiagonal systems, called cyclic reduction, which allows for vectorization, and which is optimal for the Cray-1. Software based on this algorithm is now being used in LASNEX to solve tridiagonal linear systems in the subroutines mentioned above. The new algorithm runs as much as five times faster than the standard algorithm on the Cray-1.

A. *The Basic Algorithm*

Consider a tridiagonal linear system of equations

$$b_{i-1}x_{i-1} + a_i x_i + c_i x_{i+1} = y_i, \quad i = 1, 2, \ldots, n, \tag{1}$$

with $b_0 = c_n = 0$. The standard algorithm for solving these equations is

$$d_i = (a_i - l_{i-1}c_{i-1})^{-1} \quad \text{for} \quad i = 1, 2, \ldots, n \tag{2}$$

and

$$l_i = b_i d_i \quad \text{for} \quad i = 1, 2, \ldots, n-1, \tag{3}$$

followed by

$$w_i = y_i - l_{i-1}w_{i-1}, \quad i = 1, 2, \ldots, n, \tag{4}$$

and

$$x_i = d_i(w_i - c_i x_{i+1}), \quad i = n, (n-1), \ldots 3, 2, 1. \tag{5}$$

This algorithm is recursive and cannot be vectorized, and therefore will only run about twice as fast on the Cray-1 as it did on the CDC 7600.

Note that the standard scalar algorithm is just LU decomposition.* If our original equation is written in matrix notation as $MX = Y$, with

$$M = \begin{bmatrix} a_1 & c_1 & 0 & 0 \\ b_1 & a_2 & c_2 & 0 \\ 0 & b_2 & a_3 & c_3 \\ 0 & 0 & b_3 & a_4 \end{bmatrix}, \tag{6}$$

then we may decompose M as a lower-triangular-matrix L with unit diagonal elements and an upper-triangular-matrix U; i.e., $M = LU$. Here

$$L = \begin{bmatrix} 1 & 0 & 0 & 0 \\ l_1 & 1 & 0 & 0 \\ 0 & l_2 & 1 & 0 \\ 0 & 0 & l_3 & 1 \end{bmatrix} \tag{7}$$

and

$$U = \begin{bmatrix} d_1^{-1} & c_1 & 0 & 0 \\ 0 & d_2^{-1} & c_2 & 0 \\ 0 & 0 & d_3^{-1} & c_3 \\ 0 & 0 & 0 & d_4^{-1} \end{bmatrix}. \tag{8}$$

Then we solve $LW = Y$ and $UX = W$ to find X.

*For a good discussion of LU decomposition, see Wilkinson (1965).

To obtain an algorithm which allows some vectorization we reorder the rows and columns of our matrix so that first we take all the odd multiples of 1, then all the odd multiples of 2, then all the odd multiples of 2^2, etc. Thus we apply a permutation P which takes the original ordering, $1, 2, 3, \ldots, n$ into the new ordering

$$1, 3, 5, 7, \ldots, 2, 6, 10, 14, \ldots, 4, 12, 20, 28, \ldots, 8, 24, 40, 56, \ldots,$$
$$2^q, 3 \cdot 2^q, 5 \cdot 2^q, 7 \cdot 2^q, \ldots, 2^p. \qquad (9)$$

Here p is the highest power of 2 with $2^p \leq n$.

Our original matrix equation now becomes

$$(PMP^{-1})(PX) = (PY). \qquad (10)$$

If we perform LU decomposition on this reordered matrix, we obtain a new algorithm called cyclic reduction. This algorithm is vectorizable because we are eliminating all the odd variables first (which are not coupled to each other), then all the odd-multiple-of-2 variables (which are not coupled to each other), and so on. Furthermore, the fact that the algorithm is just LU decomposition on a permuted matrix tells us that it is just as stable as the standard scalar algorithm, a virtue not shared by other methods for vectorizing tridiagonal systems such as recursive doubling.

Applying the standard formulas for LU decomposition, and eliminating those terms which are zero because of the particular sparsity pattern of our reordered matrix, we obtain

$$\tilde{l}_i^0 = b_i, \quad \tilde{u}_i^0 = c_i, \quad \text{and} \quad \tilde{d}_i^0 = a_i, \quad i = 1, \ldots, (n-1), n. \qquad (11)$$

Then for each $q = 0, 1, 2, 3, \ldots, p$ we take

$$\begin{aligned}
d^q_{(2i-1)} &= 1/\tilde{d}^q_{(2i-1)}, & 1 &\leq 2i-1 \leq r(q,n), \\
l^q_{(2i-1)} &= \tilde{l}^q_{(2i-1)} d^q_{(2i-1)}, & 1 &\leq 2i-1 < r(q,n), \\
l^q_{(2i)} &= \tilde{l}^q_{(2i)}, & 2 &\leq 2i < r(q,n), \\
u^q_{(2i-1)} &= \tilde{u}^q_{(2i-1)}, & 1 &\leq 2i-1 < r(q,n), \\
u^q_{(2i)} &= \tilde{u}^q_{(2i)} d^q_{(2i+1)}, & 2 &\leq 2i < r(q,n), \\
\tilde{d}_i^{q+1} &= \tilde{d}^q_{2i} - l^q_{(2i-1)} u^q_{(2i-1)} - l^q_{(2i)} u^q_{(2i)}, & 1 &\leq i \leq r(q+1, n), \\
\tilde{l}_i^{(q+1)} &= -l^q_{2i} l^q_{(2i+1)}, & 1 &\leq i < r(q+1, n), \\
\tilde{u}_i^{(q+1)} &= -u^q_{2i} u^q_{(2i+1)}, & 1 &\leq i < r(q+1, n),
\end{aligned} \qquad (12)$$

where $r(q, n)$ is the largest integer r such that $r \cdot 2^q \leq n$.

For the solve, we let

$$w_i^0 = y_i, \quad i = 1, 2, \ldots, n.$$

Then for each $q = 0, 1, 2, \ldots, (p-1)$ we take

$$w_i^{(q+1)} = w_{(2i)}^q - l_{(2i-1)}^q w_{(2i-1)}^q - u_{(2i)}^q w_{(2i+1)}^q$$
$$\text{for} \quad i = 1, 2, \ldots, r(q+1, n). \tag{13}$$

We set $x_1^p = d_1^p w_1^p$, and then for each $q = (p-1), (p-2), \ldots, 4, 3, 2, 1, 0$ we take

$$x_{(2i)}^q = x_i^{(q+1)}, \quad 2 \le 2i \le r(q, n), \tag{14}$$

and

$$x_{(2i-1)}^q = d_{(2i-1)}^q [w_{(2i-1)}^q - l_{(2i-2)}^q x_{(2i-2)}^q - u_{(2i-1)}^q x_{(2i)}^q], \quad 1 \le 2i - 1 \le r(q, n). \tag{15}$$

Finally, $x_i = x_i^0$ for $i = 1, \ldots, n$.

Clearly, all the above operations at the q level of reduction are vectorizable with vector length $r(q+1, n)$. Thus we now have a vectorizable algorithm with vectorization on lengths $n/2, n/4, n/8, \ldots, 1$.

B. *Implementation on the Cray-1*

The numbers $r(q, n)$ are easily calculated from $r(0, n) = n$ and

$$r(q+1, n) = \text{SHIFTR}[r(q, n), 1], \tag{16}$$

where $\text{SHIFTR}(n_1, n_2)$ is n_1 (in binary notation) shifted right n_2 positions with the rightmost n_2 bits of n_1 lost off the end. $\text{SHIFTR}(r, 1) = r/2$, but on the Cray-1 the shift operation is much faster than the integer divide.

It is important to give consideration to storage layout and possible memory bank conflicts. Consider, for example, the forward sweep of the solve. The "natural" storage layout is to store $w_i^{(q+1)}$ in the $[i \cdot 2^{(q+1)}]$ element of the Y array and just keep overwriting $w_{(2i)}^q$ with $w_i^{(q+1)}$. The problem with this is that vector reads from and writes to memory in increments of 2^q on the Cray-1 cause bank conflicts and loss of speed if $q > 2$. For $q = 0, 1, 2$ we can read or write one word per clock period. For $q = 3$ this degrades to one word every two clock periods, and for $q \ge 4$ the performance degrades to one word every four clock periods. To eliminate this problem we use the following storage scheme.

$w_i^0 = y_i$ is stored in the first $n = r(0, n)$ elements of the Y array. w_i^1 is stored in the next $r(1, n)$ elements of the Y array, and so on. Thus

$$w_i^q = y_{[s(q, n) + i]}, \tag{17}$$

where $s(0, n) = 0$ and $s(q+1, n) = s(q, n) + r(q, n)$. x_i^q is stored in the same locations as w_i^q. The matrix elements l_i^q and $\tilde{l}_i^q$ are stored in $b_{[s(q, n) + i]}$; $\tilde{u}_i^q$ and

u_i^q are stored in $c_{[s(q,n)+i]}$; d_i^q and $\tilde{d}_i^q$ are stored in $a_{[s(q,n)+i]}$. With this storage scheme all memory reads and writes are done in increments of 1 and 2, and all bank conflicts are eliminated. Since

$$\sum_{q=0}^{p} r(q,n) \leq \sum_{q=0}^{p} \frac{n}{2^q} \leq 2n, \tag{18}$$

the arrays a, b, c, and y must now be $2n$ elements long instead of n elements long, and so we need twice as much storage as is required for the standard scalar algorithm. If this increase in storage is unacceptable, then it is possible to shuffle the elements in place and have no increase in storage for y and a but at the expense of the extra time required for the shuffle. For b and c (l and u) the storage must double because both $l_{(2i)}^q$ and $l_{(2i-1)}^q$ and $u_{(2i)}^q$ and $u_{(2i-1)}^q$ are needed for the solve. Thus on the forward sweep of the solve, after calculating w_i^1, with $i = 1, 2, \ldots, r(1, n)$, we could store w_i^1 in the first $r(1, n)$ words of y_i and move (shuffle) $w_{(2i-1)}^0$ to the last $(n - r(1, n))$ words of y_i. Then w_i^2 is stored in the first $r(2, n)$ words of y_i while $w_{(2i-1)}^1$ is moved to the next $(r(1, n) - r(2, n))$ words of y_i, etc. (The storage for the d_i^q array may be similarly arranged.) With this scheme the storage requirement will be only 50% greater instead of 100%.

C. Performance

This algorithm was hand-coded for the Cray-1 by R. E. von Holdt for both the symmetric and nonsymmetric matrix cases. In Fig. 1 the hand-coded vector cyclic reduction algorithm is compared with the Cray-1 FORTRAN CFT-compiled standard scalar algorithm. N is the problem dimension and (t_s/t_v) is the relative execution time of the scalar and vector algorithms. The

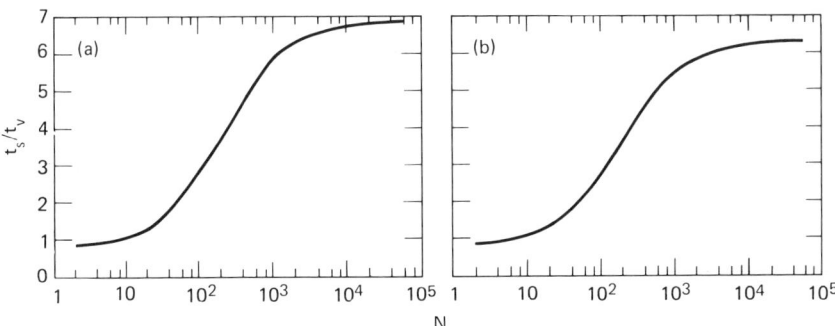

FIG. 1. Ratio of scalar to vector execution times for the solution of a tridiagonal matrix of length N. For both symmetric (a) and asymmetric (b) matrices, the asymptotic rate of the vector cyclic reduction algorithm is more than six times the scalar rate, but this speedup can be achieved only with relatively long vectors.

cyclic reduction algorithm was also coded with CFT and it was asymptotically about 4.5 times faster than the CFT scalar algorithm; thus the hand coding made only a 50% improvement over CFT.

II. AN INCOMPLETE CHOLESKY CONJUGATE GRADIENT (ICCG) ALGORITHM FOR THE CRAY-1 COMPUTER

In LASNEX, the two-dimensional radiation transport, electron thermal conduction, ion thermal conduction, and neutron transport all use ICCG to solve the transport equations. We have been using the ICCG method (Kershaw, 1978) to solve the diffusion equation with a nine-point coupling (Kershaw, 1981) scheme on the CDC 7600. In going from the CDC 7600 to the Cray-1, a large part of the algorithm consists of solving tridiagonal linear systems on each L line of the Lagrangian mesh in a manner which is not vectorizable. Therefore a direct translation from the 7600 to the Cray would not give much increase in running speed because the vectorization potential of the Cray cannot be used. We have developed an alternate ICCG algorithm for the Cray-1 which utilizes a block form of the cyclic reduction algorithm described in the previous chapter. This new algorithm allows full vectorization and runs as much as five times faster than the old algorithm on the Cray-1. It is now being used in Cray LASNEX to solve the two-dimensional diffusion equation in all the physics subroutines mentioned above.

A. *Basic Algorithm*

We have an equation $MX = Y$, where M is positive definite and symmetric and the sparsity pattern is such that $M_{i,i}, M_{i,(i\pm 1)}, M_{i,(i\pm \text{KMAX})}, M_{i,(i\pm \text{KMAX} \pm 1)}$ are the only nonzero elements.

The ICCG method consists in finding an approximate Cholesky decomposition for $M \approx LDL^T$ and then using the conjugate gradient algorithm. Let $r_0 = Y - MX_0$ and $p_0 = (LDL^T)^{-1}r_0$; then

$$a_i = [r_i, (LDL^T)^{-1}r_i]/(p_i, Mp_i),$$
$$X_{i+1} = X_i + a_i p_i,$$
$$r_{i+1} = r_i - a_i Mp_i,$$
$$b_i = [r_{(i+1)}, (LDL^T)^{-1}r_{(i+1)}]/[r_i, (LDL^T)^{-1}r_i],$$
$$p_{i+1} = (LDL^T)^{-1}r_{(i+1)} + b_i p_i$$

for $i = 0, 1, 2, \ldots$.

Vector Solution of Linear Systems on the Cray-1

This algorithm is trivially vectorizable except for the evaluation of the approximate inverse on the residual, $(LDL^T)^{-1}r_i$. With the choice for the approximate Cholesky decomposition given in Kershaw (1978) this is a recursive operation which can only go at scalar speeds on the Cray-1. To improve it we must use a different approximate decomposition which allows for vectorization.

To see how this is done we first write the exact Cholesky decomposition in block form, in which our matrix M can be written

$$M = \begin{bmatrix} A_1 & B_1^T & 0 & 0 & \\ B_1 & A_2 & B_2^T & 0 & \\ 0 & B_2 & A_3 & B_3^T & \\ 0 & 0 & B_3 & A_4 & \ddots \\ & & & \ddots & \ddots \end{bmatrix},$$

where the A_i and B_i are tridiagonal matrices of dimension KMAX, and $i = 1, 2, 3, \ldots,$ LMAX. The standard block Cholesky decomposition may now be written $M = LDL^T$, where

$$L = \begin{bmatrix} L_1 & 0 & 0 & 0 & \\ C_1 & L_2 & 0 & 0 & \\ 0 & C_2 & L_3 & 0 & \\ 0 & 0 & C_3 & L_4 & \\ & & & \ddots & \ddots \end{bmatrix},$$

and

$$D = \begin{bmatrix} D_1 & 0 & 0 & 0 & \\ 0 & D_2 & 0 & 0 & \\ 0 & 0 & D_3 & 0 & \\ 0 & 0 & 0 & D_4 & \\ & & & & \ddots \end{bmatrix}.$$

Here the L_i are lower triangular matrices whose diagonal elements are all 1, the D_i are diagonal matrices, and L_i, C_i, D_i are given recursively by

$$C_i = B_i L_i^{-T} D_i^{-1} \tag{19}$$

and

$$L_i D_i L_i^T = A_i - C_{(i-1)} D_{(i-1)} C_{(i-1)}^T = \tilde{A}_i \qquad (20)$$

for $i = 1, 2, 3, \ldots,$ LMAX. We use the notation $L^{-T} = (L^T)^{-1}$.

If we write the unknown and right-hand-side vectors in block form,

$$X = \begin{bmatrix} X_1 \\ X_2 \\ X_3 \\ \vdots \end{bmatrix}, \quad Y = \begin{bmatrix} Y_1 \\ Y_2 \\ Y_3 \\ \vdots \end{bmatrix},$$

where X_i, Y_i are vectors of length KMAX, then $LDL^T X = Y$ may be solved by a forward sweep,

$$W_i = L_i^{-1}[Y_i - B_{(i-1)} L_{(i-1)}^{-T} D_{(i-1)}^{-1} W_{(i-1)}], \qquad (21)$$

for $i = 1, 2, 3, \ldots,$ LMAX, and a backward sweep,

$$X_i = L_i^{-T} D_i^{-1}[W_i - L_i^{-1} B_i^T X_{(i+1)}], \qquad (22)$$

for $i =$ LMAX, (LMAX $- 1$), ..., 3, 2, 1.

To obtain an algorithm which allows vectorization we again apply the cyclic reduction permutation, but we apply it only to the KMAX-by-KMAX blocks and within a block we retain the standard ordering.* Thus the blocks are now ordered $L = 1, 3, 5, 7, \ldots, 2, 6, 10, 14, \ldots, 2^p$. Here p is the highest power of 2 with $2^p \leq$ LMAX. Within a block the elements are still ordered $K = 1, 2, 3, 4, \ldots,$ KMAX.

As in Section I our equation $MX = Y$ becomes

$$(PMP^{-1})(PX) = (PY), \qquad (23)$$

and we now perform block LU decomposition of this block permuted matrix.

Applying the standard formulas for block LU decomposition and retaining only the nonzero blocks, we obtain the following algorithm: Let

$$B_i^0 = B_i, \quad A_i^0 = A_i, \quad i = 1, 2, \ldots, \text{LMAX}.$$

* Another possibility is to apply the cyclic reduction permutation only within each block, but let the blocks retain the standard ordering. Thus within each block we have $K = 1,3,5,7,\ldots, 2, 6, 10, 14, \ldots 2^p$, but the blocks are still ordered $L = 1, 2, 3, 4, \ldots,$ LMAX. This alternate vector algorithm is described in detail in Lawrence Livermore National Laboratory (1979). It was found to run at roughly the same speed as the algorithm presented in this article on most of our test problems, but in this author's opinion it is considerably less elegant and much harder to understand and program.

Then for each $q = 0, 1, 2, 3, \ldots, p$ we take

$$L^q_{(2i-1)} D^q_{(2i-1)} (L^q_{(2i-1)})^T = A^q_{(2i-1)},$$
$$1 \leq 2i - 1 \leq r(q, \text{LMAX}), \quad (24)$$

$$C^q_{(2i-1)} = B^q_{(2i-1)} (L^q_{(2i-1)})^{-T} (D^q_{(2i-1)})^{-1},$$
$$1 \leq 2i - 1 < r(q, \text{LMAX}), \quad (25)$$

$$C^q_{(2i)} = (D^q_{(2i+1)})^{-1} (L^q_{(2i+1)})^{-1} B^q_{(2i)},$$
$$2 \leq 2i < r(q, \text{LMAX}), \quad (26)$$

$$A_i^{(q+1)} = A^q_{(2i)} - C^q_{(2i-1)} D^q_{(2i-1)} (C^q_{(2i-1)})^T - (C^q_{(2i)})^T D^q_{(2i+1)} C^q_{(2i)},$$
$$1 \leq i \leq r(q+1, \text{LMAX}), \quad (27)$$

$$B_i^{(q+1)} = -C^q_{(2i+1)} D^q_{(2i+1)} C^q_{(2i)},$$
$$1 \leq i < r(q+1, \text{LMAX}), \quad (28)$$

where $r(q, \text{LMAX})$ is the largest integer r such that $r \cdot 2^q \leq \text{LMAX}$. This completes the decomposition.

For the solve we let

$$\tilde{W}_i^0 = Y_i, \quad i = 1, 2, 3, \ldots, \text{LMAX}.$$

Then for each $q = 0, 1, 2, 3, \ldots, (p-1)$ we take

$$W^q_{(2i-1)} = (L^q_{(2i-1)})^{-1} \tilde{W}^q_{(2i-1)}, \quad 1 \leq 2i - 1 \leq r(q, \text{LMAX}), \quad (29)$$

$$\tilde{W}_i^{(q+1)} = \tilde{W}^q_{(2i)} - C^q_{(2i-1)} W^q_{(2i-1)} - (C^q_{(2i)})^T W^q_{(2i+1)},$$
$$1 \leq i \leq r(q+1, \text{LMAX}), \quad (30)$$

and finally $W_1^p = (L_1^p)^{-1} \tilde{W}_1^p$. This completes the forward sweep.

For the backward sweep we set

$$X_1^p = (L_1^p)^{-T} (D_1^p)^{-1} W_1^p,$$

and then for $q = (p-1), (p-2), \ldots, 3, 2, 1, 0$ we take

$$X^q_{(2i)} = X_i^{(q+1)}, \quad 1 \leq i \leq r(q+1, \text{LMAX}), \quad (31)$$

followed by

$$X^q_{(2i-1)} = (L^q_{(2i-1)})^{-T} ((D^q_{(2i-1)})^{-1} W^q_{(2i-1)} - C^q_{(2i-2)} X^q_{(2i-2)} - (C^q_{(2i-1)})^T X^q_{(2i)}),$$
$$1 \leq (2i-1) \leq r(q, \text{LMAX}), \quad (32)$$

and finally $X_i = X_i^0$ for $i = 1, 2, 3, \ldots, \text{LMAX}$. This completes the solve.

Note that the $L_i^q D_i^q (L_i^q)^T$ decomposition in Eq. (24) is uniquely determined by requiring that the diagonal elements of the KMAX-by-KMAX lower triangular matrices L_i^q all be equal to 1. The D_i^q are diagonal KMAX-by-KMAX matrices and the Cholesky decomposition theorem assures us that all of the diagonal elements of D_i^q are positive.

Now let us examine the sparsity pattern of A_i^q, L_i^q, and C_i^q for the case of exact Cholesky decomposition. A_i^0 is tridiagonal, so $L_{(2i-1)}^0$ will be lower bidiagonal. Since B_i^0 is tridiagonal, Eqs. (25) and (26) imply that $(C_{(2i-1)}^0)_{jk}$ will be nonzero for $k + 1 \geq j$ and $(C_{(2i)}^0)_{jk}$ will be nonzero for $j + 1 \geq k$. Equations (27) and (28) thus imply that A_i^1 and B_i^1 are dense matrices (all elements nonzero) and so $L_{(2i-1)}^1$ is dense lower triangular and in general A_i^q and B_i^q are dense matrices. Thus we have extensive fill-in.

We seek an approximate LDL^T decomposition which has no fill-in. To accomplish this, in Eqs. (25) and (26) we throw away all but the tridiagonal part of C_i^q. In evaluating $(L_{(2i+1)}^q)^{-1}B_{(2i)}^q$ and $(B_{(2i-1)}^q)(L_{(2i-1)}^q)^{-T}$, we neither compute nor store any elements outside the tridiagonal bands. Then in Eqs. (27) and (28) we throw away (i.e., neither compute nor store) all but the tridiagonal part of $C_{(2i-1)}^q D_{(2i-1)}^q (C_{(2i-1)}^q)^T$,

$$(C_{2i}^q)^T D_{(2i+1)}^q C_{(2i)}^q \quad \text{and} \quad C_{(2i+1)}^q D_{(2i+1)}^q C_{(2i)}^q.$$

Thus we assure that A_i^q and B_i^q retain the same sparsity pattern as A_i and B_i and we obtain an incomplete block LDL^T decomposition.

Having obtained the incomplete LDL^T decomposition, we perform the forward and backward sweep of the solve exactly as in Eqs. (29)–(32) except that in Eqs. (30) and (32) $C_{(2i-1)}^q$ is replaced by $B_{(2i-1)}^q(L_{(2i-1)}^q)^{-T}(D_{(2i-1)}^q)^{-1}$ and C_{2i}^q is replaced by $(D_{(2i+1)}^q)^{-1}(L_{(2i+1)}^q)^{-1}B_{(2i)}^q$. These are no longer equivalent because, for example, $C_{(2i-1)}^q$ is only the tridiagonal part of

$$B_{(2i-1)}^q(L_{(2i-1)}^q)^{-T}(D_{(2i-1)}^q)^{-1}$$

which is dense upper triangular. This has several advantages. First, the Cs are now only temporary quantities used during the course of the incomplete decomposition and need not be saved. Thus we save memory space. Second, using $B_{(2i-1)}^q(L_{(2i-1)}^q)^{-T}(D_{(2i-1)}^q)^{-1}$ is closer to the exact LDL^T decomposition than using the tridiagonal $C_{(2i-1)}^q$, and so we expect (and numerical experiment confirms this expectation) to converge in fewer conjugate gradient iterations than if we had used the Cs. The only price we pay is that evaluating $(Y - CX)$ requires 3 adds and 3 multiplies per vector element whereas $(Y - (BL^{-T}D^{-1})X)$ requires 4 adds and 5 multiplies per element.

Since we are doing an incomplete decomposition there is no longer any guarantee that in Eq. (24) the pivots $(D_{(2i-1)}^q)_{jj}$ will be >0. Therefore, in evaluating $D_{(2i-1)}^q$ we take $(D_{(2i-1)}^q)_{jj} = |Z|$, where

$$Z = (A_{(2i-1)}^q)_{jj} - ((A_{(2i-1)}^q)_{(j-1)j})^2 / (D_{(2i-1)}^q)_{(j-1)(j-1)}, \tag{33}$$

instead of the usual $(D^q_{(2i-1)})_{jj} = Z$. This assures that our incomplete LDL^T will be positive definite.

Clearly, all the operations at the qth level of reduction are vectorizable with vector length $r(q + 1, \text{LMAX})$. Thus we now have a vectorizable algorithm on vector lengths LMAX/2, LMAX/4, LMAX/8, ..., 1.

B. Array Storage

For our memory usage we simply generalize the storage layout described in Section I [see Eqs. (16)–(18)]. On input X_i, Y_i, $M_{(i+\text{KMAX}+1)i}$, $M_{(i+\text{KMAX})i}$, $M_{(i+\text{KMAX}-1)i}$, $M_{(i+1)i}$, and M_{ii} with $i = 1, 2, 3, \ldots, (\text{KMAX} * \text{LMAX})$ are stored as 7 vectors, each of length $(2 * \text{KMAX} * \text{LMAX})$, except for Y_i, which has length $(\text{KMAX} * \text{LMAX})$. On input, X_i has the initial guess for the solution vector. In these vectors the input numbers are stored in the first $(\text{KMAX} * \text{LMAX})$ words, while the second $(\text{KMAX} * \text{LMAX})$ words are scratch space. Then as we proceed through the decomposition, $(D^q_i)_{jj}$ is stored in M_{kk}, where

$$k = \text{KMAX} * s(q, \text{LMAX}) + i * \text{KMAX} + j, \quad (34)$$

and $s(q, \text{LMAX})$ is defined after Eq. (17), $(L^q_i)_{(j+1)j}$ is stored in $M_{(k+1)k}$, and $(B^q_i)_{jj}$, $(B^q_i)_{(j\pm 1)j}$ are stored in $M_{(k+\text{KMAX})k}$, $M_{(k+\text{KMAX}\pm 1)k}$, respectively. As we proceed through the solve, $(W^q_i)_j$ and $(X^q_i)_j$ are stored in X_k. Clearly, since i is the index on which we vectorize, we have avoided all bank conflicts on the Cray so long as KMAX is odd. If KMAX is not odd we simply let KMAX' = KMAX + 1 and solve a new matrix equation $M'X' = Y'$, where if $i = n * \text{KMAX}'$ ($n = 1, 2, \ldots, \text{LMAX}$), then $M'_{ij} = M'_{ji} = \delta_{ij}$ and if i and j are not multiples of KMAX', then $M'_{ij} = M_{ij}$. Note also that we store $(D^q_i)^{-1}$ instead of (D^q_i) since only $(D^q_i)^{-1}$ appears in the course of the solve, and multiplies are much faster than divides on the Cray-1.

C. Minimization of Bandwidth

It was shown in Kershaw (1978) that if KMAX > LMAX, faster convergence of the ICCG algorithm can be obtained by first transposing the grid so we have KMAX blocks of dimension LMAX instead of the initial choice of LMAX blocks each of dimension KMAX. This is true because by making the dimension of (A^q_i) as small as possible, we minimize the error we make when we throw away those elements of (A^q_i) outside the tridiagonal bands. With our new vectorizable scheme, transposing the grid if KMAX > LMAX has the added advantage that $(\text{LMAX}/2^{(q+1)})$ which is the length of all our vector operations, is made as large as possible.

D. Incomplete Cyclic Reduction

A further improvement of the algorithm may be obtained by early termination of the cyclic reduction. It has been shown (Heller, 1976) that in many cases as we proceed through the decomposition algorithm [Eqs. (24)–(28)] the elements of $B_i^{(q+1)}$ become negligible relative to the elements of $A_i^{(q+1)}$ as q increases. This suggests that instead of continuing the algorithm until $q = p$, we perform (24)–(28) for $q = 0, 1, 2, 3, \ldots, (S - 1)$ (where $S < p$) and then simply assume that $B_i^S = 0$. Then we take

$$L_i^S D_i^S (L_i^S)^T = A_i^S, \quad 1 \leq i \leq r(S, \text{LMAX}),$$

and we are done with the decomposition.

For the forward sweep of the solve we perform (29) and (30) for $q = 0, 1, 2, \ldots, (S - 1)$ and then

$$W_i^S = (L_i^S)^{-1} W_i^S, \quad 1 \leq i \leq r(S, \text{LMAX}).$$

For the backward sweep of the solve we do

$$X_i^S = (L_i^S)^{-T} (D_i^S)^{-1} W_i^S, \quad 1 \leq i \leq r(S, \text{LMAX}),$$

and then perform (31) and (32) for $q = (S - 1), (S - 2), \ldots, 3, 2, 1, 0$.

Thus we can often obtain almost as good an approximate inverse with considerably less computational effort. Furthermore, we have eliminated just those computations which have vector lengths of $(\text{LMAX}/2^q)$, where $q > S$, and these are just the very short vector computations which will run at not much better than scalar speeds on the Cray-1. Finally, if our assumption that $B_i^S = 0$ is wrong, it will only mean that LDL^T is not as good an approximate inverse as if we had continued the cyclic reduction until $q = p$. This will only make the conjugate gradient algorithm take more iterations until convergence is reached, and the correct answer will still be obtained. In practice one needs to experiment with one's typical mix of problems until the best compromise between fewer levels of cyclic reduction and fewer conjugate gradient iterations is found.

E. Performance

The new algorithm has been tried out on a wide variety of test problems, including the model problem from Kershaw (1978), a typical real problem taken from a laser fusion simulation and a set of M matrices whose elements were randomly generated and whose mean degree of diagonal dominance was adjustable to obtain matrices with different degrees of ill conditioning.

Incomplete cyclic reduction worked very well. For all the matrices we tried it was true that the "B" blocks decreased very rapidly (in fact they decreased quadratically as predicted in Heller (1976) relative to the "A" blocks. Even for the stiffest problems it was never necessary to go beyond $S = 3$ to get the fastest possible running time (see Section II.D for the definition of S).

Figure 2 shows relative running times for the new vectorizable algorithm and the scalar algorithm given in Kershaw (1978). KMAX was held fixed at KMAX = 5 and LMAX was varied from 4 to 8000. The matrices were generated by setting each subdiagonal element $M_{(i+1)i}$, $M_{(i+\text{KMAX})i}$, $M_{(i+\text{KMAX}-1)i}$, and $M_{(i+\text{KMAX}+1)i}$ to a different random number between -1 and 0.

This fixes the superdiagonals since M is symmetric. Then we set

$$M_{ii} = \alpha * \text{RANF} - \sum_j M_{ji},$$

where the summation is over all the off-diagonal elements in the ith column, RANF is a random number between 0 and 1, and α is a stiffness parameter. By construction M is a diagonally dominant M matrix and so is positive definite (Plemmons, 1977). We chose a random X vector, found Y from

$$Y = MX,$$

and then solved $Y = MX$ using ICCG until $\|X^n - X\|/\|x\| \leq 10^{-7}$, where X^n was the ICCG solution for X after n iterations and $\|x\|$ was the Euclidean norm. Both the old and new algorithms were written in FORTRAN, compiled with CFT, and run on the Cray-1 computer. The new algorithm was

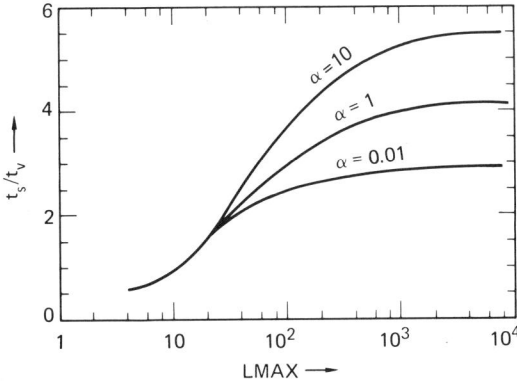

FIG. 2. Ratio of scalar to vector execution times for the solution of a 9-point coupling, block tridiagonal matrix. LMAX is the number of blocks; α is a stiffness parameter.

completely vectorized. The old algorithm was completely vectorized except for the incomplete factorization

$$LDL^T \approx M$$

and the incomplete solve

$$(LDL^T)^{-1} r_i.$$

These operations are recursive and cannot be vectorized. The ratio of the CPU times is shown in Fig. 2 for three different values of α. Asymptotically the new algorithm runs 3 to 5 times faster than the old. As the problem got stiffer, the new algorithm tended to take a few more iterations than the old. This is why the speed increase is less for small α.

III. CYCLIC REDUCTION ON FUTURE MACHINES

On the Cray-1 computer the algorithms presented in Sections I and II run 3–7 times faster than their scalar counterparts. The Cray-1 is a pipeline machine with no multiprocessing. There is only one vector add unit, one vector multiply unit, etc. As soon as computing machines become available with true parallel processing our algorithms can take immediate advantage of them *with no modifications to the algorithms*. Thus on a machine with four parallel vector pipelines for each arithmetic operation, each of our vector operations could split into four pieces each of which went to one of the four pipelines. Assuming the memory was properly designed so it could feed the four pipelines as fast as they can compute, the algorithms presented in this article would then run 12–28 times faster than their scalar counterparts. In general, if we have P parallel processors, then our vector algorithm for the tridiagonal linear system will run in $\sim (N/Q) + \log_2(Q)$ units of time [where N is the order of the linear system and $Q = \mathrm{MIN}(N, P)$], as compared to $\sim N$ units of time for the single pipeline ($P = 1$) machine (such as the Cray-1). Similarly the vector block tridiagonal algorithm will run in $\sim \mathrm{KMAX} * ((\mathrm{LMAX}/Q) + \log_2(Q))$ units of time as compared to $\sim \mathrm{KMAX} * \mathrm{LMAX}$ for the Cray-1 (here $Q = \mathrm{MIN}(\mathrm{LMAX}, P)$).

Thus even though these algorithms already have a distinct speed advantage on current computing machines, their ultimate potential will only be realized on future machines where the ratio of scalar algorithm to vector algorithm execution times will be $\sim N/(\log_2 N)$ for a triadiagonal system of order N and a machine with P parallel processors with $P \geq N$.

ACKNOWLEDGMENTS

I should like to thank Garry Rodrigue and Alex Friedman for a number of helpful discussions and Alex Friedman for his able programming of the algorithm presented in Section II.

REFERENCES

Heller, D. (1976). Some aspects of the cyclic reduction algorithm for block tridiagonal linear systems. *SIAM J. Numer. Anal.* **13**, 484.

Kershaw, D. S. (1978). The ICCG method for the iterative solution of systems of linear equations. *J. Comput. Phys.* **26**, 43.

Kershaw, D. S. (1980). Differencing of the diffusion equation in Lagrangian hydrodynamics codes. *J. Comput. Phys.* **39**, 375 (1981).

Lawrence Livermore National Laboratory (1979). Laser Program Annual Report. Rep. UCRL-50021-79, Sect. 3, p. 79, Lawrence Livermore National Laboratory, Livermore, California.

Plemmons, R. J. (1977). M matrix characterizations. *Linear Algebra Appl.* 18, 175.

Wilkinson, J. H. (1965). "The Algebraic Eigenvalue Problem," Chap. 4, Sect. 36. Oxford University Press, London and New York.

An Implicit Numerical Solution of the Two-Dimensional Diffusion Equation and Vectorization Experiments

Garry Rodrigue
Chris Hendrickson
Mike Pratt

Computation Department
Lawrence Livermore National Laboratory
University of California
Livermore, California

I. INTRODUCTION

The diffusion equation

$$h(\mathbf{x}) \frac{\partial \phi}{\partial t}(\mathbf{x}, t) = \nabla \cdot \Lambda(\mathbf{x}) \nabla \phi(\mathbf{x}, t) - s(\mathbf{x})\phi(\mathbf{x}, t) \qquad t \geq 0, \quad \mathbf{x} \in \Omega, \quad (1)$$

is solved when studying heat conduction or transport problems (Pomraning, 1973; Tarter and Westbrook, 1975; Weinberg and Wigner, 1958). Associated with the diffusion equation are the boundary conditions

$$\alpha(\mathbf{x})\phi(\mathbf{x}, t) + \beta(\mathbf{x})\langle \nabla \phi(\mathbf{x}, t), \mathbf{\eta}(\mathbf{x}) \rangle = \gamma(\mathbf{x}), \qquad \mathbf{x} \in \Gamma, \quad t \geq 0,$$

---

* Work performed under the auspices of the U.S. Department of Energy by the Lawrence Livermore National Laboratory under contract number W-7405-ENG-48.

where Γ is the boundary of Ω, $\boldsymbol{\eta}(\mathbf{x})$ is a direction normal to Γ at $\mathbf{x}$ and $\langle \cdot, \cdot \rangle$ denotes the ordinary Euclidean dot product. We also assume

(1) h, Λ are strictly positive in $\Omega \cup \Gamma$ and

(2) $\iint [\nabla \cdot \Lambda(\mathbf{x}) \nabla \psi - s(\mathbf{x})\psi]\psi \, d\Omega < 0$ (2)

for all twice differentiable functions ψ;

(3) $\beta(\mathbf{x})$, $\alpha(\mathbf{x}) \geq 0$ and $\beta(\mathbf{x}) + \alpha(\mathbf{x}) > 0$, $\mathbf{x} \in \Gamma$.

The associated initial condition is

$$\phi(\mathbf{x}, 0) = g(\mathbf{x}), \quad \mathbf{x} \in \Omega. \tag{3}$$

In this chapter, the numerical solution of the diffusion equation is studied within the context of the flow of a compressible fluid. Here, temperature differences can be considerable from one point to another, and the transfer of energy by thermal conduction may have a significant effect on the motion. This is accounted for by coupling the heat diffusion equation (1) to the hyperbolic equations of fluid dynamics and the two phenomena are calculated concurrently. The equations themselves can be very complex, so that any analysis of the finite difference approximation in space and time is difficult to perform; however, for vector processors, explicit time differencing schemes are preferred since they give rise to highly vectorizable formulas. In fact, for nonviscous fluid flows in which the effects of temperature is minimal, explicit time differencing schemes can be used very effectively with the time step controlled by the Courant-Friedricks-Levy condition (Richtmyer and Morton, 1967). However, with the introduction of heat through the diffusion equation, the time step can be forced to be excessively small if an explicit method is used on this parabolic equation. This problem is typically circumvented by "splitting" off the diffusion equation from the other equations and calculated at a different time level with an altogether different time-differencing scheme. Generally, this is an implicit method where unconditional stability prevails. When this is done, the time step for the entire problem is governed not only by the Courant-Friedricks-Levy condition for the hyperbolic fluid equations, but also by the physical changes occurring within the problem itself, e.g., steep energy gradients. This leads to a considerable variation in the time steps for which the differencing scheme used for the diffusion equation must be stable. For small time steps, no problem is posed as there are a wide variety of implicit-explicit time-differencing schemes to choose from, e.g., Crank-Nicholson. However, for large time steps, the dissipation of the solution of the diffusion equation forces the solution of the finite difference equations to behave the same way. This requirement, as well as that imposed by the "stiffness" of the diffusion equation itself, necessitate the use of full implicit time schemes, such as the backward Euler method. The major disadvantage of these schemes is the requirement of the solution of a large matrix equation, a very time-consuming calculation on any computer.

Consequently, the method used to solve these equations is extremely important.

For two-dimensional problems the number of unknowns in the equation is large ($\sim 10{,}000$), so that direct matrix inversion techniques are impractical and, instead, iterative methods are used. Iterative methods are especially attractive for vector processors since the majority of the operations in the calculation involve vectors. However, iterative methods can be a problem since slow convergence of first-degree methods such as successive overrelaxation for diffusion problems with widely varying diffusion coefficients often necessitates acceleration by second-degree schemes, e.g., the conjugate gradient method (Concus et al., 1976). Here, a penalty is paid in that second-degree schemes require extra computer storage. Despite this problem, the conjugate gradient method has been shown to be an effective iterative method for solving matrix equations arising from the finite differencing of the diffusion operator on a Cartesian coordinate system (Kershaw, 1978). In this situation, the conjugate gradient method can be used since the important M-matrix and positive definiteness properties of the matrices are easily established (Varga, 1962).

Convergence behavior of the conjugate gradient method depends, as in first-degree methods, on the particular preconditioning matrix that is used within the iterative process. However, in contrast to first-degree methods, convergence rates of the different versions have not been established mathematically and comparisons have to be made by computer timings. As a result, the particular computer executing the algorithm plays an important role in choosing the appropriate version of the conjugate gradient algorithm to be used in the solution of the implicit equations (Greenbaum, 1981). This is especially true when vector processors are involved since the algorithms that will exhibit the best performance on these machines are those that make optimal use of the high-speed vector arithmetic units, i.e., algorithms that maximize vector activity and minimize scalar activity.

Compressible fluid flows are characterized by moving boundaries and internal discontinuities, such as shocks, so that it becomes practical numerically to solve the equations in a Lagrangian or material coordinate system. Here, a given differential mass is labeled by its Cartesian coordinates at an initial time, and the dependent variables of the fluid equations are defined as functions of this initial labeling. The equations of motion then describe the properties and position of this differential mass as the fluid changes with time. The numerical approximation must, of course, behave in the same way. That is, the mass equation is defined on an initial grid of points and the positions of the grid points change with time. Consequently, the grid on which the diffusion equation is to be solved is nonorthogonal and, in fact, can be highly distorted. However, the grid must maintain a quadrilateral structure or else the calculation becomes meaningless. The spatial difference approximations to the diffusion are carried out by first mapping the distorted grid to an orthogonal grid and then carrying out the approximations. This mapping introduces nonzero cross-derivative terms into the diffusion equation and the corresponding matrix equation no longer possess the important M-matrix properties. Moreover, positive definite properties for these particular finite difference operators have not been established and application of the conjugate gradient method to the matrix equations has to be verified heuristically.

Another anomaly imposed on the diffusion equation by the fluid equation is the position on the grid at which the diffused quantity ϕ is to be solved. The diffusion equation is generally coupled to the energy equation via a specific heat term [i.e., the function h in (1)] and on a two-dimensional grid; the accuracy of the calculation dictates that energy (and consequently ϕ) be determined at the center of a grid zone rather than at the grid intersection points (Richtmyer and Morton, 1967). This complicates the difference approximations of the diffusion equation at material interfaces and boundaries where the differencing must be performed so as to preserve the important definiteness properties of the diffusion difference matrix.

With the foregoing scenario in mind, this chapter will study the performance of different versions of the conjugate gradient method for solving, on a vector processor, an implicit system of equations arising from a given finite difference formula. The chapter begins by first expressing the diffusion equation (1) in terms of an arbitrary coordinate system, thereby modeling a Lagrangian framework. Line integral techniques are used to generate a spatial finite difference scheme. The region Ω is assumed to be a rectangle in order to ease the application of the boundary conditions (2). It is then possible to establish heuristically symmetry and negative definiteness of the corresponding diffusion matrix. Using the method of lines, a linear system of ordinary differential equations is generated and solved by a fully implicit, first-order-accurate, time-differencing scheme based on the backward Euler method. This time differencing scheme is shown to be stable and have the appropriate time dissipative properties. Different data structures for the implicit matrix are discussed with emphasis on their appropriateness for vector processors. It is then possible to describe vectorizable versions of the conjugate gradient method and to discuss memory requirements and computational complexities for inverting the particular preconditioning matrices on a vector processor. A convergence rate estimate for the different versions of the conjugate gradient method is derived that indicates the behavior of the different versions for small time steps. Finally, as a computational example, each of the different versions of the conjugate gradient method are tested on the implicit equations that arise from the differencing of the two-dimensional heat equation on a highly distorted mesh. A time-step control is defined for the

II. SPATIAL DIFFERENCING

We first express the diffusion operator in terms of an arbitrary coordinate system. Let $S_{kl} = \{0 \le k \le N_k\} \times \{p \le l \le N_l\}$, where N_k, N_l are arbitrary positive integers. Let $\mathbf{R}: S_{kl} \to \tilde{\Omega} \subset \Omega$ be a twice differentiable function given by

$$\mathbf{R}(k, l) = \begin{bmatrix} r \\ z \end{bmatrix} = \begin{bmatrix} r(k, l) \\ z(k, l) \end{bmatrix}. \tag{4}$$

Here, (r, z) are the natural Cartesian coordinates. The Jacobian of $\mathbf{R}$ is

$$J(k, l) = \begin{bmatrix} \partial r/\partial k & \partial r/\partial l \\ \partial z/\partial k & \partial z/\partial l \end{bmatrix}, \tag{5}$$

where in order to assure that $\mathbf{R}$ is a one-to-one positively oriented transformation, we assume $j(k, l) = \text{determinant}[J(k, l)] > 0$ for all $(k, l) \in S_{kl}$. Thus, $\mathbf{R}$ defines a mapping of an orthogonal coordinate into a curvilinear coordinate system on $\tilde{\Omega}$. By the chain rule,

$$\begin{bmatrix} \partial/\partial k \\ \partial/\partial l \end{bmatrix} = \begin{bmatrix} \partial r/\partial k & \partial z/\partial k \\ \partial r/\partial l & \partial z/\partial l \end{bmatrix} \begin{bmatrix} \partial/\partial r \\ \partial/\partial z \end{bmatrix} = J^t(k, l) \begin{bmatrix} \partial/\partial r \\ \partial/\partial z \end{bmatrix},$$

so that

$$\begin{bmatrix} \partial/\partial r \\ \partial/\partial z \end{bmatrix} = [J^t(k, l)]^{-1} \begin{bmatrix} \partial/\partial k \\ \partial/\partial l \end{bmatrix}$$

$$= j^{-1}(k, l) \begin{bmatrix} \partial z/\partial l & -\partial z/\partial k \\ -\partial r/\partial l & \partial r/\partial k \end{bmatrix} \begin{bmatrix} \partial/\partial k \\ \partial/\partial l \end{bmatrix}$$

$$= j^{-1}(k, l) \left\{ \mathbf{N}_l \frac{\partial}{\partial k} - \mathbf{N}_k \frac{\partial}{\partial l} \right\},$$

$$\mathbf{N} = \begin{bmatrix} z \\ -r \end{bmatrix}, \quad \mathbf{N}_l = \frac{\partial \mathbf{N}}{\partial l}, \quad \mathbf{N}_k = \frac{\partial \mathbf{N}}{\partial k}.$$

The effect of this mapping on the diffusion operator is as follows:

$$\nabla \cdot \Lambda \nabla \phi(k, l) = j^{-1}(k, l) \left\{ \frac{\partial}{\partial k} a(k, l) \frac{\partial \phi}{\partial k} + \frac{\partial}{\partial l} b(k, l) \frac{\partial \phi}{\partial l} \right.$$

$$\left. - \frac{\partial}{\partial k} c(k, l) \frac{\partial \phi}{\partial l} - \frac{\partial}{\partial l} c(k, l) \frac{\partial \phi}{\partial k} \right\}, \tag{6}$$

where

$$a(k, l) = \mathbf{j}^{-1}(k, l)\Lambda(k, l)\langle \mathbf{N}_l, \mathbf{N}_l \rangle,$$
$$b(k, l) = \mathbf{j}^{-1}(k, l)\Lambda(k, l)\langle \mathbf{N}_k, \mathbf{N}_k \rangle,$$
$$c(k, l) = \mathbf{j}^{-1}(k, l)\Lambda(k, l)\langle \mathbf{N}_k, \mathbf{N}_l \rangle.$$

We are now in a position to difference the equation (1) with the diffusion term given by (6). Introduce the grid G of points $\{i, j\}$, $i = 0, 1, \ldots, N_k$, $j = 0, 1, \ldots, N_l$ on S_{kl}. This defines an orthogonal lattice on S_{kl} and the image $\mathbf{R}(G)$ is a lattice (not necessarily orthogonal) on $\tilde{\Omega}$, i.e.,

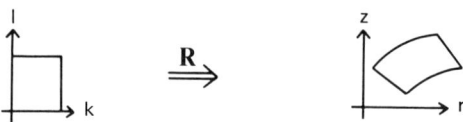

We assume that the coordinates of $\mathbf{R}(G)$ are given and attempt to use this information to difference (1).

Associate with each point $\{i - \frac{1}{2}, j - \frac{1}{2}\}$, $i = 1, \ldots, N_k$, $j = 1, \ldots, N_l$, a closed region $A_{i-1/2, j-1/2}$ bounded by the lines $k = i - 1$, $k = i$, $l = j - 1$, and $l = j$. Let $B_{i-1/2, j-1/2}$ be the image of $A_{i-1/2, j-1/2}$ under $\mathbf{R}$. Now, integrating (1) (Varga, 1962)

$$\iint h \frac{\partial \phi}{\partial t} d\tilde{\Omega} = \iint \nabla \cdot \Lambda \nabla \phi \, d\tilde{\Omega} - \iint s\phi \, d\tilde{\Omega}. \tag{7}$$

Summing the integrals over the different regions $\{B_{i-1/2, j-1/2}\}$ to achieve the calculation of (7) we arrive at

$$\iint h \frac{\partial \phi}{\partial t} dB = \iint \nabla \cdot \Lambda \nabla \phi \, dB - \iint s\phi \, dB, \tag{8}$$

where B is any of the regions $B_{i-1/2, j-1/2}$. Using the midpoint rule on the left of (8) (recalling area of $A_{i-1/2, j-1/2} = 1$),

$$\iint h \frac{\partial \phi}{\partial t} dB = \iint h(k, l) \frac{\partial \phi}{\partial t}(k, l) \mathbf{j}(k, l) \, dA \cong \left[h \frac{\partial \phi}{\partial t} \mathbf{j} \right]_{i-1/2, j-1/2}$$

[here we use the notation $f_{a,b} = f(a, b)$.] Now, the first term on the right of (8) is

$$\iint \nabla \cdot \Lambda \nabla \phi \, dB = \iint \nabla \cdot \Lambda \nabla \phi(k, l) \mathbf{j} \, dA.$$

Implicit/Diffusion Equation and Vectorization

Let C be the boundary of A. Then by (6) and Green's theorem,

$$\iint_A \frac{\partial}{\partial k} a(k,l) \frac{\partial \phi}{\partial k} dk\, dl = \int_C a(k,l) \frac{\partial \phi}{\partial k} dl \qquad (9)$$

$$\iint_A \frac{\partial}{\partial l} b(k,l) \frac{\partial \phi}{\partial l} dk\, dl = \int_C b(k,l) \frac{\partial \phi}{\partial l} dk, \qquad (10)$$

$$\iint_A \frac{\partial}{\partial l} c(k,l) \frac{\partial \phi}{\partial k} dk\, dl = \int_C c(k,l) \frac{\partial \phi}{\partial k} dk, \qquad (11)$$

$$\iint_A \frac{\partial}{\partial k} c(k,l) \frac{\partial \phi}{\partial l} dk\, dl = \int_C c(k,l) \frac{\partial \phi}{\partial l} dl. \qquad (12)$$

Central difference approximation to the spatial derivatives will be used and we adopt the following notation:

$$\begin{aligned}
&\delta_k f_{i,j} = f_{i+1/2,j} - f_{i-1/2,j}; \quad \delta_{2k} f_{ij} = (f_{i+1,j} - f_{i-1,j})/2; \\
&\delta_l f_{i,j} = f_{i,j+1/2} - f_{i,j-1/2}; \quad \delta_{2l} f_{ij} = (f_{i,j+1} - f_{i,j-1})/2; \\
&\Delta_k f_{i,j} = \tfrac{1}{2}[f_{i+1/2,j+1/2} + f_{i+1/2,j-1/2} - f_{i-1/2,j+1/2} - f_{i-1/2,j-1/2}]; \\
&\Delta_l f_{i,j} = \tfrac{1}{2}[f_{i+1/2,j+1/2} + f_{i-1/2,j+1/2} - f_{i+1/2,j-1/2} - f_{i-1/2,j-1/2}].
\end{aligned} \qquad (13)$$

Note that

$$\begin{aligned}
(\Delta_k + \Delta_l) f_{i,j} &= f_{i+1/2,j+1/2} - f_{i-1/2,j-1/2}, \\
(\Delta_k - \Delta_l) f_{i,j} &= f_{i+1/2,j-1/2} - f_{i-1/2,j+1/2}
\end{aligned} \qquad (14)$$

Consider now the interior points $\{i - \tfrac{1}{2}, j - \tfrac{1}{2}\}$, $i = 2, \ldots, N_k - 1$, $j = 2, \ldots, N_l - 1$ of S_{kl}.

If the midpoint rule is used on (9) and (10), we obtain the following difference formulas:

$$\begin{aligned}
\int_C a \frac{\partial \phi}{\partial k} dl &\cong \left[a \frac{\partial \phi}{\partial k}\right]_{i, j-1/2} - \left[a \frac{\partial \phi}{\partial k}\right]_{i-1, j-1/2} \\
&\cong a_{i, j-1/2} \delta_k \phi_{i, j-1/2} - a_{i-1, j-1/2} \delta_k \phi_{i-1, j-1/2} \\
&= a_{i, j-1/2}[\phi_{i+1/2, j-1/2} - \phi_{i-1/2, j-1/2}] \\
&\quad - a_{i-1, j-1/2}[\phi_{i-1/2, j-1/2} - \phi_{i-3/2, j-1/2}] \qquad (15)
\end{aligned}$$

$$\begin{aligned}
\int_C b \frac{\partial \phi}{\partial l} dk &\simeq \left[b \frac{\partial \phi}{\partial l}\right]_{i-1/2, j} - \left[b \frac{\partial \phi}{\partial l}\right]_{i-1/2, j-1} \\
&\cong b_{i-1/2, j} \delta_l \phi_{i-1/2, j} - b_{i-1/2, j-1} \delta_l \phi_{i-1/2, j} \\
&= b_{i-1/2, j}[\phi_{i-1/2, j+1/2} - \phi_{i-1/2, j-1/2}] \\
&\quad - b_{i-1/2, j-1}[\phi_{i-1/2, j-1/2} - \phi_{i-1/2, j-3/2}]. \qquad (16)
\end{aligned}$$

Using the trapezoidal rule on (11) and (12) we obtain

$$\int_C c \frac{\partial \phi}{\partial k} dk \cong \tfrac{1}{2}\left[\left(c\frac{\partial \phi}{\partial k}\right)_{i-1, j-1} + \left(c\frac{\partial \phi}{\partial k}\right)_{i, j-1}\right]$$
$$- \tfrac{1}{2}\left[\left(c\frac{\partial \phi}{\partial k}\right)_{i, j} + \left(c\frac{\partial \phi}{\partial k}\right)_{i-1, j}\right]$$
$$= \tfrac{1}{2}[(c \Delta_k \phi)_{i-1, j-1} + (c \Delta_k \phi)_{i, j-1}]$$
$$- \tfrac{1}{2}[(c \Delta_k \phi)_{i, j} + (c \Delta_k \phi)_{i-1, j}] \tag{17}$$

$$\int_C c \frac{\partial \phi}{\partial l} dl \cong \tfrac{1}{2}\left[\left(c\frac{\partial \phi}{\partial l}\right)_{i-1, j} + \left(c\frac{\partial \phi}{\partial l}\right)_{i-1, j-1}\right]$$
$$- \tfrac{1}{2}\left[\left(c\frac{\partial \phi}{\partial l}\right)_{i, j-1} + \left(c\frac{\partial \phi}{\partial l}\right)_{i, j}\right]$$
$$= \tfrac{1}{2}[(c \Delta_l \phi)_{i-1, j} + (c \Delta_l \phi)_{i-1, j-1}]$$
$$- \tfrac{1}{2}[(c \Delta_l \phi)_{i, j-1} + (c \Delta_l \phi)_{i, j}]. \tag{18}$$

Combining the above gives

$$\int_C c \frac{\partial \phi}{\partial k} dk + c \frac{\partial \phi}{\partial l} dl \cong \tfrac{1}{2}[c_{i-1, j-1}(\Delta_k + \Delta_l)\phi_{i-1, j-1}$$
$$+ c_{i, j-1}(\Delta_k - \Delta_l)\phi_{i, j-1} - c_{i, j}(\Delta_l + \Delta_k)\phi_{i, j}$$
$$- c_{i-1, j}(\Delta_k - \Delta_l)\phi_{i-1, j}].$$

Using the identities in (13), we obtain

$$\iint_{A_{i-1/2, j-1/2}} j\nabla \cdot \Lambda \nabla\phi \, dk \, dl$$
$$\cong r^E_{i-1/2, j-1/2}[\phi_{i+1/2, j-1/2} - \phi_{i-1/2, j-1/2}]$$
$$- r^W_{i-1/2, j-1/2}[\phi_{i-1/2, j-1/2} - \phi_{i-3/2, j-1/2}]$$
$$+ r^{NE}_{i-1/2, j-1/2}[\phi_{i+1/2, j+1/2} - \phi_{i-1/2, j-1/2}]$$
$$- r^{SW}_{i-1/2, j-1/2}[\phi_{i-1/2, j-1/2} - \phi_{i-3/2, j-3/2}]$$
$$+ r^N_{i-1/2, j-1/2}[\phi_{i-1/2, j+1/2} - \phi_{i-1/2, j-1/2}]$$
$$- r^S_{i-1/2, j-1/2}[\phi_{i-1/2, j-1/2} - \phi_{i-1/2, j-3/2}]$$
$$+ r^{SE}_{i-1/2, j-1/2}[\phi_{i+1/2, j-3/2} - \phi_{i-1/2, j-1/2}]$$
$$- r^{NW}_{i-1/2, j-1/2}[\phi_{i-1/2, j-1/2} - \phi_{i-3/2, j+1/2}] \tag{19}$$

(the schematic

Implicit/Diffusion Equation and Vectorization

NW	N	NE
W	$(i - \frac{1}{2}, j - \frac{1}{2})$	E
SW	S	SE

is used), and we define

$$r^W_{i-1/2,j-1/2} = [\rho_1 \langle \delta_l N, \delta_l N \rangle]_{i-1,j-1/2},$$
$$r^E_{i-1/2,j-1/2} = [\rho_1 \langle \delta_l N, \delta_l N \rangle]_{i,j-1/2},$$
$$r^{SW}_{i-1/2,j-1/2} = -[\sigma \langle \delta_{2k} N, \delta_{2l} N \rangle]_{i-1,j-1},$$
$$r^{NE}_{i-1/2,j-1/2} = -[\sigma \langle \delta_{2k} N, \delta_{2l} N \rangle]_{i,j},$$
$$r^S_{i-1/2,j-1/2} = [\rho_2 \langle \delta_k N, \delta_k N \rangle]_{i-1/2,j-1},$$
$$r^N_{i-1/2,j-1/2} = [\rho_2 \langle \delta_k N, \delta_k N \rangle]_{i-1/2,j},$$
$$r^{NW}_{i-1/2,j-1/2} = [\sigma \langle \delta_{2k} N, \delta_{2l} N \rangle]_{i-1,j}, \quad (20)$$
$$r^{SE}_{i-1/2,j-1/2} = [\sigma \langle \delta_{2k} N, \delta_{2l} N \rangle]_{i,j-1},$$
$$D_{ij} = (\mathbf{j}^{-1} \Lambda)_{i,j},$$
$$(\rho_1)_{ij} = \tfrac{1}{2}(D_{i-1/2,j} + D_{i+1/2,j}),$$
$$(\rho_2)_{ij} = \tfrac{1}{2}(D_{i,j-1/2} + D_{i,j+1/2}),$$
$$\sigma_{ij} = \tfrac{1}{8}(D_{i-1/2,j-1/2} + D_{i-1/2,j+1/2} + D_{i+1/2,j-1/2} + D_{i+1/2,j+1/2}),$$
$$\mathbf{j}_{ij} = [\langle \Delta R_k, \Delta N_l \rangle]_{ij}.$$

Focusing on the points $\{i - \frac{1}{2}, j - \frac{1}{2}\}$ adjacent to the boundary of S_{kl}, the portion of the line integrals that enter into the calculation of the boundary difference formulas are the lines that coincide directly with the boundary. Green's theorem is used to convert the volume integral to a line integral as before and the boundary conditions (2) are substituted when appropriate. To illustrate the idea, only the portion of the boundary given by $i = 0, 0 < j \leq N_l$, will be considered. As before,

$$\int_{j-1}^{j} a \frac{\partial \phi}{\partial k} dl \cong \left[a \frac{\partial \phi}{\partial k} \right]_{0, j-1/2}.$$

In this chapter only the following two cases will be considered:

(i) $\beta_{0, j-1/2} = 0$. Then $\alpha_{0, j-1/2} > 0$, and

$$\left[a \frac{\partial \phi}{\partial k} \right]_{0, j-1/2} = 2 a_{0, j-1/2} [\phi_{0, j-1/2} - \phi_{1/2, j-1/2}]$$

$$\simeq r^W_{0, j-1/2} \left[\left(\frac{\gamma}{\alpha} \right)_{0, j-1/2} - \phi_{1/2, j-1/2} \right].$$

(ii) $\beta_{0,j-1/2} > 0$. Then $\alpha_{0,j-1/2} = 0$, and

$$\left[a\frac{\partial\phi}{\partial k}\right] \simeq r^W_{0,j-1/2}\left[\frac{\tau - \alpha\phi}{\beta}\right]_{0,j-1/2}.$$

Since the values of $\phi_{0,j-1/2}$ are known boundary conditions, these quantities remain fixed throughout the solution of the problem and, in effect, represent numerical source terms in the calculation.

Finally, the midpoint rule is used on the second term of the right side of (8) to obtain

$$\iint s\phi \, dB = \iint s(k,l)\phi(k,l)\mathbf{j}(k,l) \, dA = [s\phi\mathbf{j}]_{i-1/2,j-1/2}.$$

III. MATRIX FORMULATION

The spatial differencing defined in Section II gives rise to a system of ordinary differential equations

$$C\frac{d\mathbf{\phi}}{dt} = A\mathbf{\phi} + \mathbf{b} - S\mathbf{\phi}, \tag{21}$$

where C, A, S are $N_{kl} \times N_{kl}$ matrices, $N_{kl} = N_k * N_l$, and $\mathbf{\phi}$ is an N_{kl}-dimensional vector of spatial approximations to ϕ. Different orderings of the dependent quantities $\{\phi_{l-1/2,k-1/2}\}$ give rise to different data structures associated with the difference approximations of (7). Orderings are defined by labeling each zone of the grid G with a distinct integer value ranging between 1 and N_{kl}. For example, the "natural" ordering (Dubois and Rodrigue, 1977) is given by

$2N_k + 1$		
$N_k + 1$	$N_k + 2$	
1	2	3

In this case,

$$\mathbf{\phi}(t) = \begin{bmatrix} \phi_1(t) \\ \phi_2(t) \\ \vdots \\ \phi_{N_{kl}}(t) \end{bmatrix},$$

Implicit/Diffusion Equation and Vectorization

$$C = \begin{bmatrix} \ddots & & 0 \\ & \mathbf{h} & \\ 0 & & \ddots \end{bmatrix}, \quad \mathbf{h} = \begin{bmatrix} (hj)_1 \\ (hj)_2 \\ \vdots \\ (hj)_{N_{kl}} \end{bmatrix},$$

$$S = \begin{bmatrix} \ddots & & 0 \\ & \mathbf{s} & \\ 0 & & \ddots \end{bmatrix}, \quad \mathbf{s} = \begin{bmatrix} (sj)_1 \\ (sj)_2 \\ \vdots \\ (sj)_{N_{kl}} \end{bmatrix}.$$

The matrix A is a banded matrix whose mth row vector (where the mth zone of G is not a boundary zone) is given by

$$\underbrace{0 \cdots 0 \, r_m^{SW} r_m^{S} r_m^{SE} \overbrace{0 \cdots 0}^{N_k} r_m^{W}}_{N_{kl}/2}, \; \sum_m r_m^{E} \overbrace{0 \cdots 0}^{N_k} r_m^{NW} r_m^{N} r_m^{NE} \, 0 \cdots 0,$$

$$\sum_m = -[r^{SW} + r^{S} + r^{SE} + r^{W} + r^{NW} + r^{N} + r^{NE}]_m.$$

When the mth zone is a boundary zone, the mth row vector still has the same structure as above but some of the r_ms will be zero.

Since we shall be solving the differential equation (21) on a vector processor, the vectors available within the equation must be isolated and identified. For the quantities ϕ, C, and S this is obvious. For the matrix A, two possible vector configurations are available. They are the *full zone* form

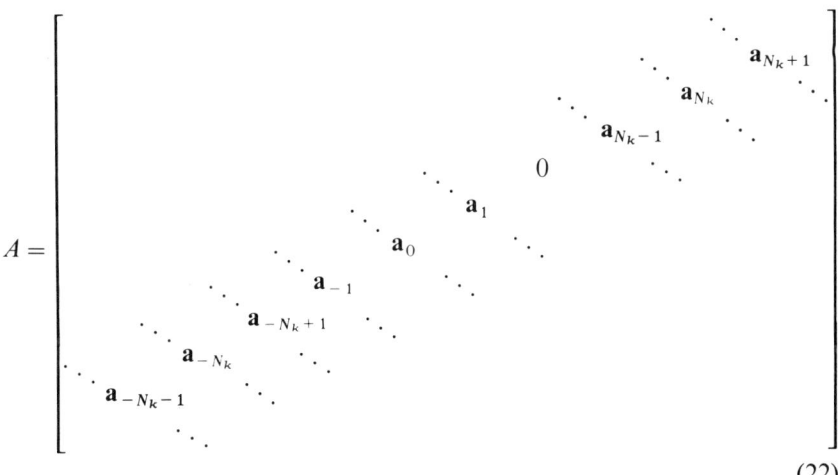

(22)

where the vectors $\mathbf{a}_i$ are $(N_{kl} - |i|)$-dimensional vectors and the *line* form

$$A = \begin{bmatrix} Y_1 & Z_1 & & & \\ X_2 & Y_2 & Z_2 & & \\ & \ddots & \ddots & \ddots & \\ & & & & Z_{N_l-1} \\ & & & X_{N_l-1} & Y_{N_l} \end{bmatrix}, \qquad (23)$$

where the submatrices X_i, Y_i, Z_i are $N_k \times N_k$ tridiagonal matrices. The pertinent vector lies within the tridiagonal systems as

$$\begin{bmatrix} \ddots & & & & 0 \\ & \ddots & & \mathbf{t}_1 & \\ & & \mathbf{t}_0 & & \ddots \\ & \mathbf{t}_{-1} & & \ddots & \\ 0 & & \ddots & & \end{bmatrix}.$$

Some useful notation to be used later is

$$\text{diag}(A) = \begin{bmatrix} \ddots & & 0 \\ & \mathbf{a}_0 & \\ 0 & & \ddots \end{bmatrix},$$

$$\text{bdiag}(A) = \begin{bmatrix} Y_1 & & & \\ & Y_2 & & \\ & & \ddots & \\ & & & Y_{N_l} \end{bmatrix},$$

where $\mathbf{a}_0$ and the Y_i are as in (22) and (23).

The vector **b** is an N_{kl}-dimensional vector containing the numerical approximations on the boundary. Hence, it contains mostly zero components except possibly at the components corresponding to boundary zones.

Another ordering that proves quite useful for vector processors is the "odd-even line" ordering (Heller, 1977). In this situation the zones of the grid G are numbered in the following manner:

N_k	$N_k + 1$		
$N_{kl}/2 + 1$	$N_{kl}/2 + 2$		
1	2	3	

For this ordering, the structure of the matrices C and S and the vectors $\boldsymbol{\phi}$ and $\mathbf{b}$ remain the same as before (however, their particular entries may have been shuffled around), but the structure of the matrix A is given by its *two-cyclic* form

$$\begin{bmatrix} Y_1 & & & & & Z_1 & & & & \\ & Y_3 & & & & X_3 & Z_3 & & & \\ & & \ddots & & & & X_3 & \ddots & & \\ & & & Y_{N_l-1} & & & & & X_{N_l-1} & Z_{N_l-1} \\ \hline Z_2 & X_2 & & & & Y_2 & & & & \\ & Z_4 & X_4 & & & & Y_4 & & & \\ & & \ddots & \ddots & & & & \ddots & & \\ & & & X_{N_l} & & & & & & \\ & & & Z_{N_l} & & & & & & Y_{N_l} \end{bmatrix}.$$

IV. PROPERTIES OF THE MATRIX A

Two important mathematical properties of the matrix A that are necessary for developing time stepping techniques are symmetry and definiteness. Formally, a matrix Q is said to be

(i) symmetric if $\langle Q\mathbf{y}, \mathbf{z}\rangle = \langle \mathbf{y}, Q\mathbf{z}\rangle$ for all vectors $\mathbf{y}, \mathbf{z}$;
(ii) negative definite (positive definite) if $\langle Q\mathbf{y}, \mathbf{y}\rangle < 0\ (>0)$ for all vectors $\mathbf{y} \neq 0$;

The approximations in Section II allow a heuristic establishment of these properties for the diffusion matrix A. To begin we have by Green's first identity that

$$\iint (\nabla \cdot \Lambda \nabla \psi)\psi\, d\tilde{\Omega} = -\iint \Lambda \langle \nabla\psi, \nabla\psi\rangle\, d\tilde{\Omega}$$
$$+ \int_{\tilde{\Gamma}} \mathbf{n} \cdot \Lambda \psi\, \nabla\psi\, d\tilde{\Gamma}, \quad (24)$$

where $\tilde{\Gamma}$ is the boundary of $\tilde{\Omega}$ and ψ is a sufficiently differentiable real-valued function. Now, if $\mathbf{y} = (\mathbf{y}_i)$ is an arbitrary N_{kl}-dimensional vector, let the function ψ be defined so that it is zero on $\tilde{\Gamma}$ and takes on the value y_i at the ith zone center of the grid G. Then, the line integral in (24) vanishes and, by assumptions (2),

$$\iint (\nabla \cdot \Lambda \nabla \psi)\psi\, d\tilde{\Omega} = -\iint \Lambda \langle \nabla\psi, \nabla\psi\rangle\, d\tilde{\Omega} < 0. \quad (25)$$

Using the same rectangular region $A_{i-1/2,\,j-1/2}$ (or A_m in the natural order index) as in Section II,

$$\iint (\nabla \cdot \Lambda \nabla \psi)\psi \, d\tilde{\Omega} = \sum_m \iint (\nabla \cdot \Lambda \nabla \psi)\psi \, dB_m$$

$$\cong \sum_m \psi_m \iint (\nabla \cdot \Lambda \nabla \psi) \mathbf{j} \, dA_m \quad \text{(midpoint rule)}$$

$$= \sum_m y_m \iint (\nabla \cdot \Lambda \nabla \psi) \mathbf{j} \, dA_m.$$

Using the approximation (18),

$$\sum_m y_m \iint (\nabla \cdot \Lambda \nabla \psi) \mathbf{j} \, dA_m \cong \langle A\mathbf{y}, \mathbf{y} \rangle.$$

The inequality in (25) is then invoked to establish heuristically that A is negative definite. Symmetry of A follows by the same argument as above; i.e.,

$$\langle A\mathbf{y}, \mathbf{z} \rangle = \iint (\nabla \cdot \Lambda \nabla \psi_1)\psi_2 \, d\tilde{\Omega} = \iint \Lambda \langle \nabla \psi_1, \nabla \psi_2 \rangle \, d\tilde{\Omega}$$

$$= \iint (\nabla \cdot \Lambda \nabla \psi_2)\psi_1 \, d\tilde{\Omega}$$

$$\cong \langle \mathbf{y}, A\mathbf{z} \rangle.$$

Note that by (2), the matrix $A - S$ is also symmetric and negative definite.

V. METHOD OF LINES

A simple form of the differential equation (21) is the equivalent

$$\frac{d\boldsymbol{\phi}}{dt} = T\boldsymbol{\phi} + \mathbf{q}, \tag{26}$$

where $T = C^{-1}(A - S)$, $\mathbf{q} = C^{-1}\mathbf{b}$. Conceptually, the solution of (26) is a set of N_{kl} curves in space each emanating from a zone center of the grid G. Formally, the solution of (26) is

$$\boldsymbol{\phi}(t) = e^{tT}[\boldsymbol{\phi}(0) + T^{-1}\mathbf{q}] - T^{-1}\mathbf{q}, \tag{27}$$

where

$$e^{tT} = \sum_{i=0}^{\infty} (i!)^{-1}(tT)^i. \tag{28}$$

Implicit/Diffusion Equation and Vectorization

It follows from the symmetry and negative definiteness of the matrix A that the eigenvalues of T are real and negative. Hence, by (27), ϕ has the dissipative property

$$\lim_{t \to \infty} \phi(t) = -T^{-1}\mathbf{q}. \tag{29}$$

Time-stepping techniques are generated by substituting a rational matrix polynomial $E(tT)$ for the exponential e^{tT} in (27). This yields the numerical formula

$$\psi(t) = E(tT)[\phi(0) + T^{-1}\mathbf{q}] - T^{-1}\mathbf{q}. \tag{30}$$

For stability reasons (Gear and Shampine, 1979), $E(tT)$ is required to have the property

$$\max |\text{eigenvalue}[E(tT)]| < 1$$

for all $t \geq 0$. From (29), it is also necessary that $E(tT)$ satisfy

$$\lim_{t \to \infty} \psi(t) = -T^{-1}\mathbf{q}.$$

That the Euler backward difference operator

$$E(tT) = [I - tT]^{-1} \tag{31}$$

satisfies these properties is easy to verify.

The local accuracy in time of this operator is

$$\|\psi(t) - \phi(t)\| \leq \|E(tT) - e^{tT}\| \, \|\phi(0) + T^{-1}\mathbf{q}\|$$
$$\leq \{\tfrac{1}{2}t^2\|T^2\| + O(t^3)\}\|\phi(0) + T^{-1}\mathbf{q}\|;$$

i.e., it is first-order accurate in time and the subsequential numerical method for solving (26) is given by

$$[I - tT]\psi(t) = \phi(0) - t\mathbf{q},$$

or the more computationally attractive form

$$[C - tQ]\psi(t) = C\phi(0) - tC\mathbf{q},$$
$$Q = A - S. \tag{32}$$

Note that the matrix $V(t) = C - tQ$ must be inverted to carry out the computation in (32). By the argument put forth in Section IV, A can be assumed to be symmetric and negative definite, so that $V(t)$ is symmetric and positive definite for $t \geq 0$. Hence the generalized conjugate gradient method

Concus et al., 1976) can be used to solve the system of equations. The algorithm is as follows:

Step 1:

(i) Pick $M(t)$—an arbitrary symmetric positive definite matrix.
(ii) Let $\mathbf{x}^{(0)} = \boldsymbol{\phi}(0)$ and $\mathbf{b} = C\boldsymbol{\phi}(0) - tC\mathbf{q}$.
(iii) Set $\mathbf{p}^{(-1)} = 0$ and $k = 0$.
(iv) Compute $\mathbf{r}^{(0)} = \mathbf{b} - V(t)\mathbf{x}^{(0)}$.

Step 2: Compute

(i) $M(t)\mathbf{z}^{(k)} = \mathbf{r}^{(k)}$,
(ii)
$$\beta_k = \begin{cases} 0, & k = 0 \\ \dfrac{\langle \mathbf{z}^{(k)}, \mathbf{r}^{(k)} \rangle}{\langle \mathbf{z}^{(k-1)}, \mathbf{r}^{(k-1)} \rangle}, & k \geq 1 \end{cases}$$

(iii) $\mathbf{p}^{(k)} = \mathbf{z}^{(k)} + \beta_k \mathbf{p}^{(k-1)}$.
(iv) $\mathbf{q}^{(k)} = V(t)\mathbf{p}^{(k)}$.
(v) $\alpha_k = \langle \mathbf{z}^{(k)}, \mathbf{r}^{(k)} \rangle / \langle \mathbf{p}^{(k)}, \mathbf{q}^{(k)} \rangle$.
(vi) $\mathbf{x}^{(k+1)} = \mathbf{x}^{(k)} + \alpha_k \mathbf{p}^{(k)}$.
(vii) $\mathbf{r}^{(k+1)} = \mathbf{r}^{(k)} - \alpha_k \mathbf{q}^{(k)}$.

Step 3:

(i) Compute $\text{res}^2 = \langle \mathbf{r}^{(k+1)}, \mathbf{r}^{(k+1)} \rangle / \langle \mathbf{b}, \mathbf{b} \rangle$.
(ii) If $\text{res} < 10^{-6}$, terminate; else set $k = k + 1$ and go to step 2.

The following important facts can be established (Concus et al., 1976) letting $\boldsymbol{\varepsilon}^{(k)} = \boldsymbol{\psi}(t) - \mathbf{x}^{(k)}$:

$$\mathbf{r}^{(k)} = \mathbf{b} - V(t)\mathbf{x}^{(k)} = V(t)\boldsymbol{\varepsilon}^{(k)},$$

$$\langle V(t)\boldsymbol{\varepsilon}^{(k)}, \boldsymbol{\varepsilon}^{(k)} \rangle \leq 4\left(\frac{\sqrt{\kappa}-1}{\sqrt{\kappa}+1}\right)^2 \langle V(t)\boldsymbol{\varepsilon}^{(k-1)}, \boldsymbol{\varepsilon}^{(k-1)} \rangle, \qquad (33)$$

where κ, the condition number of $[M(t)^{-1}V(t)]$ is given by

$$\kappa = \frac{\max |\lambda_i|}{\min |\lambda_i|}, \qquad \lambda_i \text{ an eigenvalue of } [M(t)^{-1}V(t)]. \qquad (34)$$

Since Q is negative definite,

$$\langle C\boldsymbol{\varepsilon}^{(k)}, \boldsymbol{\varepsilon}^{(k)} \rangle \leq \langle V(t)\boldsymbol{\varepsilon}^{(k)}, \boldsymbol{\varepsilon}^{(k)} \rangle,$$

so that if C is the identity matrix, (33) also gives a bound for mean square of the error.

VI. THE GENERALIZED CONJUGATE GRADIENT ALGORITHM

With the exceptions of the inner products and the solution of the system $M(t)\mathbf{z}^{(k)} = \mathbf{r}^{(k)}$, all of the operations in the conjugate gradient algorithm are highly amenable to vector processors (Heller, 1977). Since inner products are typically hardware instructions, "vectorizing" the conjugate gradient algorithm amounts to selecting a suitable matrix $M(t)$ with the following properties:

1. $M(t)$ should increase the rate of convergence of the algorithm which, according to (34), means the eigenvalues of $M(t)$ should be good approximations to those of $V(t)$;
2. $M(t)$ should be "computationally easy" to invert on a vector processor.

It is difficult to determine when property (1) holds for general systems $V(t)$ as most of the known results are for systems arising from the solution of Laplace's equation (Chandra, 1978). Property (2), on the other hand, is approachable when "computationally easy" on a vector processor and is equivalent to using the operation of matrix multiplication as much as possible within the inversion process (Karush et al., 1976). Conceptually, this amounts to determining the "explicitness" available within the "implicit" equations $V(t)$. In light of this, the following choices for $M(t)$ will be compared:

1. Point-Jacobi (PJ): Let

$$M(t) = C - t \operatorname{diag}(Q). \tag{35}$$

Then $M(t)$ is a diagonal matrix and

$$M(t)^{-1} V(t) = I + O(t). \tag{36}$$

The relation (36) indicates the behavior of the condition number κ in (34) with respect to the step size t. Here, no extra computer memory is required and the inversion process is a simple vector division by an N_{kl}-dimensional vector. (When the inversion process involves only vectors of length N_{kl}, then it is said to be *full zone*.)

2. Line-Jacobi (LJ): Let

$$M(t) = C - t \operatorname{bdiag}(Q). \tag{37}$$

Then $M(t)$ is a block-diagonal matrix with blocks consisting of tridiagonal matrices. Also $[M(t)^{-1} V(t)] = I + O(t)$. Inverting $M(t)$ consists of inverting a sequence of tridiagonal matrices in parallel for which several techniques exist (Heller, 1977). Typically, these techniques require additional memory to accommodate two vectors of length N_{kl}. All of the techniques have in common a decomposition phase and a back-substitution phase. For the conjugate gradient method the decomposition phase can be performed before the algorithm begins and the back-substitution phase is performed during each iteration. The decomposition phase involves vectors of length N_k. (The process is thus called a line process.) On the other hand, the more time-consuming back-substitution phase is a full-zone process.

3. Symmetric Gauss–Seidel (Young, 1971) (SGS): Let $Q = \text{bdiag}(Q) - L - L^t$, where

$$L = \begin{bmatrix} 0 & & & & \\ L_1 & 0 & & & \\ & L_2 & 0 & & \\ & & \ddots & \ddots & \\ & & & L_{N_l-1} & 0 \end{bmatrix}$$

and the L_i are tridiagonal matrices. Define

$$M(t) = [C - t\,\text{bdiag}(Q) - tL][C - t\,\text{bdiag}(Q)]^{-1}[C - t\,\text{bdiag}(Q) - L^t]. \tag{38}$$

Then $M(t)^{-1}V(t) = I + O(t^2)$, indicating that the convergence rate should be better than the point- and line-Jacobi versions for small values of t. The inversion of $M(t)$ is a 3-stage process:

1. inversion of $C - t\,\text{bdiag}(Q) - tL$,
2. multiplication by $C - t\,\text{bdiag}(Q)$,
3. inversion of $C - t\,\text{bdiag}(Q) - tL^t$.

$[C - t\,\text{bdiag}(Q) - tL]$ has the form

$$\begin{bmatrix} X_1 & & & & 0 \\ Y_2 & X_2 & & & \\ & Y_3 & \ddots & & \\ & & \ddots & \ddots & \\ 0 & & & Y_{N_l} & X_{N_l} \end{bmatrix},$$

where the X_i and Y_i are tridiagonal matrices. The solution of the system $[C - t\,\text{bdiag}(Q) - tL]\mathbf{z} = \mathbf{w}$ is given by the first-order vector recursion (a line process)

$$\begin{aligned} \mathbf{z}_1 &= X_1^{-1}\mathbf{w}_1, \\ \mathbf{z}_i &= X_i^{-1}[\mathbf{w}_i - Y_i\mathbf{z}_{i-1}], \quad 2 \le i \le N_l. \end{aligned} \tag{39}$$

The multiplication by $[C - t\,\text{bdiag}(Q)]$ is a full-zone process and the inversion of $[C - t\,\text{bdiag}(Q) - tL^t]$ is done with a first-order vector recursion similar to (39). The vector recursion requires the inversion of each of the blocks in $C - t\,\text{bdiag}(Q)$ and, as in the line-Jacobi method, their decompositions can be done before the conjugate gradient algorithm begins and back substitution is performed during each iteration. However, because of the recursiveness of (39), the back substitution is a line process. The memory requirement is two vectors of length N_{kl}.

4. *Two-Cyclic Symmetric Gauss–Seidel (TCSGS):* For this version, it is assumed that the matrix A has been permuted to its two-cyclic form. Then $V(t)$ is also in two-cyclic form. Define $M(t)$ as in the symmetric Gauss–Siedel version. Then the matrix $[C - t\,\text{bdiag}(Q) - tL]$ has the form

$$\begin{bmatrix} X_1 & 0 \\ \hline Y & X_2 \end{bmatrix},$$

where

$$X_i = \begin{bmatrix} W_1^{(i)} & & & 0 \\ & W_2^{(i)} & & \\ & & \ddots & \\ 0 & & & W_{N_l/2}^{(i)} \end{bmatrix},$$

$$Y = \begin{bmatrix} G_1 & H_1 & & & \\ & G_2 & H_2 & & 0 \\ & & \ddots & \ddots & \\ 0 & & & & H_{N_l/2-1} \\ & & & & G_{N_l/2} \end{bmatrix}$$

and the $W_j^{(i)}, G_j, H_j$ are tridiagonal matrices. The solution of $[C - t\,\text{bdiag}(Q) - tL]\mathbf{z} = \mathbf{w}$ is given by the simple two-term vector recursion

$$\mathbf{z}_1 = X_1^{-1}\mathbf{w},$$
$$\mathbf{z}_2 = X_2^{-1}[\mathbf{w}_2 - Y\mathbf{z}_1].$$

In this case, the inversion of X_1 and X_2 involves the inversion of $N_l/2$ tridiagonal matrices. Again, the decomposition of these matrices is done before starting the conjugate gradient algorithm and the back substitution is performed done during each iteration. However, in this case the back-substitution process involves vectors of length $N_{kl}/2$. The memory requirement is two vectors of length N_{kl}.

5. *Incomplete Cholesky (IC):* The basic motivation behind the incomplete Cholesky version of the conjugate gradient algorithm is that a symmetric positive definite matrix can always be expressed as a product LL^t, a Cholesky decomposition, where L is a lower triangular matrix. The matrix L is calculated by the Gaussian elimination process and, even though the original matrix may have many zero terms, L can be densely nonzero. The matrix $M(t)$ is generated by performing the Gaussian elimination algorithm on $V(t)$, purportedly to calculate its Cholesky decomposition $L(t)L(t)^t$, but instead processes elements in $L(t)$ that are in the same matrix position as nonzero elements of $V(t)$ while the other elements are set immediately to zero (Martin *et al.*, 1974). $M(t)$ is then defined by

$$M(t) = \tilde{L}(t)\tilde{L}(t)^t, \tag{40}$$

where matrix $\tilde{L}(t)$ has the form

$$\tilde{L}(t) = \begin{bmatrix} X_1 & & & \\ Y_1 & X_2 & & \\ & \ddots & \ddots & \\ & & Y_{N_l-1} & X_{N_l} \end{bmatrix}$$

and the X_i are bidiagonal matrices having nonzero diagonal and subdiagonal elements and the Y_i are tridiagonal matrices. The solution of the system $\tilde{L}(t)\mathbf{z} = \mathbf{w}$ is given by the first-order vector recursion relation (39) (Greenbaum and Rodrigue, 1977). Unfortunately the matrix $M(t)$ does not bear any resemblance to $V(t)$ and memory must be allocated to accomodate the storage of the matrix $\tilde{L}(t)$, i.e., five vectors of length N_{kl}.

VII. COMPUTATIONAL EXAMPLE

The diffusion equation

$$\frac{\partial \phi}{\partial t} = \Lambda \nabla \cdot \nabla \phi, \qquad \Lambda = 47.56,$$

is solved on the grid given in Fig. 1. The boundary conditions are

$$\frac{\partial \phi}{\partial z}(t, r, 0) = \frac{\partial \phi}{\partial z}(t, r, 1) = 0,$$

$$\phi(t, 1, z) = 3,$$

$$\phi(t, 0, z) = 0,$$

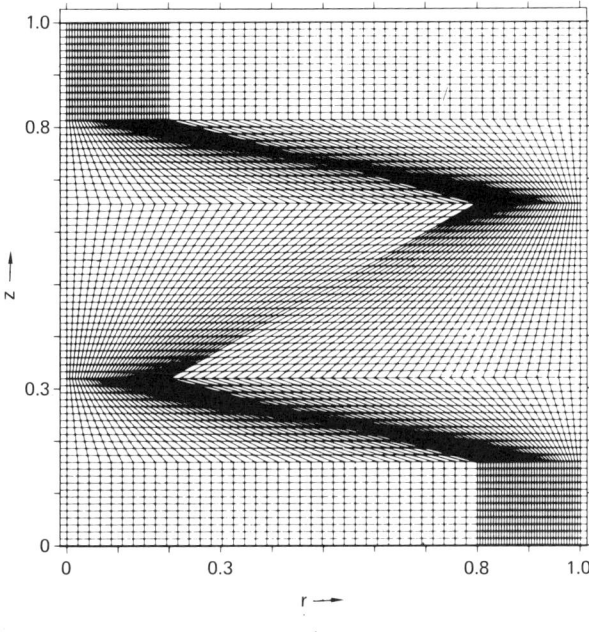

Fig. 1

and the initial conditions

$$\phi(0, r, z) = 0, \quad r \neq 1,$$
$$\phi(0, 1, z) = 3.$$

The method of lines is used to generate the ordinary differential equation

$$\frac{d\mathbf{\phi}}{dt} = A\mathbf{\phi} + \mathbf{q},$$

where the matrix A is given by the formulas (19), (20). For a given time step Dt, the method (32) is used to update the dependent variable $\mathbf{\phi}$ to the next time level.

The time steps are given by

$$Dt^{(0)} = 10^{-11},$$
$$Dt^{(n+1)} = \beta Dt^{(n)},$$

where

$$\beta = 0.65 + 0.7/(1 + \alpha^2)$$
$$\alpha = 6.65 \frac{\|\mathbf{\phi}^{(n+1)} - \mathbf{\phi}^{(n)}\|_\infty}{\|\mathbf{\phi}^{(n)}\|_\infty}$$

(here $\|v\|_\infty = \max |v_i|$). Note that

$$\alpha = 0 \Rightarrow \beta = 1.35,$$
$$\alpha \to \infty \Rightarrow \beta = 0.65,$$
$$\alpha = 1 \Rightarrow \beta = 1.0.$$

The first two implications indicate that if change in ϕ is small, then the step size is increased by a factor of 1.35, and if the change is large then the step size is decreased by a factor of 0.65. The last implication indicates that a constant step size is maintained provided the relative change in ϕ is no greater than 15%. Figure 2 indicates some of the isolines at steady state using the point-Jacobi version of the conjugate gradient algorithm as the linear system solver. Figure 3 indicates the behavior of the step size Dt during the computer run.

Figure 4 illustrates the behavior of the iteration count per cycle for each of the versions of the conjugate gradient method described in Section VI. When the algorithm required the solution of tridiagonal systems of equations, e.g., line-Jacobi, the decomposition phase was done in the usual scalar recursive Gauss elimination process and the back substitution was done using the recursive doubling algorithm (Dubois and Rodrigue, 1977). The reason for using a scalar algorithm for the decomposition of a sequence of tridiagonal matrices, rather than simply decomposing the matrices in parallel, is that the CDC-STAR-100 computer does not perform well on vector algorithms involving noncontiguous vectors. This, of course, is not the case for the

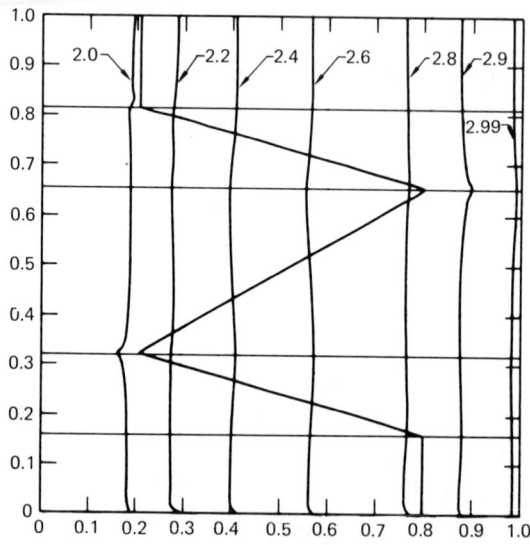

Fig. 2

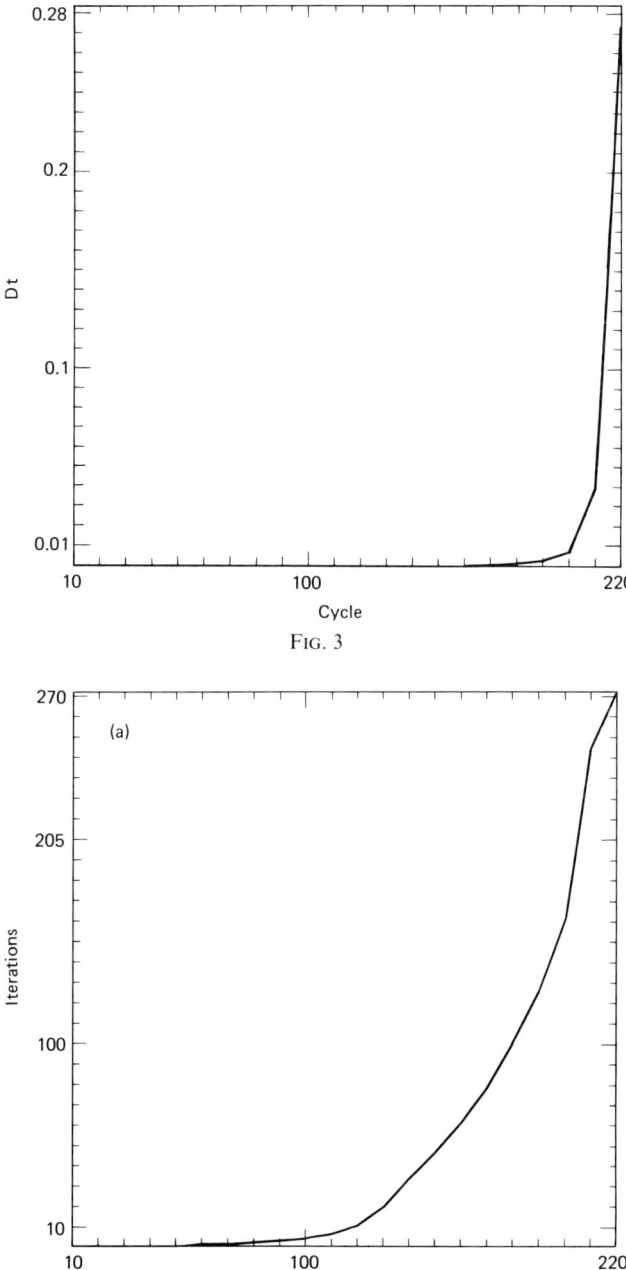

FIG. 3

FIG. 4a

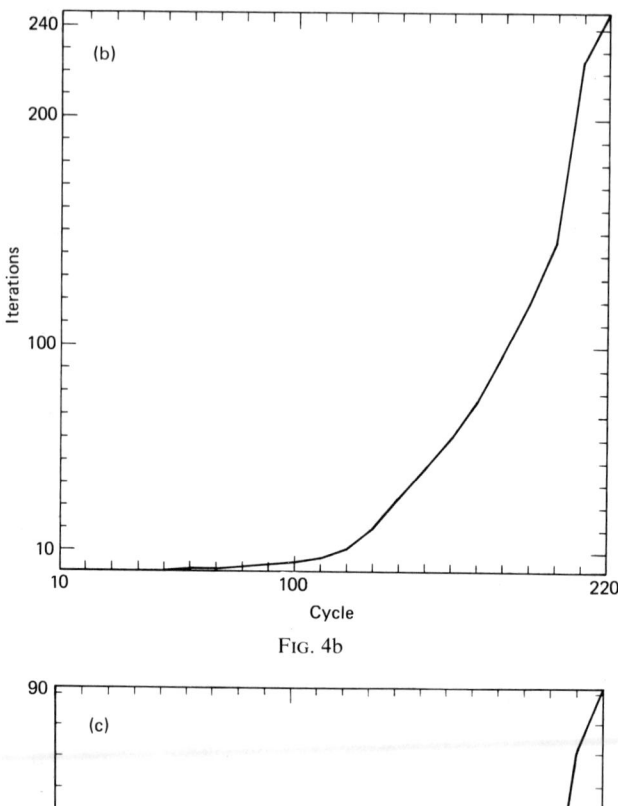

Fig. 4b

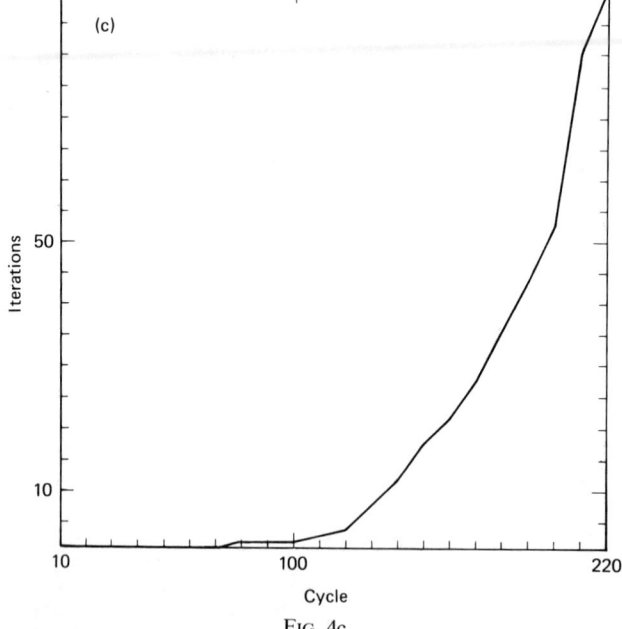

Fig. 4c.

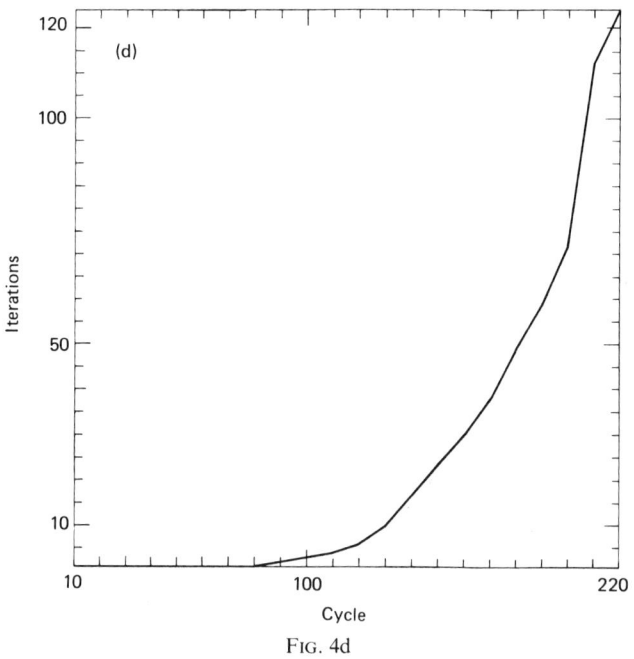

FIG. 4d

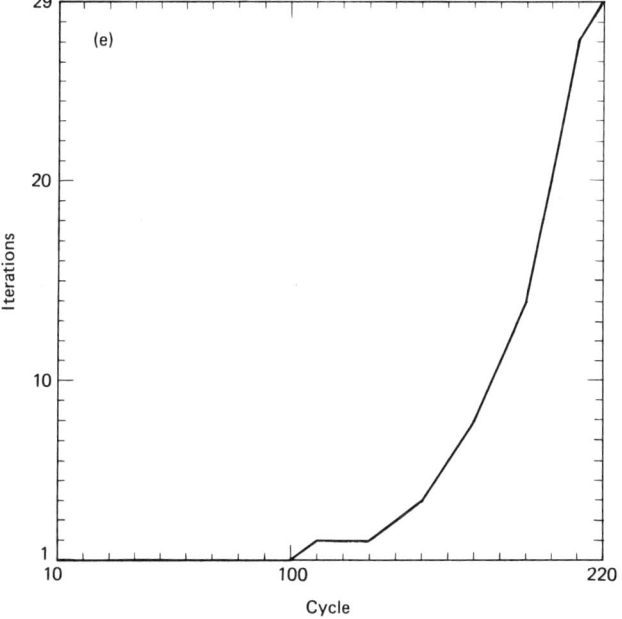

FIG. 4e

TABLE I

Method	Time (sec)	S
Point-Jacobi	170.6	1.65
Line-Jacobi	318.7	0.88
Symmetric Gauss–Seidel	616.4	0.46
Two-cyclic symmetric Gauss–Seidel	247.8	1.13
Incomplete Cholesky	281	1

Cray-1 machine. However, since the decomposition need be performed only once during the conjugate gradient algorithm, the insertion of a scalar decomposition method into the process does not significantly degrade the execution time.

The first column of Table I lists the total computing time expended on a CDC-STAR-100 for each of the versions of the conjugate gradient algorithm. The second column lists the speed-ups, where

$$S = \frac{\text{(time for the IC version)}}{\text{(time for other version)}}.$$

The IC version was chosen as the normalizer since it achieved the least amount of overall iterations/cycle. All versions were coded in LRLTRAN with vector extensions (Martin et al., 1974).

VIII. COMMENTS AND CONCLUSIONS

As can be seen from Fig. 4 and Table I, iterative algorithms yielding the least number of iterations need not be computationally the fastest. The point-Jacobi version, although its iteration count per time cycle far exceeds the other version, is still computationally faster on a vector machine such as the CDC-STAR-100. The reason for this is clear in that the point-Jacobi method is a highly vectorizable full-zone method, an ideal setting for a machine such as the STAR, whereas the other schemes are only line methods. The timing results are, of course, machine dependent, and may vary when the algorithms are performed on a different vector computer such as the Cray-1, which has a better capability of dealing with line methods than the STAR. For other machines, other versions of the conjugate gradient may be better, and a series of tests such as those elaborated in Section I in this paper must be performed to again obtain the optimal one. However, the results also yield information regarding the use of explicit and implicit time stepping schemes on vector computers. Although the full implicit backward Euler method was used for numerically solving the differential equation (26), the most efficient version of the conjugate gradient method is the point-Jacobi, the version that has the most explicitness within it, i.e., the version that makes the most use of the full zone matrix multiplication operation. This naturally leads to the question whether more explicit time stepping techniques for numerically solving the differential equation (26) would be better.

The basic problem with explicit schemes is their inherent conditional stability condition. Although an explicit time step can be executed quite rapidly on a vector processor, it is not difficult to generate an equation (26) with enough "stiffness" in it to force the time step to be so small that it becomes computationally unattractive. Hence, some form of implicitness is required in the time differencing scheme in order to maintain a reasonable computational step size. This leads to the possibility of combining an explicit with an implicit scheme in the same manner as predictor–corrector schemes are used in the solution of ordinary differential equations (Berger *et al.*, 1980). Whatever the direction that is taken, it is clear from the results of Section VII that explicit schemes can be used to a higher degree on vector computers than they have been on scalar computers.

References

Ames, W. G., Buning, P. G., Calahan, D. A., Orbits, D. A., and Sesek, E. J. (1979). Sparse matrix and other high-performance algorithms for the CRAY-1. SEL Rep. No. 124, January, University of Michigan, Ann Arbor.

Berger, M., Oliger, S., and Rodrigue, G. (1980). Predictor-corrector methods for the solution of time-dependent parabolic problems on parallel processors. *Proc. Conf. Elliptic PDE Methods, Santa Fe, New Mexico, July.*

Chandra, R. (1978). Conjugate gradient methods for partial differential equations. Ph.D. Thesis, Dept. of Computer Science, Yale University, New Haven, Connecticut.

Concus, P., Golub, G., and O'Leary, D. (1976). A generalized conjugate gradient method for the numerical solution of elliptic partial differential equations. Stanford Computer Science Rep. STAN-CS-76-533, Stanford University, Palo Alto, California.

Crowley, W. P., Hendrickson, C. P., and Rudy, T. E. (1978). The SIMPLE code. Rep. UCID-17715, Lawrence Livermore Laboratory, Livermore, California.

Dubois, P. F., and Rodrigue, G. (1977). An analysis of the recursive doubling algorithm. *Proc. Symp. High Speed Comput. Algorithm Organ., University of Illinois, Urbana, April,*

Dubois, P. F., Greenbaum, A., and Rodrigue, G. (1979). Approximating the inverse of a matrix for use in iterative algorithms on vector processors. *J. Comput.* **22**, 257–258.

Gear, C. W., and Shampine, L. F. (1979). A user's view of solving stiff ordinary differential equations. *SIAM Rev.* **21**, 1–17.

Greenbaum, A. (1981). Comparison of splittings used with the conjugate gradient algorithm. *Numer. Math.*

Greenbaum, A., and Rodrigue, G. (1977). The incomplete Cholesky conjugate gradient method for the STAR (5-point operator). Rep. UCID-17574, Lawrence Livermore Laboratory, Livermore, California.

Heller, D. (1977). Direct and iterative methods for block tridiagonal linear systems. Ph.D. Thesis, Dept. of Computer Science, Carnegie-Mellon University, Pittsburgh, Pennsylvania.

Karush, J., Madsen, N., and Rodrigue, G. (1976). Matrix multiplication by diagonals on a vector/parallel processor. *J. Inform. Process Lett.* June, 1976.

Kershaw, D. (1978). The incomplete Cholesky-conjugate gradient method for the iterative solution systems of linear equations. *J. Comput. Phys.* **26**, 43–65.

Martin, J. T., Solbeck, S., and Zwakenberg, R. (1974). LRLTRAN language used with the CHAT and STAR compilers. Rep. M-026, Lawrence Livermore Laboratory, Livermore, California.

Pomraning, G. C. (1973). "The Equations of Radiation Hydrodynamics." Pergamon, Oxford.

Richtmyer, R. D., and Morton, K. W. (1967). "Difference Methods for Initial Value Problems." Wiley (Interscience), New York.

Schreiber, R. Implementation of the conjugate gradient method on the CDC-STAR-100. Personal communication.

Tarter, C. B., and Westbrook, C. K. (1975). On protostellar evolution. *Astrophys. J.* **200**, 48–60.

Varga, R. (1962). "Matrix Iterative Analysis." Prentice-Hall, Englewood Cliffs, New Jersey.

Weinberg, A. M., and Wigner, E. P. (1958). "The Physical Theory of Neutron Chain Reactors." University of Chicago Press, Chicago, Illinois.

Young, D. M. (1971). "Iterative Solution of Large Linear Systems." Academic Press, New York.

Swimming Upstream: Calculating Table Lookups and Piecewise Functions

Paul F. Dubois

Mathematics and Statistics Division
Lawrence Livermore National Laboratory
University of California
Livermore, California

I. INTRODUCTION TO TABLE LOOKUP

Locating a given value in an increasing table of values is like trying to swim upstream since it almost seems that a vector processor is designed to do everything except the table lookup. In designing algorithms for a vector processor, you will be able to exploit its increased competence in familiar mathematical operations, but you will notice all the difficulties of an ordinary computer remain for table lookup. This is essentially because a *decision* about what to do is being made in table lookup, instead of actually *doing* a lot of things.

Table lookup is an extremely common calculational operation in modern computational physics. By table lookup, we mean locating a given value in an

---

* This work was performed under the auspices of the U.S. Department of Energy by Lawrence Livermore National Laboratory under Contract W-7405-Eng-48.

increasing array or "table." This is almost always part of the evaluation of a function which has been defined in a piecewise manner or tabulated at a finite set of values. At Lawrence Livermore National Laboratory (LLNL), the most important table lookup problem is the equation-of-state problem, a piecewise function of two variables, which is described in more depth in Section V. In order to describe the problems of this calculation, we begin our discussion of algorithms for performing table lookup and other piecewise-defined functions with the simpler case of one-dimensional table lookup.

A. *One-Dimensional Table Lookup*

Given a value x and a table of increasing values $t_1 < t_2 < \cdots t_n$, the table lookup problem in one dimension is to find $i = i(x)$ so that i is the least integer such that $x \leq t_i$. The most common situation in which this arises is when we have pairs (t_i, f_i), $i = 1, N$, of values of a function f, that is, $f_i = f(t_i)$. If we wish to evaluate f at some point x, then we need to know $i(x)$, no matter what interpolation scheme we use. It is quite common for f to represent physical data taken at different times or different spatial locations. For example, in an atmospheric computer program, or computer "code," t_i may represent a height and f_i some property at that height, such as sound speed. Since the sound speed of the real atmosphere varies in no simple way with height, a tabular representation may be used, with values given at certain heights and interpolation used at intermediate values.

The lookup operation is frequently bypassed. This is done, for example, by fitting an analytic model to the data—a process which may be difficult and time consuming. A faster table lookup algorithm can eliminate the need for this fitting process in the cases where the principal reason for doing the fitting is to permit rapid function evaluation. We investigate various strategies for this one-dimensional table lookup problem in Section IV.A.

B. *Vector Arguments*

The second major category of problems we describe are those in which x is a vector, usually of considerable length (200–20,000). In computational physics, x usually represents values at the points of a computational mesh, and it is necessary to evaluate a function at all of the points. The important thing about the problem in which x is a vector is that the values of different components of x are almost never random. Rather, there is usually a high degree of correlation between different components. On the simplest level, the probability that $i(x_{j+1}) = i(x_j)$ is usually fairly high. Thus an approach that fails to take advantage of this is probably inferior to one that does. Some ideas for the case where x is a vector are discussed in Section IV.B.

C. Equation-of-State Calculations

A history of equation-of-state calculations as performed on the STAR and the Cray-1 at LLNL is given in Section V. This is the problem on which the most effort has been expended since it is the one that has the largest payoff in time savings. At LLNL, an improvement by a factor of 10 in the EOS algorithm saves more than a million dollars worth of computer time in a year. As discussed in Section V, equation-of-state tables are one of the principal connections between codes and the properties of materials in the real world.

II. EVALUATING ALGORITHMS ON VECTOR PROCESSORS

There is no longer any simple, unified way to evaluate the performance of algorithms on vector or parallel processors. The standard method of evaluating algorithms by counting the asymptotic number of operations is now at best only a rough guideline. We shall see examples of its failure below and in Section IV. As yet, no unified framework has been developed which can help us decide between two algorithms.

There are, however, a few concepts that do not work in evaluating algorithms. First, tradition has it that numerical stability is an absolute must. I still think so, but at least one person has told me they are using the recursive doubling algorithm despite its instability, as shown in Dubois and Rodrigue (1977), and with good success.

Tradition also has it that one can count only the multiply/divide operations and get a good estimate of the speed of an algorithm. Purists may insist on counting the additions and subtractions as well. Memory reference time is ignored as "negligible." Yet it is precisely memory reference time which dominates the behavior of the Cray-1 and STAR-100. Reducing memory traffic is the soul of the clever but rather bizarre algorithm of Fong and Jordan (1977) for full linear systems. (I feel one is allowed to call an algorithm bizarre which starts off by fetching columns of a matrix from the bottom up!)

Third, one cannot use a higher-level language to get a "feel" for the running time and assume that the speedup obtainable by hand coding is uniform across algorithms. Sometimes the difference can be an order of magnitude. Indeed, some of the interesting algorithms are not even expressible in most higher-level languages, as we shall see.

A. How Operation Counts Can Be Misleading

Here is an example of the inadequacy of "operation counts." Suppose we wish to compute $C(I) = A(I) + B(I)$ for $I = 1, 2, \ldots, N$. There are clearly N addition operations. The old implicit assumptions were

(1) Time used = (number of operations) × (cost per operation).
(2) Cost per operation is a fixed feature of the hardware.
(3) Therefore, time used is proportional to the number of operations.

This argument does not hold on vector processors where the cost per operation is *variable*. Consider the above calculation as performed on the Cray-1. If done in a DO loop, the operations are

(a) Fetch $A(I)$
(b) Fetch $B(I)$
(c) Add $A(I) + B(I)$
(d) Store in $C(I)$

This takes 28 cycles, of which only 8 are the addition (see Table I for details). The storing of C does not inhibit the loop, so we ignore its completion time. Thus the total time will be close to $28N$. More sophisticated strategies for this calculation have been evolved on scalar machines to try to "hide" some of the memory time (McMahon et al., 1972).

Table II shows a detailed analysis of the same loop performed in vector mode. However, such analysis is not necessary if we use T. Jordan's "chime" concept (Fong and Jordan, 1977): a chime is a unit of time equal to one vector operand of length 64 passing through a functional unit. The different functional units have different times, but since they range around 10 cycles, 1 chime is roughly $64 + 10 = 74$ cycles, or just short of 1 μsec (80 cycles = 1

TABLE I

CYCLE ANALYSIS OF AN ADDITION LOOP[a]

Operation	Cycle number at start	Cycle number at finish
Fetch A	0	11
Fetch B	2	13
Add	13	21
Store C	21	(33)
Loop back	23	28

[a] It takes 28 cycles per element to add two arrays on the Cray-1.

Calculating Table Lookups and Piecewise Functions

TABLE II
Cycle Analysis of a Vector Addition of Length 64[a]

Operation	Cycle number at start	Cycle number at end	"Chime" number
Fetch A	0	75	1
Fetch B	68	77	2
Add $A + B$	82	149	2
Store result	149	218	3

[a] The fetch of B begins when the memory unit is free, at cycle 68. The addition chains to this fetch. No chaining is allowed for a store. Note that 3 chimes is approximately 3 μsec, or 240 cycles.

μsec). If several vector operations "chain" together the total time remains around 1 "chime." Thus we can analyze the vector add as

Chime 1: Fetch A
Chime 2: Fetch B and add
Chime 3: Store C and loop

The total time is thus 3 chimes/loop, for $N/64$ loops, or

$$\text{time per element} = \frac{(80 \text{ cycles/chime})(N/64 \text{ loops})(3 \text{ chimes/loop})}{N \text{ elements}}$$

$= 3(\frac{80}{64})$, or about 4 cycles/element.

For a simple add operation, 80 cycles/chime is 10% too high, but this method gives us a pretty good idea without a lot of detailed counting.

The end result is that a "plus" can cost from 3.5 to 28 cycles, depending on how it is done. If the operation had been $C(I) = (A(I) + B(I)) * 3.14$, then the cost of the scalar loop would be 37 cycles/element, but the vector computation would remain less than 4 cycles/element. Therefore a scalar calculation can be an order of magnitude slower than a vector calculation.

Much is made in traditional analysis of the virtues of an algorithm that is order of N, $o(N)$, instead of $o(N \log_2 N)$, or $o(N^2)$. But, since the cost of an operation may vary greatly, such distinctions can be very misleading over a realistic range for N.

In concepts like the "chime," we see the beginnings of a framework for an evaluation of algorithms for the parallel and vector processors. Lacking such a framework, each algorithm must be carefully analyzed for a given machine, with timing tests on assembly language implementations being our only final judge.

III. BASIC PROCESSES ON VECTOR PROCESSORS

Here are some concepts which can help us design vector algorithms in general, and table lookup algorithms in particular. These basic processes are *conceptual processes* that have different implementations, and hence different relative speeds, on different types of machines. We describe these processes here to avoid interrupting the discussion later; the experienced reader may wish to skip ahead to Section IV.

A *bit array* is an array Z whose elements $Z(I)$ are bit quantities. Bit arrays arise frequently from comparisons. We might set $Z = X$ .LT. Y, where X and Y are conformable arrays. Then $Z(I)$ would be 1 where $X(I)$ .LT. $Y(I)$, and 0 elsewhere.

A. Description of the Processes

A *compress* of X into Y on a bit array Z has as its arguments Z and the array X, of the same length. The output Y is an array whose length is the number of 1 bits in Z and consists of those elements $X(I)$ for which $Z(I) = 1$, in order. The length of Y is known after the operation. We write $Y = X(Z)$.

Expand is the opposite: $X(Z) = Y$ spreads the elements of Y into the elements of X for which $Z(I) = 1$.

Scatter and *gather* are another pair of basic processes. We are given two arrays A and B, not necessarily of the same length, and an index vector I of the same length as A. The entries of I are integers between 1 and the length of B. To *gather* A from B by I means to set $A(J) = B(I(J))$ for $J = 1, \ldots,$ length(A). We write $A = B(I)$. To *scatter* is to set $B(I(J)) = A(J)$, and we write $B(I) = A$.

These four concepts are all hardware instructions on the CDC STAR 100 and its descendants. On the STAR, compress/expand is very fast and gather/scatter is very slow. Gather/scatter collectively are called Transmit Indexed List (TIL) instructions, and avoiding them was an early watchword on the STAR. CDC has been working on improving such instructions for future versions.

Surprisingly, compress is not necessarily expensive on other machines. Usually, the bit pattern of Z is not random but consists of relatively long patches of all ones or all zeros. This allows some use of the vector registers on Cray, for example.

The final process to be described is what I call the *cache loop*. The idea behind a cache loop is that there are registers, possibly including vector registers, within which one can operate very quickly as long as no arguments need to be fetched from memory to the registers (we call the registers the "cache"). On the Cray and STAR there are also mechanisms for mass transfers of data from memory at high data rates. By a cache loop, we mean doing a calculation staying entirely within the registers. For example, the calculation

```
      C(1) = 1
      DO 100 I = 2, 1000
      C(I) = A(I) + B(I)
100   IF (C(I).LT. A(I) * B(I)) C(I) = 2. * C(I-1)
```

might be performed on the Cray by fetching blocks of A and B and forming $A + B$ and $A * B$ in vector registers, forming a bit vector $(A + B).LT.(A * B)$ in the vector mask register, and then changing $A + B$ where required by pulling elements over to the scalar registers, performing the required arithmetic, and replacing the result before storing C. This avoids many time-wasting processes which a typical FORTRAN compiler would go through if presented with this loop. Much arithmetic can be overlapped, no shuffling of temporary results to memory takes place, and the vector registers are used where possible. Cache loops are of particular interest for table lookup problems, calculating array extremes, and array norms, since such operations are not normally vectorizable.

IV. ONE-DIMENSIONAL PROBLEMS

A. *Functions with Scalar Arguments*

A piecewise function f of one variable on a scalar argument x is given by

$$f(x) = f_i(x), \qquad t_{i-1} < x \leq t_i,$$

where $t_1 < t_2 < \cdots < t_M$ and $t_1 < x \leq t_M$. In practice M is usually a modest number, say in the range 5–500, and we shall imagine it so here. In saying that x is a scalar, we mean to imply that we have no information about $i = i(x)$. The intervals (t_{i-1}, t_i) are often referred to as "boxes," and $i(x)$ is sometimes called the "index" or "box index" of x.

Such a function is usually evaluated in two steps:

(1) Table lookup: find $i(x)$;
(2) Evaluation: find $f_{i(x)}(x)$.

Usually (1) is supplied as a central user utility, but (2) is left up to the particular user. However, such a separation is not at all necessary. One can imagine algorithms in which $f_i(x)$ could be calculated for *all i at the same time as* deciding which i is correct. On multiple processors this may not be as silly as it sounds for a standard computer. However, we confine ourselves here to the traditional lookup problem in which (1) and (2) are performed separately. Specifically, we discuss various algorithms to meet the following specifications:

FUNCTION LUF $(X, T\ N)$ returns

(1) 1 if $N \leq 0$,
(2) the least J such that $X \leq T(J)$, or
(3) $N + 1$ if $X > T(N)$,

where $T(1) < T(2) < \cdots < T(N)$ is an increasing array.

In evaluating algorithms for LUF we will assume that $5 \leq N \leq 500$, and X is equally likely to fall in each box, with a small chance, say 2%, of $X < T(1)$ or $X > T(N)$. Property (1) can be useful to users in certain circumstances not of interest in this discussion.

Accepted wisdom tells us that, especially for $N \geq 10$, the bisection method is the best algorithm for table lookup. We shall see that for the Cray-1, this accepted wisdom is always wrong. We shall compare three methods: bisection, vector scan, and M-section.

1. Bisection

In roughly $\log_2 N$ iterations, $i(x)$ is found by testing a middle element that falls between the current upper and lower bounds. How fast can we do bisection on the Cray? At each iteration we must

(1) Determine the next J
(2) Fetch $T(J)$
(3) Compare with X
(4) Decide whether to change the lower or the upper bound
(5) Repeat until the upper and lower bounds differ by 1.

Between successive iterations we might overlap (5) with (1) and (2), but (2)–(4) have to be done essentially in sequence. The fetch takes 11 cycles. The comparison, a subtract, takes 6. The decision takes 12 cycles in our implementation of the bisection method, and it is difficult to imagine it taking much less since a jump cannot start until 2 cycles after the subtract completes, and since it takes 5 cycles at the least if the branch is taken. Table III shows an analysis of our implementation that takes 35 cycles per iteration after the first. The loop jump is hidden by the fetch operation.

Calculating Table Lookups and Piecewise Functions

TABLE III

Cycle Analysis of Bisection[a]

	Assembly language			Cycle number	
Label	Instruction	Operand	Comment	At start	At end
loop	s0	s6 − fs7	$t(j) − x$	0	6
	a6	a5	low = j	1	3
	a7	a1	high	2	4
	jsm	newlo	branch if $t(j) < x$	8	13
	a6	a4	low	10	12
	a7	a5	high = j	11	13
newlo	a5	a6 + a7	low + high	13	15
	a1	a7	new high	14	16
	a7	a6 + 1		15	17
	s5	a5		16	18
	s5	s5 > 1	(low + high)/2	13	21
	a0	a1 − a7	is low + 1 = high?	19	21
	a5	s5	new j	21	22
	a7	a5 + a2	.loc. $t(j)$	22	24
	s6	,a7	read $t(j)$	24	*35
	a4	a6	new low	26	28
	jan	loop		27	30

[a] The loop begins with a low index in a4, a high index (LUF) in a1, x in s7, an intermediate index j in a5, and $t(j)$ in s6. It requires 35 cycles before the instruction at "loop" can execute again.

Since we have to do $\log_2 N$ such iterations, our required time is $35 \log_2 N$, plus some overhead.

2. Linear Vector Scan

Since the Cray's vector registers hold 64 elements, 64 comparisons of X with $T(1)–T(64)$ can all be done simultaneously. A vector mask can be formed on $T(I) − X < 0$, and the number of on bits is the desired index—unless they are all on, indicating that $X > T(64)$. In that case, another 64 elements must be read.

A timing analysis made on one iteration is straightforward. Reading, subtracting, and vector-mask forming all are chaining processes, so an iteration takes a chime. Using the vector mask and deciding what to do next probably take 10–15 cycles, so in round numbers we are now looking at 100 cycles per iteration. The number of iterations on the average is $(1 + N/64)/2$— since we get to stop when we succeed—for a total of $50(1 + N/64)$. Vector

TABLE IV

Vector Linear Scan versus Bisection (Cycles)[a]

Length of table, N	Bisection $35 \log_2 N$ (cycles)	Vector linear scan (approximate) (cycles)
8	105	45
16	140	55
32	175	75
64	210	100
128	245	150
256	280	250
512	315	450
1024	350	850

[a] The number of cycles for a total subroutine call would be greater by about 80 in each case.

linear scan will be faster than bisection over most of the range of interest for N, as we see in Table IV.

3. M-Section

M-section is a vector analog of bisection. In the linear vector scan algorithm, we formed $T(I) - X$, where $I = 1, \ldots, 64$. However, we could just as well have formed $T(I) - X$ for $I = 1, 1 + L, 1 + 2L, \ldots, 1 + (M - 1)L$, where $M \leq 64$ and $1 + (M - 1)L \leq N$. That is, we pick M evenly spaced elements of T and locate the piece in which X lies. We repeat until $L = 1$, roughly $\log_M N$ times.

Each of these iterations takes an "M chime" plus some considerable overhead. For $M = 33$, our implementation yielded 93 cycles/iteration. In this case, $\log_M N \approx \log_2 N/5$, so the total time is about $19 \log_2 N$, compared to $35 \log_2 N$ for bisection. Table V shows a FORTRAN equivalent of M-section, and Table VI shows a cycle analysis of our implementation for $M = 33$. The M-section algorithm could include a test at each stage, so that if $N \leq 64$, it finishes in one read, i.e., changes M to N. We did not include this improvement in order to keep the discussion simpler.

Table VII shows comparative times, in microseconds per element, for bisection and M-section. The implementation for 64-section is different than for 33-section. We note that 33-section is universally 30–40% faster than bisection, as predicted. Those who believe operation counts tell the story are invited to count.

TABLE V

THE M-SECTION ALGORITHM IN FORTRAN

```
      FUNCTION LUF(X,T,NBIG)
      DIMENSION T(NBIG)
      PARAMETER (M=33)
C
      N=NBIG
      LUF=1
      IF(N.LE.0) RETURN
C
    1 L=(N-1)/(M-1)
      IF (L.EQ.0) GO TO 2
C PICK UP EVERY L'TH ELEMENT STARTING AT T(LUF)
C GET M ELEMENTS.  FIND K=NUMBER OF THEM .LE. X
      K=0
      DO 100 I=LUF,N,L
  100 IF(T(I).LE.X) K=K+1
C
      J=1+(K-1)*L
      IF(K.EQ.0) THEN
         RETURN
      ELSE
         LUF=LUF+J
      ENDIF
      IF(K.EQ.M) THEN
         N-N-J
      ELSE
         N=L-1
      ENDIF
      IF(N.GT.0) GO TO 1
      RETURN
C SPECIAL CASE FOR L=0, I.E. N .LE. M
    2 K=0
      DO 200 I=LUF,N,1
  200 IF(T(I).LE.X) K=K+1
      LUF=LUF+K
      RETURN
      END
```

[a] Note that in FORTRAN it is difficult even to express the basic concept of the algorithm as imagined originally. The coding in loop 100 is ridiculous, for example. We choose M one more than a power of 2, so that the calculation of L can be done with a shift operation.

TABLE VI
Cycle Analysis of M-Section for $M = 33$[a]

Assembly language instruction	Operand	Comment	Cycle number at start	Cycle number at end or chains	Timing notes
A3	A2 − 1	$N - 1$	0	2	
	(Calculate $L = (N - 1)/32$)		1	8	
VL	A3	Vector length 33	8	9	
JSZ	LAB2	To $L = 0$ case	10	15	
A4	S1	L	12	13	
A0	A1 + A5	Address T(LUF)	13	15	
V1	,A0 ,A4	Read M elements	15	24	Chime 1
	(Calculation of $1 + (M - 1)L$ hidden here)				Free
V2	S7 − FV1	$X - T$	24	32	Chain
	(More arithmetic for loop hidden here)				Free
VM	V2, P	Form mask	32	71	End 1
	(And more)				Free
S2	VM	Get mask	71	72	
S0	S2		72	73	
A6	PS2	$K =$ Count 1's	73	77	
JSZ	OUT	Done if $K = 0$	75	80	
	(Update of LUF here)		77	85	
JSN	ECASE	To $K = M$ case	86	91	
	(Jump to start of loop)		88	93	

[a] One iteration is 93 cycles. Odd cases are not shown.

TABLE VII
Average Time for One Call to LUF[a]

	Time (μsec/N)			
N	Bisection	33-section	17-section	64-section
5	2.38	1.86	1.86	1.91
15	3.10	1.98	1.98	2.03
30	3.54	2.17	2.43	2.23
40	3.74	2.45	2.76	2.34
50	3.88	2.60	2.78	2.48
100	4.34	2.99	2.82	3.48
200	4.76	3.03	2.89	3.80
300	5.02	3.06	3.10	3.90
400	5.20	3.10	3.67	3.85
500	5.32	3.13	3.39	3.92
1000	5.73	3.33	3.64	3.98

[a] 33-section is the best method for LUF.

Calculating Table Lookups and Piecewise Functions

A final note about the STAR and one-dimensional table lookup: the STAR has a hardware instruction for table lookup! The lifetime of the STAR at Livermore was short, and nobody investigated this question in any depth for scalar x. However, it is interesting to note that

(1) For $N = 10$, the hardware instruction takes over 4 µsec.
(2) For $N = 100$, it takes about 6 µsec.
(3) The process is $o(N)$.

Therefore, we can say that with the large startup time, it is terrible on short tables, and with the $o(N)$ behavior, it is terrible on long tables. But whether or not there was a better way, I do not know. The same instruction can also handle vector arguments but was never used, to my knowledge, because of these difficulties. Sometimes even a hardware instruction is a bad algorithm.

B. Functions with Vector Arguments

We consider now the same problem but with a vector argument $\mathbf{x} = (x_1, x_2, \ldots, x_L)$. We are to find $\mathbf{y} = (y_1, y_2, \ldots, y_L)$ by applying the rule $y_j = f_i(x_j)$, where $i = i(x_j)$ is the least i such that $t_i \geq x_j$. As indicated in the introduction to this chapter, the values $i(x_j)$ are rarely independent. The most frequent variations seem to be

(1) The values $x_1, \ldots, x_L$ are monotone increasing.
(2) The values $x_1, \ldots, x_L$ are autocorrelated, in the sense that $i(x_{j+1})$ is likely to be $i(x_j)$.

The first case arises because the x_i represent values obtained in discretizing a partial differential equation in some variable such as space, frequency, energy, or angle.

The second case occurs in both one- and two-dimensional space discretizations. Suppose, for example, that x_i represents temperature at the ith point on the Lagrangian grid shown in Fig. 1. There are two materials present, for example CO_2 and O_2, and we desire to calculate some tabular function of temperature, say, specific heat. Suppose the points are numbered left to right, and bottom to top. If we make a vector of all the O_2 points, perhaps by means of a compress operation using a bit vector for material, we see that neighboring x_i are usually close in space, too, and are therefore probably close in temperature and are likely to fall in the same box. Of course, real problems have many more grid points and many more materials, but the same argument applies. The values shown in Table VIII were gathered over a period of 100 cycles from a typical small hydrodynamics problem at LLNL. The 100 cycles were chosen from the middle of the problem history. There were about

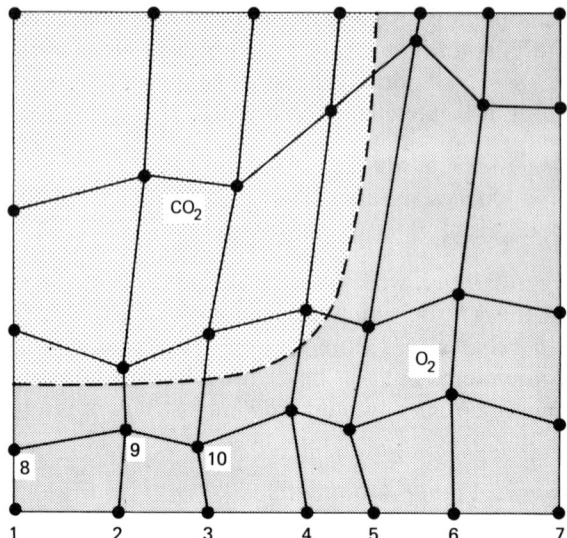

FIG. 1. Autocorrelation comes from spatial discretization. Only the first few grid points are numbered to illustrate the numbering system. The dashed line illustrates the boundary between the two materials.

1000 zones, about as coarse a spatial resolution as is ever used—and seven materials. We see that even so, the probability that $i(x_{j+1}) = i(x_j)$ in each material vector is nearly 90%.

The question of the separation of evaluation and lookup, which we dismissed in the scalar argument case, is not so easy to dismiss for vectors. The nature of the functions f_i becomes important, such as the number of coefficients there are, whether external functions are required, and whether evaluation requires a great many registers. These issues can profoundly affect

TABLE VIII

Autocorrelation in the Hydrodynamics Test Problem

Change in temperature-box index between successive points $\|i(x_{j+1}) - i(x_j)\|$	Times difference observed (%)
0	89
1	3
2–4	5
More than 4	3

algorithm choice. To discuss the vector case at all, we must settle on a problem and hope discussion of it can illuminate cases of interest to the reader. Since linear interpolation is common, we shall use it as our test problem. We shall discuss three strategies: cache loop, compress–sort, and TIL.

1. Cache-Loop Strategy

A cache loop for one-dimensional table lookup with a vector argument would be easy to design. On the Cray-1, we could read a block of 64 values for x and a beginning pair of box boundary values, say t_1, t_2. The x_j can be moved, one at a time, to an S-register and compared with the current box boundaries. We keep the required interpolation coefficients in the cache (registers), changing them only when required. We have two choices: we can either go ahead and evaluate the function or we can store the coefficients in a temporary location and do a vector evaluation later. The temporary location could be a vector register or memory. This decision is one case where the nature of the function is important, since the number of coefficients would limit our choices. If we plan to form temporary coefficient vectors, we again have a choice of stopping after every 64 to do an evaluation or using full vector-length temporaries. A scalar evaluation seems unlikely to be a good idea if we have kept in mind the expense of scalar evaluations, as discussed in Section II.

2. Compress–Sort Strategy

The compress–sort is a clever strategy invented for the STAR by LLNL's Jim Kohn. By doing vector comparisons of the form $t_i \geq x$, we form the bit vectors $b_i = t_{i-1} < x \leq t_i$.

Then the compress $y = x(b_i)$ yields a vector y whose members all lie in the same box. Because of the autocorrelation property, generally only a few boxes hold most of the components, so the vector y tends to have a fairly long length. The coefficients required are now scalars, so a vector evaluation is fast and simple. Expand operations are used for the purpose of returning the results in the correct order.

3. TIL Strategy

The TIL strategy consists of first calculating the box indices $i(x_j)$ so that one has a box index vector I, where $I_j = i(x_j)$. This can be done, for example, by a loop of the form

(1) $I = 0$,
(2)   for $i = 1, N$ let $I = I + 1$, where $t_i \leq X$.

The "controlled addition" in this loop is a natural operation on the STAR. Assuming the box coefficients have been stored in an array, say $C(1)$ to $C(N)$ the TIL gather instruction $Y(I) = C$ will form a coefficient vector to be used in the evaluation. One TIL is needed for each coefficient. There is currently no machine on which the TIL strategy can compete with the compress–sort. However, hardware improvements could change that.

Insufficient effort has been put into the one-dimensional problem with a vector argument because of the predominance of the two-dimensional equation-of-state problem. The reader is especially invited to think about this problem in the context of data-flow machines.

We offer no comparative timings of these algorithms here, although the results of the next section may aid in making an informed guess. I believe the cache loop would prove superior to the compress–sort on the Cray, and the compress–sort is the best on the STAR; I have not tried them. These judgments are formed on the basis of my experience on the equation-of-state problem.

V. TWO-DIMENSIONAL PROBLEMS: EQUATIONS OF STATE

A. *The EOS Problem*

LLNL has many codes whose running time on a production problem is 10 or more hours. Many of these codes model hydrodynamic phenomena as well as other physical processes. The basic equations of hydrodynamics can be formulated in many ways; for example,

$$\frac{\partial \rho}{\partial t} + \frac{\partial (\rho u)}{\partial x} = 0, \tag{1}$$

$$\frac{\partial (\rho u)}{\partial t} + \frac{\partial}{\partial x}(\rho u^2 + p) = 0, \tag{2}$$

$$\frac{\partial (\rho E)}{\partial t} + \frac{\partial}{\partial x}(\rho u E + pu) = 0, \tag{3}$$

where ρ is density, p is pressure, E is total energy, and u is velocity. Note that we have one more unknown than equations. The additional equation required is called an equation of state (EOS) because it, and it alone, expresses the relations between the variables, relations which are material dependent. An EOS may be simple, such as a gas law, or a complex table lookup. Many of LLNL's codes use a so-called BQL (biquadradic Lagrange) scheme which involves two-dimensional tables of, say, pressure as a function of temperature and density. For each material there is a density table $\rho_1, \rho_2, \ldots, \rho_{NR}$, and a temperature table $t_1, \ldots, t_{NT}$. The table lengths NR and NT are usually in the range 5–100. These tables define an $(NR - 1)$ by $(NT - 1)$ array of boxes, where the (i, j) box is the set of (ρ, t) such that $\rho \in (\rho_i, \rho_{i+1})$, and $t \in (t_j, t_{j+1})$. [The tables have the property that $t \in (t_1, t_{NT})$ and $\rho \in (\rho_1, \rho_{NR})$, so there is no concern about arguments out of range.] The pressure P is then given by

$$P = P(\rho, t) = \sum_{k=1}^{k=3} \sum_{m=1}^{m=3} C(i, j, k, m) \rho^{k-1} t^{m-1},$$

where $i = i(\rho)$ and $j = j(t)$. In short, on each box we have a biquadratic function of ρ and t.

Naturally, in practice ρ and t are *vectors* of length $L = 1000-20{,}000$, with perhaps 5000–10,000 being more typical. The ρ and t vectors are highly autocorrelated, as discussed in Section IV.B.

The EOS table lookup is one of the few connections with reality that the computer code has. It is how the codes know the difference between iron and air. And, for a reasonably straightforward hydrodynamics code, a scalar EOS lookup can occupy 60–70% of the computation time.

Although this calculation has always been a problem at LLNL, it became more noticeable when the STAR machines were introduced. When a major time-consuming code at the laboratory uses most of that time in one subroutine, some attention begins to be paid to it. The history of this problem at LLNL is, I believe, an important case study. It is one of the few cases of extended work on a single topic in the short history of vector algorithms, the other one of note being tridiagonal systems of equations. Also, the work was done in a practical context. The EOS subroutine must handle some exceptional cases, such as the user requesting a temperature floor (t_f, P_f) so that $P \equiv P_f$ if $t \leq t_f$, or a density multiplier wherein ρ is multiplied by a constant before lookup and evaluation. These practical constraints rule out some of the behavior seen in some academic research, such as codes that only work if N is five, codes that modify themselves on the run, etc.

B. A TIL-Based Method

The first attempt at an EOS routine on the STAR was a TIL-based method. The behavior of scalar methods was so bad that some attempt had to be made to vectorize the unvectorizable. An index vector I may be formed by the following procedure, which is related to the one mentioned in Section IV:

(1) $I = 1$
(2) For $j = 2, NT$,
    do: $I = I + 1$ where $t \geq t_j$
(3) For $j = 2, NR$,
    do: $I = I + NT - 1$ where $\rho \geq \rho_j$

I_j now contains the "box number" of (ρ_j, t_j). Nine coefficient vectors $C_{k,m}$, $k = 1, 3$, $m = 1, 3$, can be formed, by TIL instructions, from the data base. Then the evaluation can take place at full vector length.

The time to perform (2) and (3) is $NT' + NR'$ compare operations, where $NT' \le NT$ and $NR' \le NR$ because we can stop whenever, for example, all $t < t_j$; the STAR has hardware which recognizes this condition. Therefore the total time is $(NT' + NR')$ times the cost of a compare, plus 9 TIL instructions.

This algorithm is usually faster than scalar codes by a factor of two, despite the slow speed of the TIL of 20 times the length of the vector. However, after implementation of the TIL-based method, the major time-using code was still spending 50% of its time in EOS.

C. A Compress–Sort Method

By thinking about sorting cards by suit and number, Jim Kohn reinvented the "radix sort," a method of sorting Knuth (1968) explains by means of the exact same analogy. This section gives the details of Kohn's radix sort.

Let I be a vector of the same length as ρ and t, initialized to $I(i) = i$. Let b_{old} and b_{new} be bit vectors of the same length, with b_{old} initialized to "1."

For $i = 2, NR$ do:

(a) Let $b_{new} = \rho \le \rho_i$
(b) Let $b_{old} = b_{old}$ .AND. b_{new}

Thus, $b_{old} = 1$, where $\rho_{i-1} < \rho < \rho_i$.

(c) Let $\rho' = \rho(b_{old})$
$t' = t(b_{old})$
$I' = I(b_{old})$

Let $N(i)$ be the length of ρ', t', and I'. Advance the pointers for ρ', t', and I' so that the next series of compresses will place the next piece of ρ' right after the current piece, etc.

(d) Let $b_{old} = $ .NOT. b_{new}
[End i]

At the conclusion of the $i = 2, NR$ loop, we loop again on $j = 2, NT$. We denote the end of the j loop by [end j] below the end of this loop. This loop is complicated only by the need to calculate precisely the length of the set of components in the (i, j) box. The reader should picture that at this stage ρ' is ρ sorted by ρ_i value, and t and I have been sorted to match, so that

$$\rho'(j) = \rho(I'(j))$$

Calculating Table Lookups and Piecewise Functions

and

$$t'(j) = t(I'(j)),$$

and $N(i)$ is the total number of ρ in the (i, j) box summed over j.
Again, present b_{old} to "1."
Do $j = 1, k - 1$

(a) $b_{new} = t \leq t_j$
(b) $b_{old} = b_{old}$ .AND. b_{new}
(c) Compress ρ' to ρ'' on b_{old}
t' to t'' on b_{old}
I' to I'' on b_{old}

The result of this compress is that the resultants ρ'', t'' are sorted by box, the result of the compress being a vector of all elements in boxes $(1, j)$ through (L, j). If by $b_{old}(i)$ we mean those $N(i)$ bits corresponding to the part of ρ' produced by the first round of compresses, then the number of "1" bits in $b_{old}(i) = \text{pop}(b_{old}(i))$ is the length of the part of ρ'' which holds the "occupants" of box (i, j). Therefore, we can compute $P(\rho, t)$:

(d) Do $i = 1, L$. Let ρ_τ, t_τ be the portions of ρ'', t'' corresponding to the (i, j) box. The addresses of ρ_τ, t_τ can be initially set to $\rho''(1), t''(1)$ and lengths $n_i = \text{pop}(b_{old}(i))$. Then after each run through this loop the addresses can be incremented by n_i to point to the beginning of the next portion.

(i) Compute $P_\tau = P(p_\tau, y_\tau) = \sum_k \sum_m C(i, j, k, m) p_\tau^{k-1} y_\tau^{m-1}$.
(ii) Advance addresses of $\rho_\tau, y_\tau, P_\tau$.
(iii) As an added subtlety, by subtracting n_i from $N(i)$ one can keep track of the number of elements in each group $N(i)$ that have been "finished." When this number is zero, we can bypass this i on future passes of the j loop.
[End i]

(e) Advance addresses of ρ'', t'', I'' to next available location.
(f) Set $b_{old} = $ .NOT. b_{new}. [End j]

We now have $P = $ the concatenation of all the P_τ but need to restore it to its original order. Fortunately, $I(k)$ is the index of ρ and t to which $P(k)$ corresponds, so that

$$P(k) = P(\rho(I(k)), t(I(k))).$$

Therefore, by a TIL instruction we get P_{output}, so that

$$P_{output}(I) = P.$$

D. *Timing Comparison*

Without computing timing counts in detail, we can estimate the difference between the compress sort and the TIL method for reasonably large vector length. We call the methods OLD and NEW. The work done in OLD not done in NEW is roughly 8 TIL instructions plus $NT' + NR'$ controlled adds. Let N be the length of the vectors.

The work done in NEW not done in OLD is of two sorts. First, there is some extra overhead and some bit operations, such as population counts. Part of this extra overhead is the extra start-ups involved in calculating P in pieces instead of all at once. Secondly, NEW uses $3(NT' + NR')$ compresses, on vectors of length N, where NT', NR' are the number of occupied ρ and t boxes, respectively, since we can avoid doing a compress on an all-"0" bit vector. The difference, ignoring the controlled adds in OLD and the extra overhead in NEW, is $8(\text{TIL}) - 3(NT' + NR')$ (compress). On the STAR this is

$$8(124 + 20N) - 3(NT' + NR')(92 + N)$$
$$= 922 + 160N - (NT' + NR')(276 + 3N).$$

Ignoring all but terms containing N, NEW will be superior if $160N - (NT' + NR') * 3N > 0$, or if $NT' + NR' < 53$, approximately. For LLNL's EOS tables, this is almost always true. In fact, on a typical problem $NT' + NR'$ is usually in the range of 8–12.

The increase in speed was quite dramatic, as shown in Table IX. The cost of EOS lookups went from 50% of the running time to about 5%, so EOS had

TABLE IX

Timing Comparison of Various EOS Algorithms on the STAR

Method	Time per zone (μsec)
Scalar	100–200
TIL	36–79
Compress–sort	3–10

[a] The timing is problem dependent. The numbers shown are a range, with the upper number being more typical, for problems in which the vector length is 1000 or more.

ceased to be a problem (Dubois and Kohn, 1978). An inverse EOS package that relies chiefly on a similar double compress sort was also developed by S. I. Warshaw and myself (Dubois, 1978). It solves the problem $t = t(P, \rho)$, given the BQL table structure. However, when the Cray-1 arrived, suddenly everything was "back to square 1." The algorithm above had been hand coded for the STAR, so someone threw together a straightforward scalar FORTRAN version that called a scalar routine in a loop.

The Cray is good at scalar operations, but nonetheless the EOS routine was suddenly 60% of the running time again. Jim Kohn had now transferred to another assignment, so because of my inverse EOS work it was natural that I should try to develop a new algorithm for the Cray.

The inverse EOS package was written in portable vector syntax and thus was able to provide an idea of how well the compress strategy would work on the Cray. The results were startling. The inverse EOS routine ran somewhat faster than the regular EOS routine. Since the inverse package was about 3–5 times slower than the regular package on the STAR, I had an immediate estimate that a factor-of-5 savings could be gained by using the radix sort.

However, I decided to try a cache loop based upon the autocorrelation principle. Wherever possible, any branching instructions were written so that if $i(\rho_{k+1}) = i(\rho_k)$ or $j(t_{k+1}) = j(t_k)$, the result is a drop-through. The core routine assembles up to 64 coefficients and then does the evaluation in one large vector statement. While more speed could be obtained by handling an entire vector at once, assembling only up to 64 coefficients has the advantage of using only 64×9 temporary storage locations, the small amount being popular with users.

The result, after I got some advice from Harry Nelson (LLNL) on Cray assembly coding, is a cache loop that takes only 48 cycles, in the drop-through case, to store the coefficients. The full time required is a tiny average time of 1.4 μsec/element, as shown in Table X. Of this, the evaluation is about 0.2 μsec.

The following lessons are clear to me:

(1) Some of the most important and time-consuming processes in modern physics codes are simply not obviously vectorizable through "vectorizer" routines. Mathematicians and computer scientists can sometimes succeed in finding algorithms.

(2) The payoff in finding a better algorithm is usually an order-of-magnitude effect. Better coding has only a "factor-of-2" effect.

(3) Biology may or may not be destiny, but hardware certainly is not. Processes that do not fit the hardware may nonetheless be good performers, and having a hardware instruction for something does not necessarily make

TABLE X

Timing Comparison of Various
EOS Algorithms on the Cray-1

Method	μsec/zone
Scalar, FORTRAN	14
Scalar, assembly	7
Vector, compress (est.)	2–3 (?)
Vector, cache loop	1.4
Inverse EOS, vector	9.5

^a The timings are problem dependent. These times are a typical problem with 500 zones.

using it a good idea. Hardware is not destiny in another sense: the life span of a computer technology is shorter than ours, so we must be flexible.

(4) Current languages are grossly inadequate for expressing table lookup ideas. Having to write in assembly language has unfortunate, well-known consequences in the areas of maintenance, portability, and documentation.

(5) The dividends for investing in research on parallel/vector/data-flow concepts easily cover the cost of the research, as evidenced by the EOS cache-loop algorithm's saving Livermore programs an amount of computer time worth over one million dollars per year. This means much more computing time for physicists.

As a final note, let me acknowledge the dissenting view at LASL.* Calculations such as EOS lookup can be integrated into the physics calculation in which they are being used. There is no arguing that if speed is the only goal, this is a superior approach because one calculation can hide behind the delays in another. I do not agree with the integrated approach because

(1) Integrating EOS with physics requires duplicating effort both across codes and for different EOS calls within a code.

(2) Human nature leads me to believe that only a few code developers will make the effort required to really optimize the total calculation.

(3) The integrated approach drastically increases the inertia of the status quo. The division at LLNL responsible for equations of state feels BQL is not a good method for representing physical data and would like to develop a new representation. With a central utility subroutine, users do not

---

* My knowledge of the treatment of EOS in codes used at Los Alamos National Scientific Laboratory comes from conversations with Bill Buzbee (LASL) and may not be complete.

"know" about the interior representation, and therefore it can be changed simultaneously in all codes that use the EOS package without any changes in the codes. Research efforts are under way to develop better representations. If the EOS calculation is incorporated into physics modules, implementation of improvements in EOS representation becomes extremely difficult.

(4) It increases development time for new codes.

Other "unvectorizable" challenges await in legions, and almost no research is being done on them at this time: Monte Carlo methods, sparse matrices, nine-point difference matrices, to mention just a few. The processes mentioned are at the heart of modern calculational physics.

REFERENCES

Dubois, P. F. (1978). Inverse equation of state look-up. Rep. UCRL-52563, Lawrence Livermore National Laboratory, Livermore, California.

Dubois, P. F., and Kohn, J. R. (1978). Equation of state table look-up: A case study in vectorization. Rep. UCRL-81184, Lawrence Livermore National Laboratory, Livermore, California.

Dubois, P. F., and Rodrigue, G. H. (1977). In "High Speed Computer and Algorithm Organization" (D. J. Kuck, D. H. Lawrie, and A. H. Sameh, eds.), pp. 299–306. Academic Press, New York.

Fong, K., and Jordan, T. L. (1977). Some linear algebraic algorithms and their performance on Cray-1. Rep. LA-6774, Los Alamos National Laboratory, Los Alamos, New Mexico.

Knuth, D. E. (1968). "Art of Computer Programming," 7 vols. Addison-Wesley, Reading, Massachusetts.

McMahon, F. H., Sloan, L. J., and Long, G. A. (1972). Stacklibe, a vector function library of optimum stackloops for the CDC 7600. Rep. UCID-30083, Lawrence Livermore National Laboratory, Livermore, California.

Trade-Offs in Designing Explicit Hydrodynamical Schemes for Vector Computers*

Paul R. Woodward

Lawrence Livermore National Laboratory
University of California
Livermore, California

I. INTRODUCTION

In the late 1960s attempts to overcome the limitations of the speed of logic circuits and magnetic core memories led to the development of the first vector computers, the ILLIAC IV and the CDC Star-100. This development greatly increased the speed with which certain operations could be performed on a computer, while the speed at which other operations could be performed was either unaffected or reduced. Numerical methods of all kinds have since been reassessed and in some cases extensively redesigned for greater efficiency on the new machines. Although the speed of logic circuits and computer memories has continued to increase since the 1960s, the greater efficiency offered by vector computers for suitably organized computations is still attractive, and it is likely to remain so for many years to come. The purpose of this article is to acquaint the reader with the principal issues involved in

---

* Work performed in part under the auspices of the U.S. Department of Energy by the Lawrence Livermore National Laboratory under contract number W-7405-ENG-48 and by the Office of Basic Energy Sciences.

designing numerical methods for hydrodynamical calculations on vector computers. Consideration of these issues leads to recommendations for the design of future vector computers which would allow more efficient hydrodynamical computations to be performed.

II. WHY VECTORIZATION OF EXPLICIT HYDRODYNAMICAL SCHEMES SHOULD BE EASY

At the time when vector computers were first proposed, it seemed that explicit hydrodynamical calculations would be ideally suited to them. Explicit hydrodynamical calculations are easily vectorized because they are local. Consider, for example, the continuity equation

$$\frac{\partial \rho}{\partial t} + \frac{\partial}{\partial x}(\rho u) = 0. \tag{1}$$

Using a simple, first-order accurate, upwind differencing scheme we may approximate this equation by

$$\rho_i^{n+1} = \rho_i^n - (\Delta t/\Delta x)(F_{i+1/2}^n - F_{i-1/2}^n). \tag{2}$$

The superscripts n and $n+1$ give the number of the time step, and the subscripts give the number of the zone. The subscripts $i - \frac{1}{2}$ and $i + \frac{1}{2}$ denote the left- and right-hand interfaces of zone i. The upstream centering of this scheme comes from the definition of the mass flux F at a zone interface. When the velocity $u_{i+1/2}$ at interface $i + \frac{1}{2}$ is positive, we use the upstream density in $F_{i+1/2}$:

$$F_{i+1/2} = \rho_i u_{i+1/2}. \tag{3}$$

When $u_{i+1/2}$ is negative, we replace ρ_i by ρ_{i+1} in $F_{i+1/2}$. This scheme is formulated most easily if the velocity is prescribed at zone interfaces rather than at zone centers. Then $u_{i+1/2}$ is available as data and need not be constructed by interpolation.

The above difference equation for the new density can be evaluated for every zone in the grid simultaneously or in a vector stream operation. This is a result of the explicit nature of the scheme. The new density in a zone depends only upon the old density in that zone and its nearest neighbors and upon the old velocities at the zone interfaces. Each operand on the right in Eq. (2) can be written as a one-dimensional array, or vector, consisting of data known at the begining of the time step. Because all zones are updated in the same way, and no unknowns appear on the right in Eq. (2), the method vectorizes easily. A distinction between zones is made in computing the flux in Eq. (3), but this can be handled by a vector mask operation. These features which allow easy vectorization are shared by the full set of hydrodynamical difference equations for many simple difference schemes. Difficulties with the vectorization of hydrodynamics schemes only arise when we wish to use the more elaborate schemes which were developed during the past 15 years to achieve higher accuracy or to deal with more complex flow problems.

III. WHY VECTORIZATION OF EXPLICIT HYDRODYNAMICAL SCHEMES CAN BE DIFFICULT

Difficulties in vectorization arise when we no longer update all zones with the same difference equations. The hydrodynamical equations governing the flow may be identical for each zone, and we may nevertheless choose to use different difference equations in different regions of the flow. The explanation of this apparent contradiction is that terms describing viscosity and heat conduction may have been omitted from the hydrodynamical equations. In regions occupied by strong nonlinear waves, these neglected terms would be important, so it is appropriate to update these regions with a different set of difference equations. Neglecting the viscous terms in the governing differential equations is justified when the coefficient of viscosity is very small. However, the nonlinear steepening of compression waves described by the inviscid equations can lead eventually to the development of extremely large density and pressure gradients in extremely thin regions of space. The neglected viscous terms must then become important and lead to the formation of a shock. Omitting the viscous terms corresponds to the assumption that the thickness of any such shock is negligible compared to the other natural length scales in the problem. Because the shock thickness is about a collisional mean free path for a molecule in the fluid, this assumption is usually justified.

In order to ease the vectorization of the numerical method, we could allow for the development of shocks in the flow by solving the hydrodynamical equations with viscous terms included. This is the approach taken in the moving finite element method described by Gelinas *et al.* (1981). However, this can involve needless effort in the vast majority of zones which do not contain shocks. A simpler method was used by Von Neumann and Richtmyer (1950). They added to the pressure an artificial viscous pressure defined in such a way that it would be significant only in shock regions. Still simpler is the approach used in the methods of Lax (1954), Rusanov (1961), and others. In these methods truncation error terms which dissipate kinetic energy to give heat are relied upon to limit the steepening of shock waves and to stabilize the numerical methods. In all of these methods vectorization is easy because all zones are treated alike. However, a price is paid for this simplicity. When viscous terms are added to the difference equations explicitly, some unnecessary work must be performed in most of the zones. In addition these methods require that shocks be smeared out over at least three zones so that the flow in these regions can be approximated as the smooth flow of a viscous fluid. In problems where shocks and their interactions are important, this shock smearing can be the dominant source of error (see the comparison of several schemes in Woodward and Colella, 1981).

In the shock-smearing methods mentioned earlier, the needless work performed in regions without shocks is not very great. If, however, we wish to spread shocks out over no more than a single zone width, considerably more elaborate methods must be used. Such methods can be applied to smooth flow as well in order to permit easy vectorization, but then the wasted effort in smooth regions of the flow will be substantial. A family of methods of this type is based upon the ideas of Godunov. In Godunov's (1959) scheme, shocks are treated by a combination of viscous smearing and a built-in knowledge of the correct nonlinear jump conditions. In this scheme structureless zones of fluid interact at their interfaces by generating shocks and rarefaction waves. These nonlinear waves are computed at the beginning of each time step by solving Riemann's shock tube problem. This Riemann solution builds viscous effects into the method through the shock jump conditions. Additional viscosity occurs as a numerical truncation error when the structure which develops in the zones is smeared out at the end of the time step to give a new set of structureless zones. The treatment of a one-dimensional flow containing a shock using Godunov's method is illustrated in Fig. 1.

As originally formulated, Godunov's method treats all zones alike. However, for most zones the discontinuities at the zone interfaces are small, and only the simplest guess at the solution of Riemann's problem is required. Use of such a simple guess at every zone interface leads to a formulation of the method as a normal difference scheme. At shocks the discontinuities at zone interfaces are large, and the method yields better results if Riemann's problem is solved more accurately. When the method is formulated in Eulerian coordinates, this improvement reduces the width of shocks on the grid from about three zones to about half this width. If shocks and their interactions are important in the flow problem, the accuracy of the entire calculation is increased significantly. When Eulerian calculations are performed using a Lagrangian hydrodynamical step followed by a remapping to the original Eulerian grid, the improvement is smaller because of the extra dissipation introduced by the remap step. The calculations shown in Fig. 1 were done this way, and the shock width is therefore about three zones.

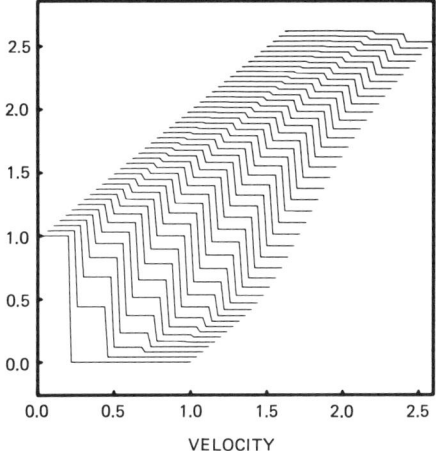

FIG. 1. Illustration of the representation of a shock by Godunov's method. A very strong shock moving to the right in a one-dimensional flow calculation is shown at each of 50 time steps. The first five zones of the grid are shown, with the display displaced slightly upward and to the right at each time step. The distributions of velocity within the zones are shown which are used in performing the computation. Because all variables are assumed constant within zones, the shock is smeared out over about three zones. Also note that the velocity rises only very gradually in time in any particular zone. The shock should entirely cross a zone in 10 steps.

Another first-order accurate method based upon the ideas of Godunov is the random choice method introduced by Glimm (1965) and reformulated as a method for practical hydrodynamics calculations by Chorin (1976), Sod (1978), and Colella (1982a). This method treats shocks as pure discontinuities. The discontinuities are forced to occur at zone interfaces, and the resulting falsification of the shock position is minimized by means of the random choice aspect of the method. The shock position jitters about its proper value, and the amplitude of the jittering is controlled by using a pseudorandom sampling technique discussed in detail in Colella (1982a). The size of the shock jump and its speed of propagation are computed by solving Riemann's problem, as in Godunov's method. However, the random choice method carries Godunov's approach to an extreme limit. The lack of any averaging in the method guarantees that the shock jump will not be smeared out, but it also eliminates the mechanism by which Godunov's method accommodates inaccuracies in the solution of Riemann's problem. In order for a shock to be well represented by the random choice method, a very accurate solution of Riemann's problem must be computed. If this is done at every zone interface, the method vectorizes easily, but a great deal of needless work is performed in smooth regions of the flow.

The random choice method is an extreme example which clarifies a general phenomenon. If a shock is to be smeared out over only a thin region on a numerical grid, considerably more labor will have to be expended in advancing the zones of that thin region in time than is required for advancing the zones in the rest of the problem. This general trend is further illustrated by the higher order schemes described by van Leer (1979), van Leer and Woodward (1979), Colella (1982b), and Woodward and Colella (1982; Colella and Woodward, 1982). In these schemes, as in Godunov's scheme, knowledge of shock jump conditions is built into the method through a Riemann solver, and some shock smearing is introduced by dissipative truncation errors. As is illustrated in Fig. 2, these schemes represent shocks as discontinuities at zone interfaces which are preceded and followed by smooth compression waves in the neighboring zones. The schemes differ in their description of the smooth compression waves, and this affects the overall thickness of the shock region as well as the formal order of accuracy of the scheme.

All of the methods discussed earlier can be implemented so that they treat all zones alike. However, the most accurate description of shocks results from adding to these methods a technique for tracking shocks explicitly on the grid. Such techniques are generally referred to as "shock fitting." Glimm et al. (1980) have adapted the random choice method to treat a very simple form of shock in this manner. Chorin (1980) has used a more elaborate technique to track flame fronts in calculations with the random choice method. His technique is a modification of the SLIC algorithm devised by Noh and Woodward (1976) to track fluid interfaces. The SLIC algorithm grew out of the earlier work of DeBar (1974) and of Sutcliffe (1973). When coupled with a Riemann solver, this algorithm could be used to track shocks in multidimensional calculations. However, such a method would be extremely difficult to implement efficiently on a vector computer. Results of the use of SLIC to track fluid interfaces are shown in Noh and Woodward (1976) and Woodward (1979). In such applications the increased accuracy of the calculation resulting from the use of SLIC is so great that the cost of implementing the method is not a concern. The astrophysical calculation in Woodward (1979) is in fact a single-fluid problem, but it would be pointless to attempt such a calculation with a standard single-fluid difference method. If SLIC tracking of shocks could give a similar increase in accuracy in problems dominated by strong shock interactions, then the difficulties of implementing the method on a vector computer would be offset.

We have been focusing upon the problems in designing efficient numerical schemes which result when shocks are an important feature of the flow. These problems, and the alternative means of handling them, are very similar for a number of other discontinuities which can arise in the flow. We restrict our

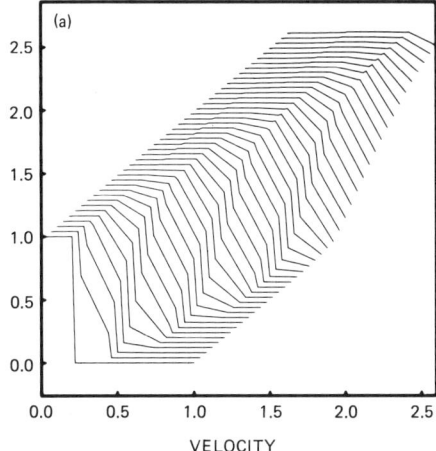

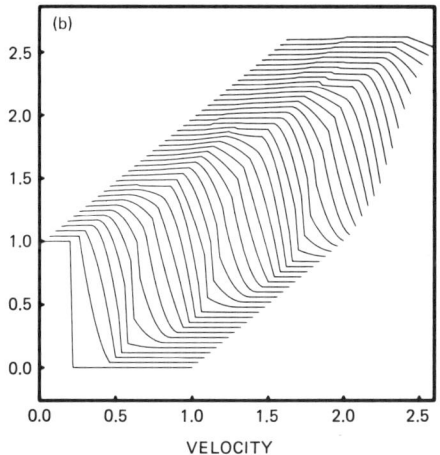

FIG. 2. Illustration of the representation of a shock by (a) MUSCL and (b) the piecewise-parabolic method. The problem and its display format are the same as in Fig. 1. The shock involves a pressure increase of more than six orders of magnitude, yet it is represented smoothly by these methods with very little smearing. The linear representation of variables within zones in MUSCL already gives a very narrow shock, so that the parabolas used by the piecewise-parabolic method give only a modest reduction in the shock thickness. The main advantage of this method over MUSCL is its improved representation of contact discontinuities. See Woodward and Colella (1981).

discussion to shocks because these are the most commonly occurring discontinuities. Examples of others which can be treated in much the same ways as shocks are contact discontinuities, flames, detonations, ionization fronts, and thermal waves driven by transport of radiation. These discontinuities arise from physical processes which operate on time or length scales which are much shorter than those characteristic of the bulk of the flow. For an ionization front, the relevant process is the ionization of atoms in the gas by high-energy photons. For a flame, chemical reactions are involved. These processes are important only in thin regions, but the simplest numerical methods compute them in every zone of the grid, as is the case for shocks. Again, the most accurate calculations are performed by methods which involve a minimum of artificial smearing out of these thin regions. In some problems even a small amount of smearing can lead to very large errors. Examples of such problems for the case of ionization fronts have been given by Hill and Marsh (1971).

IV. ALTERNATIVE APPROACHES AND THEIR COSTS ON VECTOR COMPUTERS

The schemes which update a single zone in the shortest time on present-day vector computers are generally those which employ the fewest logical operations. To see why logical operations are so costly, it is necessary to understand how they are performed by a vector computer. The simplest logical operation is the vector mask. This operation can be illustrated by the simple upwind advection scheme described in Section II. The flux $F_{i+1/2}$ is $\rho_i u_{i+1/2}$ when $u_{i+1/2}$ is positive, and it is $\rho_{i+1} u_{i+1/2}$ otherwise. $F_{i+1/2}$ can be constructed by using a vector mask. In this case the flag vector $u_{i+1/2}$ need not be separately computed, and this will save some time. We begin by computing each of the two possible fluxes for every zone interface. Then the resulting two vectors are merged, with elements of the final flux vector coming from one or the other operand vectors depending upon the sign of the flag vector $u_{i+1/2}$. For each zone interface we compute an unnecessary flux and also perform a mask operation which requires the time, on a Cray-1, for about 2 multiplies. Because we already had our flag vector, this overall operation cost us only 4 multiplies. This figure should be contrasted with the single multiply which the flux calculation would cost in an ideal world. It is obvious that vector masks are to be avoided wherever possible, at least on a Cray-1.

This flux calculation could have been performed using other vector logical operations, namely, the vector compress, merge, and demerge. These are hardware operations on a CDC-STAR-100A, but on a Cray-1 they must be implemented in software. They can be so implemented on any vector computer, with the simplest and slowest coding consisting of scalar DO loops containing IF statements. We shall describe the computation of the flux $F_{i+1/2}$ in the above example, using the software developed at the Lawrence Livermore National Laboratory by McMahon (1982) for use on the CDC-7600 and the Cray-1. We begin by constructing a bit vector. This bit vector consists of a bit for each zone i, and the bit is 1 when $u_{i+1/2}$ is positive and is 0 otherwise. Under control of this bit vector we split the velocity vector apart to form a shorter vector consisting of velocities $u_{i+1/2}$ for zones i for which our bit vector is 0 and another shorter vector for zones for which it is 1. This is a vector demerge operation. We also construct similar shorter density vectors by means of vector compress operations which select the densities ρ_i for which the ith bit of our bit vector is 1 and the densities ρ_{i+1} for which it is 0. We obtain the positive and negative fluxes by vector multiplies and then merge the resulting short vectors under control of our bit vector to give the final result. If the merge, demerge, and compress operations are not specially provided for by the computer hardware, the above process can be very expensive. On a CDC-STAR-100A these are hardware instructions and they are performed at multiply rates. On a Cray-1 they must be provided as software. When the 1 or 0 values in the bit vector are very sparse, these instructions can be performed at multiply rates. However, when this is not the case they can be up to 12 times slower. Therefore these operations are usually avoided on a Cray-1 unless a very long calculation is to be performed on a set of compressed vectors.

It should now be clear why numerical schemes which treat all zones alike and therefore use a minimum number of logical operations execute most rapidly on vector computers. These schemes are also the easiest to program. The trade-offs which must be faced in choosing which scheme of this type to use have been discussed in the previous section. Simple schemes are generally very fast but also very inaccurate. Elaborate schemes are accurate but slow. Because elaborate schemes can achieve the same accuracy with coarser grids in space and time than must be used with simple schemes, there is for any given class of problems some optimal amount of elaboration. The extensive comparison of schemes reported in Woodward and Colella (1981, 1982; Colella and Woodward, 1982) indicates that when shock interactions are important this optimal elaboration may not yet have been reached. For this class of problems the advantage of an elaborate difference scheme is often its more accurate treatment of discontinuities in the flow. In this case a simpler and faster scheme may be used in regions of smooth flow. In order for such a hybrid scheme to be efficient, the vector merge, demerge, compress, and decompress operations must be executed rapidly. In the next section a comparison of two difference schemes is presented to illustrate this point.

There is an additional trade-off in implementing difference schemes on vector computers which we have not yet discussed. Fairly simple schemes can be effective in describing smooth flow, but generate large errors near discontinuities which contaminate the entire solution (see Majda and Osher, 1977). These errors can be greatly reduced by using an adaptive grid. Adaptive grid techniques to deal with this sort of problem have been discussed in a number of contexts by several authors (Eggleton, 1971; Castor et al., 1977; Tscharnuter and Winkler, 1979; Dwyer et al., 1981; Gropp, 1980; Gelinas et al., 1981). They usually amount to refining the mesh automatically in regions of steep gradients. These methods usually employ implicit difference schemes so that the refined mesh in a region of steep gradients will not force an extremely small time step upon the rest of the calculation. On a vector computer the use of an implicit difference scheme in regions of the flow which could have been treated explicitly can slow down the calculation considerably. This extra expense results from the nonlocal nature of implicit calculations, which makes efficient vectorization of implicit algorithms extremely difficult.

In adaptive grid schemes an implicit difference scheme may only be needed near discontinuities, where the zones are made very small. If a method is devised which passes smoothly from the explicit to the implicit regime as the zones become smaller, it is possible to compress the implicit zones out of the grid and to update them in a separate, vectorized calculation. This technique would be similar to the applications of SLIC discussed earlier in that the speed of vector compress and decompress operations would be critical to its success.

V. THE EXAMPLE OF THE INTERACTION OF TWO BLAST WAVES

The trade-offs in accuracy and speed which we have discussed are most easily appreciated with the help of a specific example. A flow problem involving the generation and interaction of strong nonlinear waves in one dimension is the interaction of two blast waves. The initial conditions for this problem consist of a fluid at rest between two reflecting walls one length unit apart. The density is everywhere unity, and a region of length 0.1 next to each wall is hot. At the left the pressure is 1000, while at the right it is 100. The remainder of the fluid is cold, with pressure 0.01. The gas has a gamma-law equation of state with gamma equal to 1.4.

The evolution of this system is quite complex. Initially, strong shocks are driven into the cold gas from both sides while strong rarefactions run into the hot gas. The shocks compress the cold gas to a density of 6, creating dense slabs of gas which move toward each other. The rarefaction waves reflect off the walls and eventually interact with these dense slabs. This happens most rapidly for the left-hand slab. Its interaction with the reflected rarefaction wave gives it the appearance of a sharp density spike. A contact discontinuity is formed at the point where the slabs eventually collide. Strong reflected shocks are then sent back into the slabs and interact with the large density gradients there. Ultimately these shocks also interact with the two contact discontinuities which separate the slabs from the original hot gas.

This problem is very difficult to compute accurately. The shocks which arise are extremely strong and must be well resolved so that their interaction with steep density gradients can be computed. Also, contact discontinuities occur which are difficult to follow as they move across the grid. After the slabs interact with the rarefaction waves these contact discontinuities are very difficult to pick out from the steep decline in density in the slabs. Even harder to compute is the formation of the third contact discontinuity when the slabs collide. This requires the computation of detailed structure within the narrow density spike which results from the collision.

In order to illustrate the issues discussed in the previous sections, we have computed the evolution of these interacting blast waves using a variety of numerical methods and computational grids. Timings for the various computations will be given in microseconds per zone per time step on a Cray-1. All the codes used have been completely vectorized. The times given here include the extra computational overhead required to make these one-dimensional methods useful in the component sweeps of two-dimensional computations. All computations shown have used time steps of half the Courant limit. The methods can be run at 90% of the Courant limit, with a slight improvement of the results. When run at 10% of the Courant limit, the higher-order schemes yield still better results. However, this dependence of the accuracy upon the time step is small compared to the dependence upon the zone width. Results of the computations are shown at time 0.026, shortly before the dense slabs collide, and at time 0.038, shortly before the left-hand reflected shock emerges into the hot gas and shortly after the right-hand shock does so. The computed flows are shown by dotted lines. Circles representing the zone-averaged densities are shown for the calculations on coarse grids. For comparison, a solid line displays the solution obtained with the most accurate method available (the piecewise-parabolic method) using an extremely fine grid of 1200 zones. Because the method produces this same solution on a grid of only 400 zones, there is no doubt that this solution is correct.

In Fig. 3 results of four computations at time 0.026 are compared. Results of Godunov's method using 200 and 400 zones are shown at the bottom left

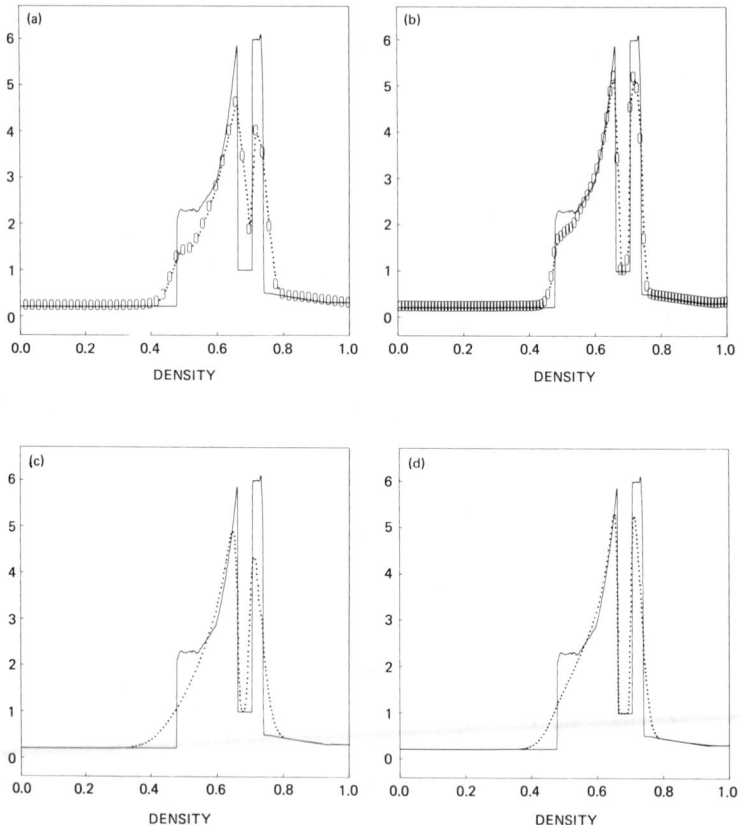

FIG. 3. Results for the blast wave interaction problem discussed in the text are plotted here at time 0.026 for four different calculations. Results obtained with the piecewise-parabolic method with (a) a grid of 50 zones and (b) 100 zones. Results obtained with Godunov's method (c) using 200 zones and (d) 400 zones. All results are indicated by dotted lines. The solid lines are results of the piecewise-parabolic method using a 1200-zone grid which represent the correct solution to this flow problem. Note the slow convergence of the solutions in the neighborhoods of discontinuities in the flow.

and right of the figure. Above them are shown results of the piecewise-parabolic method described in Woodward and Colella (1981, 1982; Colella and Woodward, 1982). These results were obtained with much coarser grids of 50 and 100 zones. Despite the coarser grids, these results are superior to those of Godunov's method, particularly in their representation of the contact discontinuities. Results of these computations at time 0.038 are displayed in Fig. 4 in the same manner. Both figures demonstrate the dramatic increase in resolution of flow structure which can result from the use of an elaborate

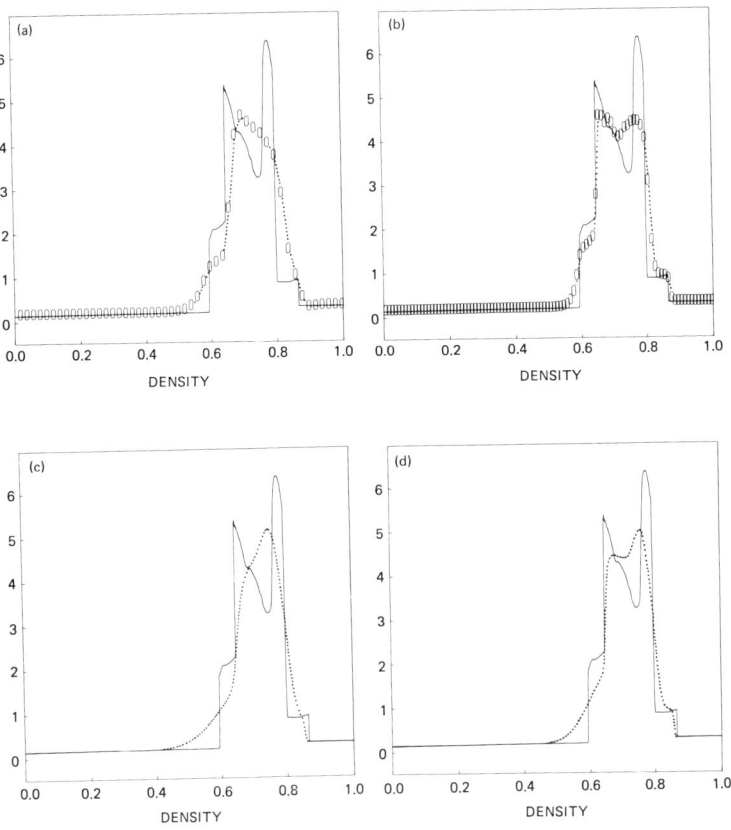

FIG. 4. Results of the same calculations presented in Fig. 3 are shown here in the same format at time 0.038. Note the poor representations of the contact discontinuity near $x = 0.75$ which formed when the two blast waves collided.

difference scheme. This increased resolution can be ascribed only indirectly to the order of accuracy of the method. The largest errors are generated by finite widths of shocks and contact discontinuities. For the piecewise-parabolic method these widths decrease linearly with the zone size. Thus despite its third-order accuracy in smooth flow, this method yields only a linear rate of convergence for this blast wave problem. For Godunov's method, which is first-order accurate in smooth flow, the rate of convergence is even slower. A mesh refinement yields very little improvement in the description of the contact discontinuities in the blast wave problem. An impractically large number of zones would be required to obtain the correct solution to this problem with Godunov's method.

The results shown in Figs. 3 and 4 indicate that Godunov's method requires more than four times as many zones as the piecewise-parabolic method in order to produce results of comparable accuracy. When we consider the computed pressure distributions, we find that the disparity in performance of these two schemes is even greater. In Fig. 5 pressure profiles at time 0.038 are compared. Results of the piecewise-parabolic method using grids of 50 and 100 zones are again shown at the top. Below them are Godunov

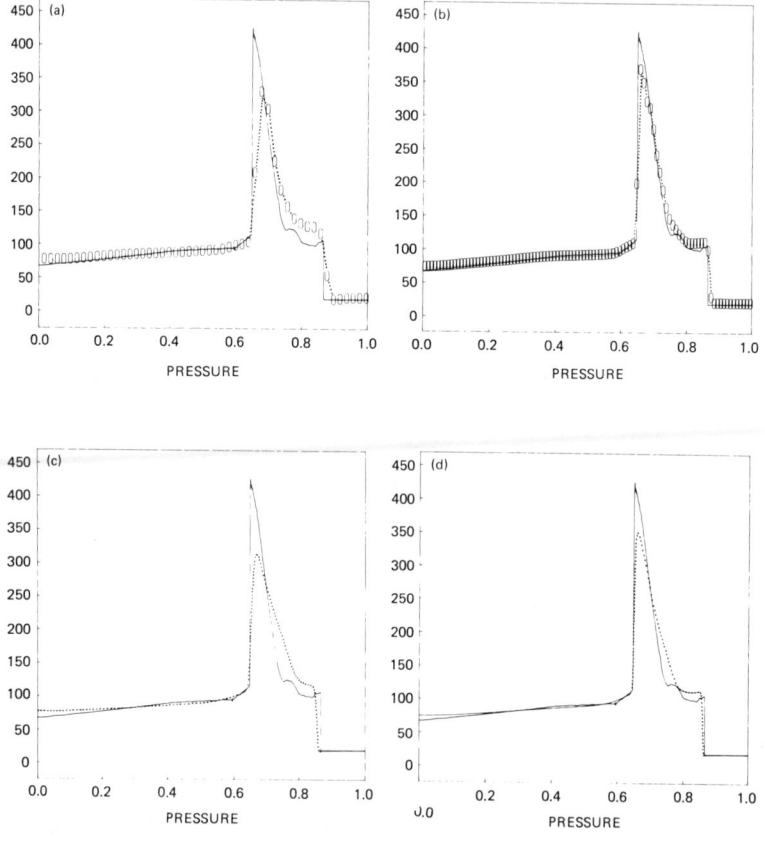

FIG. 5. Pressure profiles at time 0.038 in the blast wave interaction problem are shown compared to the 1200-zone correct result. The results at the top correspond to the density profiles at the top in Fig. 4. They have been obtained with the piecewise parabolic method using (a) 50 zones and (b) 100 zones. The pressure is much more accurately represented than the density in these calculations, and the result in (a) is quite acceptable. Results of Godunov's method using grids of (c) 300 zones and (d) 600 zones. Clearly more than six times as many zones are required by this method to achieve the same accuracy as the piecewise-parabolic method.

results using grids of 300 and 600 zones. Again, the piecewise-parabolic results are slightly superior. Thus if we desire comparable accuracy for all variables, we must use more than six times as many zones with Godunov's method as are required with the piecewise-parabolic method. Because of the Courant limit to the time step, more than six times as many time steps are required as well. To match the efficiency of the piecewise-parabolic method, Godunov's method would therefore have to run more than 36 times faster. For two-dimensional problems Godunov's method would have to run more than 216 times faster. Thus we see that the extra programming labor involved in using a more elaborate difference scheme can yield many times its cost in increased computational efficiency.

Although it is clear that Godunov's scheme cannot be more efficient than the piecewise-parabolic method, it is worthwhile to note the precise speeds of these two methods, as they depend strongly upon the method of vectorization and the type of vector computer upon which the calculations are run. When it is formulated to treat all zones alike, Godunov's scheme can update a zone in 7.5 μsec on a Cray-1. If the Riemann problems at zone interfaces are solved using a simple formula valid only for small jumps in the variables there, we obtain a much simpler difference scheme which treats strong shocks less accurately. This simpler scheme can update a zone in 4.75 μsec on a Cray-1. We can combine the advantages of both these formulations of Godunov's method by solving Riemann's problem with the full nonlinear iterative technique only for those zone interfaces at which large jumps in the variables occur. Because the necessary vector compress and decompress operations are relatively slow on a Cray-1, this hybrid formulation of the scheme also requires 7.5 μsec to update a zone. If the compress and decompress operations could be performed at multiply rates, this hybrid scheme would update a zone in about 5.3 μsec.

The piecewise-parabolic method is considerably slower than Godunov's method regardless of the way it is programmed and the machine on which it is run. Its extra two orders of accuracy inevitably require more work. However, this scheme can benefit more from a hybrid formulation because it must do an enormous amount of work in zones with steep gradients or large discontinuities at their interfaces. When the method treats every zone as if it were the most difficult case, a zone is updated in 42 μsec on a Cray-1. This is six times slower than a similar formulation of Godunov's method. Even this relatively inefficient formulation of the scheme is more cost effective than Godunov's method for one-dimensional problems. In smooth regions of the flow, the Lagrangian step of the calculation can be replaced by difference formulas accurate to third order in the zone size. These formulas involve no logical operations, and they allow a zone to be updated in a tenth of the time required for zones in which the changes in the variables are not small.

Unfortunately, little time can be saved in the remap step of the Eulerian calculation for smooth regions of the flow. Thus zones in smooth regions can only be updated about three times faster than those near discontinuities. Thus for the typical case of 10% difficult zones we would require about 17 μsec to update a zone on a Cray-1, provided that the necessary vector compress and decompress operations could be performed at multiply rates. However, for this case of 10% difficult zones these operations are performed about 12 times more slowly. Consequently the overall calculation is slowed by 30%, and 22 μsec are required to update a zone.

We stated earlier that the most accurate flow calculations result from treating discontinuities separately as jumps rather than as regions of smooth flow with large gradients. Through their use of the solutions to Riemann's problem, both Godunov's method and the piecewise-parabolic method treat shocks this way to some extent. Indeed, the shocks in Fig. 4 are merely one zone wide. However, contact discontinuities are smeared out by these methods. For Godunov's method, this smearing is disastrous, while for the piecewise-parabolic method contact discontinuities grow in width very little once they are about two zones wide. The contact discontinuity on the left in the blast wave problem is a pathological case because of the steep density gradient beside it. Nevertheless, this discontinuity is detected, although the amount of the density jump is falsified.

The versions of Godunov's scheme and the piecewise-parabolic method used here are formulated in terms of a Lagrangian step followed by a remapping to the original Eulerian grid. Because contact discontinuities are Lagrangian lines, these schemes may be trivially modified for one-dimensional problems so that they always place a zone interface at the location of a contact discontinuity. In Fig. 6 results of such modified schemes are shown for the blast wave interaction problem. Grids of 200 zones were used. The improvement in the accuracy of Godunov's method resulting from this modification is dramatic. However, it should be noted that only the density distribution is improved; the pressure and velocity distributions are unaffected by the modified treatment of contact discontinuities. The main improvement for the piecewise-parabolic method is in the description of the third contact discontinuity formed in the collision of the blast waves. Results of this method without modification on a 200-zone grid are shown in Fig. 7. These results already achieve an excellent representation of the left- and rightmost contact discontinuities.

In two dimensions this sort of contact tracking becomes much more difficult to formulate. In particular, zones containing contact discontinuities can no longer be treated in the same way as the other zones. A contact tracking algorithm must therefore make heavy use of the vector compress, decompress, and mask operations. For example, advancing a three-fluid

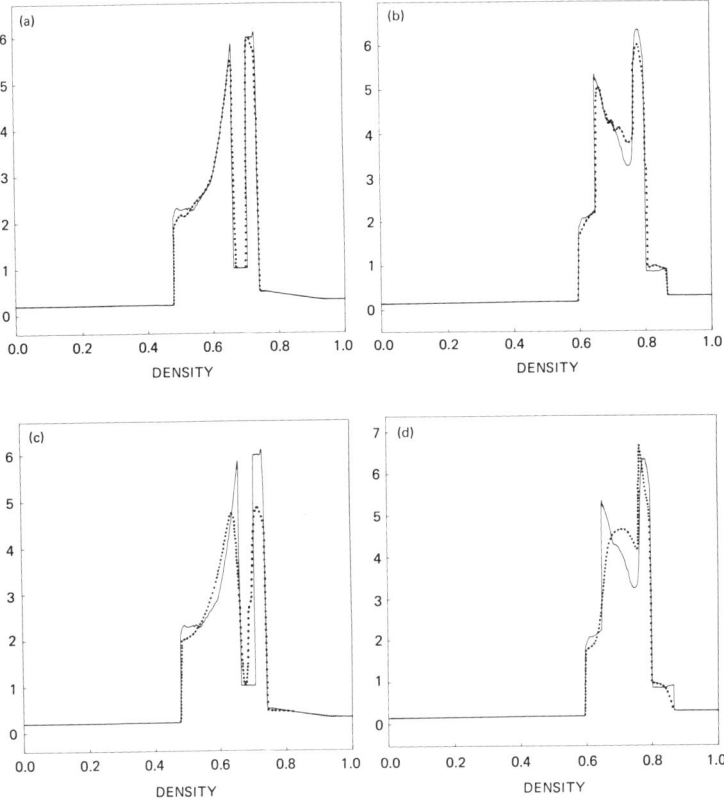

FIG. 6. Density profiles are shown here for the blast wave interaction problem at times [(a) and (c)] 0.026 and [(b) and (d)] 0.038. Results obtained by [(a) and (b)] the piecewise-parabolic method and [(c) and (d)] Godunov's method. These calculations have both used grids of 200 zones, and special modifications to the methods have been made to track contact discontinuities without any artificial smearing of the density jumps. This modification gives the most dramatic improvement when it is applied to Godunov's method. The pressure profiles for these computations are not shown, but they are essentially identical to those obtained with the unmodified schemes. The correct solutions are indicated by solid lines.

hydrodynamics problem one time step using the implementation of the SLIC algorithm described in Noh and Woodward (1976) requires the execution of 42 vector compress or decompress operations, 95 vector mask operations, and 411 vector Boolean operations. Compared to these logical instructions, the algebra performed is trivial. Nevertheless, these operations were provided for in the CDC-STAR-100A hardware, so that two-dimensional multifluid hydrodynamical calculations required only about 35 μsec to update a zone on that machine. This computation speed was only attainable

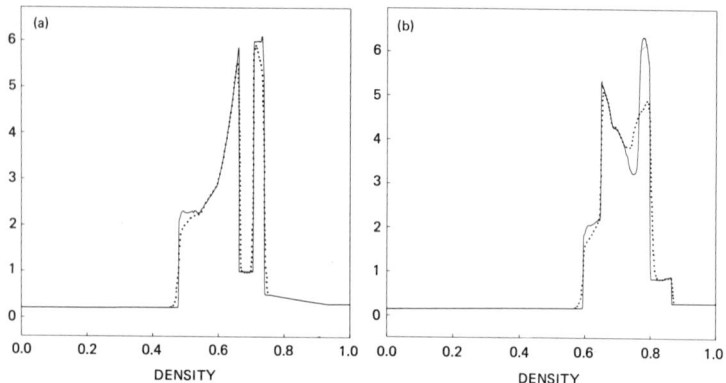

FIG. 7. Density profiles obtained with the piecewise-parabolic method unmodified for contact discontinuity tracking are shown here for comparison with Fig. 6. Results using a 200-zone grid are shown as dotted lines, and the 1200-zone results are indicated by solid lines. Only the contact discontinuity resulting from the collision of the blast waves is poorly represented. With 400 zones this method yields essentially the same result as the 1200-zone computation.

when most of the vector logical operations were performed on vectors which were much shorter than the length of the whole grid. However, even in this case this vector logic required about a third of the overall computation time on the CDC-STAR-100A.

VI. CONCLUSIONS

The example of two interacting blast waves discussed in the previous section clearly shows that the most accurate and the most efficient vectorized algorithms for computing hydrodynamical flows make fairly heavy use of vector compress, decompress, merge, demerge, and mask operations. This is a result of the fundamental tendency of hydrodynamical flows to steepen compression waves and thus to generate thin regions of the flow where special numerical techniques must be applied. Unfortunately, these vector operations are not always provided as hardware instructions on vector computers. They can always be provided as software, but then they are usually much less efficient. There is no reason, in principle, why these operations must be slow compared to vector addition; but because of their greater complexity they are usually relatively slow if they are provided at all. The blast wave example shows the very large gains in accuracy which can be achieved by algorithms for which these vector operations are essential.

Therefore vector computers designed to perform these sorting and logical operations as fast as the arithmetical operations would be of significantly greater use than the machines presently available.

REFERENCES

Castor, J. I., Davis, C. G., and Davison, D. K. (1977). Dynamical zoning within a Lagrangian mesh by use of DYN, a stellar pulsation code. Rep. LA-66640, Los Alamos National Laboratory, Los Alamos, New Mexico.
Chorin, A. J. (1976). *J. Comput. Phys.* **22**, 517.
Chorin, A. J. (1980). *J. Comput. Phys.* **35**, 1.
Colella, P. (1982a). Glimm's method for gas dynamics. *SIAM J. Sci. Statist. Comput.* **3**, 760.
Colella, P. (1982b). A direct-Eulerian MUSCL scheme for gas dynamics. Rep. LBL-14104, Lawrence Berkeley Laboratory, Berkeley, California.
Colella, P., and Woodward, P. R. (1982). The piecewise-parabolic method for compressible flow. In preparation.
DeBar, R. (1974). Fundamentals of the KRAKEN code. Rep. UCIR-760, Lawrence Livermore National Laboratory, Livermore, California.
Dwyer, H. A., Raiszadeh, F., Otey, G. (1981). *Lecture Notes Phys.* **141**, 170.
Eggleton, P. (1971). *Monthly Notices Roy. Astronom. Soc.* **151**, 351.
Gelinas, R. J., Doss, S. K., and Miller, K. (1981). *J. Comput. Phys.* **40**, 202.
Glimm, J. (1965). *Comm. Pure Appl. Math.* **18**, 697.
Glimm, J., Marchesin, D., and McBryan, O. (1980). *J. Comput. Phys.* **37**, 336.
Godunov, S. K. (1959). *Mat. Sb.* **47**, 271.
Gropp, W. D. (1980). *SIAM J. Sci. Statist. Comput.* **1**, 191.
Hill, J. G., and Marsh, M. C. (1971). *Monthly Notices R. Astronom. Soc.* **156**, 189.
Lax, P. D. (1954). *Comm. Pure Appl. Math.* **7**, 159.
McMahon, F. H. (1982). STACKLIBE, a library of fast vector functions for complete vector formulation of program logic on Cray-1. In preparation.
Majda, A., and Osher, S. (1977). *Comm. Pure Appl. Math.* **30**, 671.
Noh, W. F., and Woodward, P. R. (1976). *Lecture Notes Phys.* **59**, 330.
Rusanov, V. V. (1961). *Ž. Vyčisl. Mat. i Mat. Fiz.* **1**, 267.
Sod, G. A. (1978). *J. Comput. Phys.* **27**, 1.
Sutcliffe, W. G. (1973). BBC hydrodynamics. Rep. UCID-17013, Lawrence Livermore National Laboratory, Livermore, California.
Tscharnuter, W. M., and Winkler, K.-H. A. (1979). *Comput. Phys. Comm.* **18**, 171.
van Leer, B. (1979). *J. Comput. Phys.* **32**, 101.
van Leer, B., and Woodward, P. R. (1979). *Proc. Internat. Conf. Comput. Methods Nonlinear Mech., Austin, Texas, March.*
von Neumann, J., and Richtmyer, R. D. (1950). *J. Appl. Phys.* **21**, 232.
Woodward, P. R. (1979). Compression of interstellar clouds in spiral density-wave shocks. *In* "The Large-Scale Characteristics of the Galaxy" (W. B. Burton, ed.), IAU Symposium, No. 84, p. 159. Reidel Publ., Dordrecht, Netherlands.
Woodward, P. R., and Colella, P. (1981). *Lecture Notes Phys.* **141**, 434.
Woodward, P. R., and Colella, P. (1982). The numerical simulation of fluid flow with strong shocks. Rep. UCRL-86952, Lawrence Livermore National Laboratory, Livermore, California.

Vectorized Computation of Reactive Flow

Jay P. Boris
Niels K. Winsor [*]

Laboratory for Computational Physics
Naval Research Laboratory
Washington, D.C.

I. INTRODUCTION AND STATEMENT OF THE PROBLEM

Large-scale computational simulations of physical systems have become an extremely valuable adjunct to theoretical and experimental research efforts in the physical and chemical sciences. They are essential because the computer allows solution of problems not soluble by other methods and because the use of detailed modeling in conjunction with theory and experiment has reduced the cost and increased the effectiveness and predictability of research project management. This article discusses a number of the decisions which a modeler must make to optimize his efforts on a reactive flow simulation project. It explains what choices are available from the perspectives of program efficiency, flexibility, reliability, accuracy, and lead time to results. Our central topic is the vectorized computation of reactive flows, but the ideas and guidelines laid down here clearly apply across the breadth of computational physics endeavors and extend well beyond the limited scope of vectorized computation.

[*] Present address: GT-Devices, Alexandria, Virginia 22312.

There are five uses for modeling in research and development programs in general, and in our reactive flow modeling in particular. These uses are

(i) to evaluate concepts,
(ii) to interpret measurements,
(iii) to extrapolate parameters,
(iv) to calibrate understanding, and
(v) in engineering design.

Figure 1 depicts the interconnection of these five uses and their relation to theoretical models and experimental data schematically. The most demanding of these, and central to sound research advances, is to calibrate our understanding quantitatively by testing detailed theoretical models against reliable experimental data. From experimental observation and approximate theoretical models, we can postulate quantitative physical laws which we expect a physical effect to obey. The only way really to test one of these "laws" against reality, however, is to incorporate it in a detailed model and then make quantitative predictions of the results of a series of experiments. Because even the simplest reactive flow systems are often so complicated as to defy accurate quantitative analysis, the models are almost always numerical.

The order of the five uses listed earlier is from least to most accurate. Often a limited or oversimplified experiment or back-of-the-envelope model is quite adequate to evaluate a new concept. The interpretation of experimental measurements and the extrapolation of phenomena beyond the regimes where data are available require much better models. Excellent models are required to calibrate our understanding and to apply that

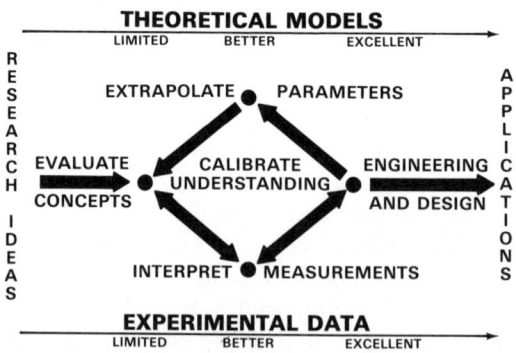

Fig. 1. Five uses of detailed computational modeling in research. Also depicted are the relations of these uses to theoretical models and experimental data.

quantitative predictive capability to engineering and design applications. The evolution of most research projects also follows this general progression.

We began our detailed modeling efforts in reactive flow with a number of numerical techniques in hand and experience in atmospheric and plasma reactive flows. Our program is advancing now on several fronts (Oran and Boris, 1981), so this Chapter pursues the optimization question from the perspective of the generic physical constraints: fast time scales (Section III), short space scales (Section IV), and physical and geometric complexity (Section V). Section II provides a rather general discussion of vectorization and optimization as a preamble to the discussion of specific techniques and algorithms in the following sections. Section VI presents a summary of programming guidelines and parallelism principles.

A. The Detailed Reactive Flow Problem

Detailed modeling, also referred to as numerical simulation, provides a description of a physical system by solving numerically the governing time-dependent conservation equations for mass, momentum, and energy. The goal is to develop a predictive model of the important controlling physical processes which can be used to evaluate concepts, to interpret measurements, to extrapolate parameters, to calibrate our understanding, and in engineering design. It is particularly important to understand quantitatively the complex interactions between the chemical kinetics and the fluid dynamics in reactive flow problems. For this goal to be attained requires accurate treatment of all the relevant physics and chemistry. Empirical components are only incorporated when the required quantities must be derived from more fundamental models and theories. This is the case for chemical rate constants, for thermal conductivity coefficients, and for other thermophysical and thermochemical data in a detailed reactive flow calculation.

There are a number of inherent numerical difficulties we face in simulating accurately the propagation of a shock or a flame front in a reacting medium. One class of problems arises as the result of wishing to reproduce faithfully the widely different time scales characteristic of the interacting fluid and chemical processes. Another set of obstacles to numerical simulation arises because conventional numerical methods are unable to resolve accurately the steep spatial gradients in pressure, density, and temperature characteristic of these phenomena. Finally, we face the twin problems of physical and geometric complexity. Because of the many species, reactions, and complex container shapes required to model a realistic system, calculation time can increase by orders of magnitude over idealized or empirical models.

Table I lists the major chemical and physical processes which have to be considered for a complete description of a complicated reactive flow system.

TABLE I

Fundamental Processes in Reactive Flow

	Single phase	Multiphase
Chemical kinetics		
Laminar hydrodynamics		
Thermal conductivity, viscosity		
Molecular diffusion	↓	
Thermochemistry		
Turbulent hydrodynamics		
Radiation		
Nucleation (soot)		
Surface effects		
Phase transitions		↓
(evaporation, condensation, . . .)		

As indicated, multiphase processes such as surface catalysis and soot formation can be important even when we are primarily interested in gas phase combustion. For most interesting systems, one finds that the basic chemical reaction scheme, the individual chemical rates, the optical opacities, or the effects of surface reactions are not known. Thus the first problem that must be solved is modeling the controlling fundamental processes separately. The equations we postulate to describe our detailed model are the time-dependent equations for conservation of mass, momentum, species concentrations, and energy (Williams, 1965; Oran and Boris, 1981):

$$\frac{\partial \rho}{\partial t} = -\nabla \cdot \rho \mathbf{v}; \tag{1}$$

$$\frac{\partial n_j}{\partial t} = -\nabla \cdot n_j \mathbf{V}_j - \nabla \cdot n_j \mathbf{v} + P_j - L_j n_j; \tag{2}$$

$$\frac{\partial (\rho \mathbf{v})}{\partial t} = -\nabla \cdot (\rho \mathbf{v} \mathbf{v}) - \nabla P + \nabla \cdot \eta \cdot \nabla \mathbf{v}; \tag{3}$$

$$\frac{\partial E}{\partial t} = -\nabla \cdot E\mathbf{v} - \nabla \cdot P\mathbf{v} + \nabla \cdot \lambda \nabla T + \left.\frac{\partial E}{\partial t}\right|_{\text{chemistry and molecular diffusion}}, \tag{4}$$

where ρ, $\rho \mathbf{v}$, E, and P are the total mass, momentum, energy density, and pressure, respectively. The $\{n_i\}$ and $\{V_i\}$ are the number density and the molecular diffusion velocities of the individual chemical species. The quantities η and λ represent the viscosity and the thermal conductivity of the gas mixture at specified $\{n_j\}$ and temperature T. The $\{P_j\}$ and $\{L_j\}$ refer to chemical production and loss processes for species j, and the last term in Eq. (4) represents the local change in energy due to molecular diffusion and chemical reactions which must be added to the fluid dynamic energy density. The diffusion velocities $\{\mathbf{V}_j\}$ are found by inverting the following matrix equation

$$\mathbf{S}_j = \sum_{\substack{k=1 \\ k \neq j}}^{M} \frac{n_j n_k}{N^2 D_{jk}} (\mathbf{V}_k - \mathbf{V}_j) \equiv \sum_{k=1}^{m} W_{jk}(\mathbf{V}_k - \mathbf{V}_j), \qquad (5)$$

where the source term $\mathbf{S}_j$ is defined by (Williams, 1965) as follows:

$$\mathbf{S}_j \equiv \nabla\left(\frac{n_j}{N}\right) - \left(\frac{\rho_j}{\rho} - \frac{n_j}{N}\right)\frac{\nabla P}{P}$$
$$- \sum_{k=1}^{M} \frac{n_j n_k}{N^2 D_{jk}} \left(\frac{D_{T_k}}{\rho_k} - \frac{D_{T_j}}{\rho_j}\right)\frac{\nabla T}{T}. \qquad (6)$$

For those elementary reactive systems where the dominant processes are fairly well understood and can be modeled accurately, there is still the problem of modeling the interactions among the processes. Modeling these interactions means finding a way to deal with the huge disparities in the time and space scales which must be represented, as well as with real physical and geometric complexity. Furthermore, the complex structure of a system may require multidimensional models with unusual boundary conditions, forcing us to handle many interacting species simultaneously. Since these diverse scales are expensive to resolve explicitly, the construction of the desired detailed models is limited primarily by the computer power available. Computationally efficient methods are required which accommodate these disparate space and time scales while maintaining the required accuracy.

The approach which we have taken at the Naval Research Laboratory (NRL) is to treat the fundamental processes of the problem individually and then to couple together these components by carefully considering the interactions among them. This "operator split" approach requires asymptotic techniques when there are short time scales which we do not wish to resolve. In contrast to this, there are global implicit approaches which attempt to deal with multiple time and space scales by solving the entire nonlinear set of equations at once. Since these global implicit techniques require iteration and the inversion of large matrices, they are computationally expensive and time consuming. Furthermore, the global implicit methods guarantee numerical stability by forward differencing, which makes the solutions first order at best in time. A major advantage of asymptotically coupling the individual modules is that it is possible to incorporate the most efficient and accurate algorithms for solving each part of the problem.

We choose to follow this modular "asymptotic" approach in this chapter. More of the reasons that mandate this choice will become clear as we present more details of the physics, chemistry, and computers. We began with the development of accurate, efficient numerical algorithms for modeling the fundamental processes governing a combustion system. The techniques which have been developed at NRL are also very flexible and have been applied in research areas ranging from ocean hydrodynamics to laser-induced fusion. They include such programs and algorithms as CHEMEQ (Young and Boris, 1977) for the solution of stiff ordinary differential equations, Flux-Corrected Transport (Boris and Book, 1976) for the solution of the convective terms in the time dependent conservation equations, and the Slow Flow (Boris, 1977; Jones and Boris, 1977) and ADINC (Boris, 1979) algorithms for handling flame propagation problems where it is too cumbersome and costly to treat sound waves explicitly. We have also developed DFLUX (Jones and Boris, 1977; Oran and Boris, 1981) for the accurate solution of coupled multispecies mass diffusion fluxes and SPLISH (Fritts and Boris, 1979), a Lagrangian two-dimensional triangular grid technique for describing flows over complicated surfaces.

In this paper, these algorithms will be discussed briefly and references to more detailed discussions will be given. With several years of development, these techniques have been coupled into detailed one- and two-dimensional models of reactive shocks and flame propagation (Jones and Boris, 1977; Oran *et al.*, 1978, 1980; Book *et al.*, 1980). In these models, special emphasis has been placed on the accurate description of the coupling between the hydrodynamic flow and the energy released or absorbed by the chemical reactions. In the multidimensional models, simplified chemical reaction schemes have been used to test the coupling of the algorithms and to see how the energy is partitioned. In the one-dimensional models, there is essentially no limit to the complexity of the chemical reaction scheme that can be used. In all cases, however, it is essential to get the coupling right.

B. *A Gedanken Flame Problem for Illustration*

To assess the impact of these various numerical simulation techniques in a real situation, a one-dimensional flame propagation problem is used as an illustrative example. Consider a premixed combustible gas which is initially at rest in a 1-m-long tube. The gas is then ignited at one end. Traveling at 100 cm/sec, the resultant flame takes 1 sec to cross the system. Unfortunately, time scales required to resolve sound waves crossing a computational cell may be 10^{-8} sec or less; chemical time scales may be on the order of 10^{-6} sec or shorter when the reactions are fast; and the characteristic diffusion times

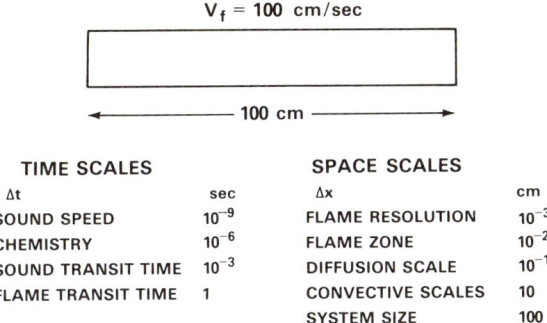

FIG. 2. Summary of the *Gedanken* flame problem.

may be 10^{-3}–10^{-4} sec. The span of important scale lengths is almost as severe. For this flame propagation problem, convective scale sizes for fluid flow may be 1–10 cm; macroscopic diffusion scales with a characteristic time of 1 sec are about 10^{-1} cm; the width of the flame front is about 10^{-2} cm; and 10^{-3} cm computational resolution is required. Figure 2 summarizes this problem.

Figure 3 displays these various time and space scales pictorially with the time scales plotted logarithmically along the abscissa and the space scales displayed logarithmically along the ordinate. The large rectangle spans the x,t region encompassed by our *Gedanken* flame problem. The smallest spatial grid resolution needed is 10^{-3} cm, about one tenth of the flame thickness. The largest spatial scale is the system size. The shortest time scale conceivably needed is the time it takes sound to cross one tenth of a cell. Assuming a maximum sound speed of 10^5 cm/sec, this CFL timestep is 10^{-9} sec. At the other end of the scale, the experiment lasts about 1 sec.

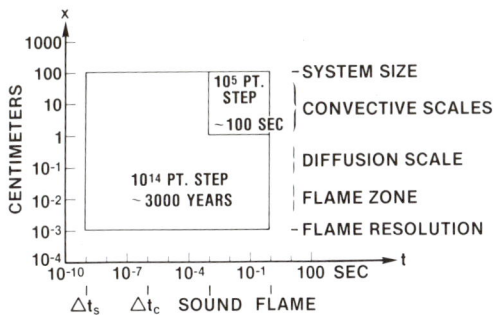

FIG. 3. Timing estimates for the *Gedanken* flame problem based on direct explicit solution of the governing equations on a uniform grid.

To estimate the cost of a given calculation we use 10^{-3} sec as the computer time needed to advance all the physical variables at one grid point by one timestep. This estimate is based on our asymptotically coupled explicit simulation models at NRL, in which roughly three quarters of the computation time is spent in the chemical kinetic calculations. The computational expense per point step is comparable when a slow flow or implicit hydrodynamics model is used to filter out sound waves and does not depend strongly on the zone size or the timestep.

The area of the large rectangle in Fig. 3 is proportional to the logarithm of the expense of performing a direct simulation in which all of these time and space scales are resolved explicitly. With 10^9 timesteps and 10^5 grid points, the calculation would last 3000 years! Our naive goal should be the small rectangle in the upper right corner of the large rectangle. This "ideal" calculation would use 1000 timesteps to follow the flame crossing 100 cells because it encompasses only the macroscopic scales and rates easily accessible to experiment. The expense of this small calculation is 100 sec, less than 2 min. In the remainder of this paper we attempt to describe techniques which allow us to model a flame such as the one described, including all the physics in the 3000-year calculation, but for the cost of the coarsely resolved macroscopic calculation.

II. VECTORIZATION AND OPTIMIZATION

Each of the computational techniques described in this chapter has been designed and developed to run efficiently on a vector computer. This section describes the principles underlying their implementation on the Texas Instruments Advanced Scientific Computer (TI ASC) and the practical methods of expressing them in FORTRAN. Within this specific context aspects of the overall system optimization problem are considered.

On fast computers, superficially minor details of the programming can change the computer time needed to solve a given problem by a large factor. Table II demonstrates some of the possibilities for these speed ranges on the ASC. A two-pipe ASC is capable of performing 1000 additions in 646 cycles (52 μsec). If the operation is coded differently, however, it requires 22,437 cycles (1800 μsec). The second version runs 35 times slower. It performs no more mathematical operations; it simply does not take advantage of "parallelism."

Parallelism is present in computers as old as the IBM 360/91 and CDC 7600. Computers which have such parallelism contain loop mode, a vector processor, a cache memory, or a fast memory buffer. The TI ASC, Cray

Research Cray-1, CDC STAR, and CYBER 200 series all make extensive use of it. All the fluid dynamics and chemical kinetics algorithms described in this chapter have benefited from program design which makes use of parallelism in these machines.

Several distinctive features are present in any computer with parallelism. It has an arithmetic unit which is faster than its memory access. In the fastest of these computers, the limit on memory access is the speed of propagation of an electrical impulse in a wire, that is, propagation at a fraction of the speed of light. In others, the limit is that the memory itself is slow. In all of these computers, access to memory is speeded up by sending several words to or from memory as a group, in a time substantially less than would be required to send them individually.

The group transfer of words to and from memory has one very large advantage—speed. This is clearly indicated in Table II. It also has two serious difficulties—complexity and causality. The vector computer is much more complex than a scalar computer. First, its arithmetic unit must process whole groups of data rather than individual words. Second, it must have a scalar computer within it since some operations cannot be vectorized. The causal problem is more subtle. The arithmetic processing of one group may require access to another group which is in transit to or from memory, and therefore unavailable. Waiting for individual words to make round trips to memory thus interrupts the group processing of data.

In Section II.A, we describe the group transfer process on several fast computers, to show how they are similar. In Section II.B we explain how this

TABLE II

COMPARISON OF INDIVIDUAL (SCALAR) AND
GROUP (VECTOR) PROCESSING TIMES[a]

	Execution time of an array of ASC additions		
		Vector time	
Number of additions	Scalar time	1 pipe	2 pipes
1	60	91	92
10	257	104	113
50	1171	149	139
100	2288	203	170
500	11,244	615	402
1000	22,437	1154	646

[a] In computer cycles (80×10^{-9} sec) for the operation $c_i = a_i + b_i$.

process is used, with FORTRAN examples. In Section II.C we discuss the impact these plus other considerations have on algorithms and explain how to avoid many potential problems by careful choice of algorithms.

A. *Speed in Hardware*

The basic limit on computing speed is the propagation speed of an electrical impulse in a circuit, approximately 50% of the speed of light. Integrated circuit elements are packed densely in the arithmetic unit and in the complete central processing unit (CPU). The fastest CPUs can now perform arithmetic operations in tens of nanoseconds or less. The physical volume of a large memory is such that the distance from its center (the logical location for its CPU) to its periphery is seldom less than 3 m, and usually greater. But 3 m of propagation implies a 20-nsec signal transit time, or a 40-nsec round trip (request for a memory word, followed by return of that word). The speed limit in such a computer is enforced by memory access delays.

Computers which bypass this speed limit are necessarily complex. This fact is best illustrated by an example. Suppose a hypothetical CPU (call it HYPPO) can complete one operation in 10-nsec, and a memory access requires 100 nsec to fetch any number of words. Then it would appear that HYPPO must fetch at least 10 words with each memory reference to keep its CPU busy. On closer examination, the number is actually much larger. Suppose HYPPO can add 10 pairs of numbers, $c_i = a_i + b_i$, in 100 nsec. Then a large number of pairs can be added without loss of CPU efficiency only if, during each 100-nsec period, the CPU is involved with 60 words in various stages of processing. Half of them (30) are being used by the CPU, while 10 c_i values are being sent to memory from the last 10 operations, and 20 a_i and b_i values are being obtained from memory for the next 10 operations. This requires many buffers, plus the associated memory drivers and logic to switch the CPU among them. This provides the other reason this hypothetical computer is named HYPPO. It is ponderous.

Most fast computers are designed so that group or pipeline access to memory occurs when arithmetic operands are stored sequentially in memory. That is, if the three arrays a_i, b_i, and c_i are arranged so that their elements are stored in consecutive memory locations, then the computer can fully utilize its CPU speed. The IBM 360/91, CDC 7600 and STAR, and TI ASC are like this. The Cray-1 provides more general memory access; it can quickly move groups of data to or from memory whenever their memory addresses have a constant displacement, e.g., every second, or every tenth word.

In each fast computer, there is an "overhead" or startup time associated with group processing. As a result, groups must have a minimum length, typically 5–20, in order to make group processing more efficient than processing individual elements. This minimum size varies significantly among computers.

On fast computers, many choices must be made in the development of an efficient program. The data layout must be carefully organized. How (or whether) to use group operations often must be decided by global properties of the algorithm being implemented. In the worst case, each programmer is faced with the necessity of spending several years learning the intricacies of very specialized hardware. Fortunately, computer manufacturers are now offering a viable alternative to that extreme in the form of high-level language compilers (FORTRAN, PASCAL, etc.), which will automatically or semi-automatically provide access to this hardware speed. Use of such a compiler is the subject of the next subsection.

B. *Speed in FORTRAN*

Very sophisticated compilers are now available. Many FORTRAN compilers are adept at recognizing group actions, such as DO loops, and taking special action to execute them efficiently. For the ASC, this means generating "vector" instructions. The degree of sophistication required for such processes is best illustrated by examples. Let us see how this generation is related to the FORTRAN source text.

We shall begin with the addition of 100 pairs of numbers, $c_i = a_i + b_i$. The FORTRAN text reads:

```
        DO 1 I = 1, 100
        C(I) = A(I) + B(I)
    1   CONTINUE
```

A traditional "scalar" computer would execute about five assembly language instructions 100 times. It would perform two memory fetches (a_i and b_i), one addition, one store to memory (for c_i), and one instruction that increments a counter, tests, and branches back to load the next pair of input operands. Thus 500 scalar instructions are executed to add arrays A and B.

The ASC can generate such "scalar" code (we shall see why in the next section) or it can generate "vector" code, which executes very differently. Vector code for adding the 100 pairs of operands consists of a single assembly language instruction and an associated table, built by the FORTRAN compiler. The table contains the starting address in central memory of the input and output arrays and the increments for stepping through the arrays (in this case, unity). A vector instruction executes by continuously streaming operands from central memory into the central processor, where the addition takes place, and continuously streaming answers back to central memory. The vectorized addition, in this example, may be thought of as 100 additions simultaneously occurring on the 100 pairs of input operands. Actually, during execution, some elements of A and B are being read from central memory, some elements of A and B are undergoing addition in the CPU, some answers (array C) are in output buffers, and some C are being written to memory.

Table II lists the time that it takes to execute a DO loop which performs various numbers of additions on the ASC in scalar mode and in vector mode. The particular ASC at the Naval Research Laboratory has two arithmetic units, (AUs), also called pipelines or pipes. The additions can be performed by one of them or the A, B, C arrays can be split in half, and each AU can do half the additions. Vectorized execution times are given for utilization of one or both AUs. Times are given in CP clock cycles; the basic CPU cycle is 80 nsec.

Notice that, for scalar execution, the time per addition remains nearly constant whether 10 or 1000 additions are performed. In two-pipe vector mode, the time per addition decreases from 11 units to 0.65 units as the array size increases from 10 to 1000 elements. This timing pattern is characteristic of all vector operations on the ASC. When arrays are large, the benefit from vectorization and from using two arithmetic units becomes very substantial. Thus for most reactive flow codes, increasing array sizes or refining grid resolution is much less costly in computing time if the code is run on a vector computer than if the code is run on a scalar computer.

For calculations involving multiply dimensioned arrays, the ASC differs from other vector computers. On it, doubly or triply subscripted arrays in loops nested two or three levels deep also may be collapsed into a single vector instruction. The loop

```
      DO 2 K = 1, 10
      DO 2 J = 1, 10
      DO 2 I = 1, 50
      C(I, J, K) = A(I, J, K) + B(I, J, K)
    2 CONTINUE
```

executes as a single vector instruction which adds 5000 pairs of numbers. No other computer can make such a triple loop into a single instruction, unless the upper limits of the loops are equal to the corresponding dimensions of all the arrays, so that the triple loop is really equivalent to a single loop. Most other fast computers will optimize or vectorize only the inner loop.

Useful computations are not restricted to simple additions. The typical DO loop in a program contains many operations, both arithmetic and logical. Here again the compiler should recognize (and the ASC compiler

TABLE III

Partial List of the More than 100 Vector Operations Implemented in the ASC Hardware[a]

*Add (Algebraic)	*Min
*Add (Magnitude)	*Multiply
*Compare ($<, <=, >, >=, \neq$)	Normalize
*Divide	*OR
*Dot Product	Order
*Exclusive OR	Peak Pick (Max and Min)
Fix-to-Float	*Replace
Float-to-Fix	Search
*Map	*Select
*Max	*Subtract (Algebraic)
Merge	*Subtract (Magnitude)

[a] Starred entries have four implementations: halfword, integrer, real, and double-precision arguments.

usually does) that the more complex operation is a series of array operations. Thus the DO loops

```
    DO 3 J = 1, M
    DO 3 I = 1, N
    Y(I,J) = A(I,J)*X(I,J)**2 + B(I,J)*X(I,J) + C(I,J)
    P(I+J) = Q(I+5*J-4) + R(I,J)
3   S(I,J) = AMIN1(T(I,J),U(I,J))
```

can become 7 vector instructions (or 14, if they are split over two AUs). Needless to say, the compiler which recognizes this has a complexity comparable with the complexity of the hardware.

An arithmetic expression must meet two criteria to be vectorizable. First, its dependence on the DO loop indices must be linear. Second, the hardware must have the desired operation implemented (see Table III). There are two circumstances, however, where the arrays are not independent, and where conditional execution (IF statements) is involved. These cases are discussed in Section III.C.

C. *Speed in Algorithms*

The remainder of this chapter is devoted to specific algorithms and techniques which have been efficiently implemented for use in reactive flow calculations. A few general words about algorithm speed are appropriate at this point. Fast hardware and good high-level compilers (FORTRAN) certainly have been a part of the computational physics revolution of the last two decades, but improvements in algorithms deserve at least an equal share of the credit for the advances which have been made. Three-dimensional calculations in the 1980s will be about as accurate as two-dimensional calculations were in the 1970s and one-dimensional calculations were in the 1960s. As the essential notions of accuracy, stability, flexibility, and computational expense have been successively refined in various disciplines, the relative merits of the various algorithms and algorithm implementations have become better understood. Different problems, being solved under different external and project constraints, still demand different solution methods, however.

Optimization of detailed computer models requires striking a balance among the often conflicting goals of short development time, code robustness, result accuracy, model flexibility, and code efficiency. These disparate goals all influence our choice of algorithms and our implementation, as indicated in Fig. 4. The five goals listed above must all be weighed subjectively and tend to shift in importance as a computational project passes through various stages of development. It has been our experience, however, that people time is more precious (up to a point!) than computer time. For example, typically 20–25% of our effective budgets usually go to computing and related services. Therefore code efficiency usually takes a backseat initially to concerns for reducing the lead time to first reliable results. The trick is to select algorithms and associated implementations which are "idiotproof" on initial implementation and yet may be easily upgraded in stages to a much more efficient and robust model for production calculations.

Codes should be flexible enough that minor conceptual modifications do not require significant revisions of all segments of the model. Efficiency clearly must suffer somewhat to keep the individual elements of the overall algorithm well separated and identifiable, but a short lead time for adapting

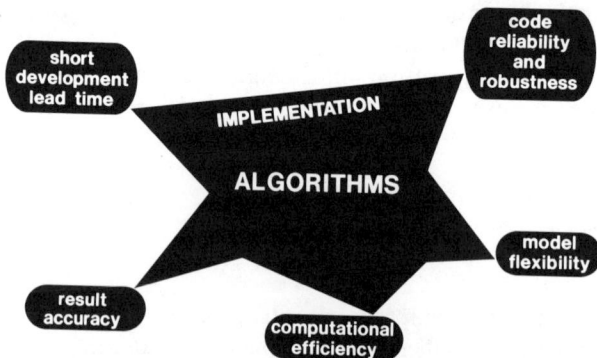

FIG. 4. The optimization of a computational model, a dynamic balance between a number of conflicting goals, is depicted schematically. Both the basic algorithms and their cladding, the computational implementation, are influenced in a complex way by the conflicting demands.

a good computational simulation model to its next application is very valuable. People time is so valuable, in fact, that considerations such as future code readability must always be a major factor in project management. For example, to allow for greater machine independence and readability, specific machine language programming should be kept to a minimum.

Well-known optimization techniques can often be applied without severely constraining the simulation. For example, requiring the number of grid points to be a power of two allows application of the most readily available highly optimized fast Fourier transform packages. In addition to CPU time, optimization also includes a consideration of data accuracy and external device timings. It is often costly to calculate a quantity more accurately than the value it is going to update. Similarly, if it is necessary to store intermediate quantities on an external device, there is unnecessary programmer effort in optimizing internal calculations to so great a degree that the computation time is dominated by the external data transfer rate.

Logical simplicity has great value in a computational algorithm to be used in conjunction with a vector computer. The newer pipelined and vector processors thrive on simplicity; their performance is seriously degraded by elegant but logically or structually cumbersome algorithms which do not treat most of the grid points in an identical way. The specific algorithms described below have been effective for use on the TI ASC, a machine with roughly a 16:1 speed improvement attendant to vectorization.

III. TECHNIQUES FOR MODELING FAST TIME SCALES

A. *Implicit and Asymptotic Techniques for Stiff Equations*

Some phenomena to be modeled have characteristic times of variation shorter than the timestep one can afford. Such phenomena are usually called temporally "stiff." There are two rather distinct approaches one can use to construct detailed models when one or more of the competing processes is temporally stiff. The first, often called "global implicit," is essentially mathematical in nature. To use it, the complete set of coupled nonlinear equations for the system is cast in a partially or fully implicit forward-differenced form. The second approach is founded much more fundamentally on the specific physics or chemistry of the problem being solved. This is the "asymptotic" approach which allows timestep splitting and modular programming. We shall discuss both of these approaches with respect to accuracy, versatility, and computational cost.

In the global implicit methods, the nonlinear terms are usually linearized about the solutions obtained numerically at the previous timestep, a process which is only valid when the values of the physical variables change very little during a timestep. A rigorously correct treatment of the nonlinear terms requires iteration and the inversion of matrices at each timestep in order to guarantee stability. The Gear method (Gear, 1971) for chemical kinetics is a restricted example of this "global implicit" approach. When neighboring points in space are coupled by fluid dynamics and transport phenomena, the use of this method requires inverting a tridiagonal matrix of matrices at each timestep.

In contrast, the asymptotic approach puts minimum strain on the computer budget but demands much more physical and chemical insight. Convergence of solutions is again easy to test, but timestep accuracy control can be difficult. Since reducing the timestep can make an asymptotic treatment of a "stiff phenomenon" less accurate rather than more accurate, the coupling algorithms for the various timestep-split phenomena must be carefully chosen to change smoothly from explicit to asymptotic form as the timestep is varied. The nonconvergence of any particular solution is usually easy to spot, however, because the calculation usually fails catastrophically. For example, lack of conservation of either mass or atoms in a kinetics calculation signals inaccuracy rather clearly.

A fundamental advantage of the asymptotic approach is the timestep splitting. This permits the development of modular simulation models. The various physical and chemical processes are tied neatly into individual packages which can be tested separately and used directly in totally different physical problems. This modularity is usually very helpful in vectorization as well because the components of the overall algorithm are broken out into easy-to-optimize "kernels" of dense computation.

Timestep splitting is most powerful when it can separate the complete simulation into parts corresponding to distinct physical processes. For example, the species density equations are separated into spatial derivative terms representing convection, those representing compression, chemical production, and loss terms. At each timestep, the partial differential fluid dynamic terms are split from the ordinary differential terms modeling chemical kinetics. The overall solution is a result of coupling the fractional steps, accomplished by providing the initial conditions for one step with the results of the solution for the other step.

This timestep-splitting approach gives accurate solutions to the equations as long as all the phenomena being modeled are slowly varying on the scale of a timestep. Accuracy is tested by decreasing the timestep and by increasing the resolution, and then checking for variations in the answers.

In general, we cannot afford 10^9 timesteps as required for sound wave resolution or even 10^7 timesteps to resolve some of the stiff chemical reactions in the flame example of Section I. Thus when stiffness is encountered in a model which is being solved by timestep splitting, the stiff terms or phenomena must be coupled "asymptotically" rather than implicitly. Timestep splitting procedures are ruled out in "global implicit" treatments of stiff phenomena since implicit coupling precludes the splitting.

The differences between these two approaches are significant. The global implicit approach puts maximum strain on the computer and minimum strain on the researcher. It is easy to test the convergence of the solutions computed by this approach by varying the grid size and timestep. Unfortunately the nonconvergence of any particular calculation is hard to spot because numerical stability is usually achieved via a strong numerical damping. This artificial smoothing changes the profiles nonphysically, so quickly detected qualitative errors are often absent or are suppressed.

The flexibility allowed by asymptotics exacts its price in the need to treat carefully all the couplings between the individual physical terms and effects. Using the asymptotic approach, one cannot sit back and turn a massive mathematical crank to get the answer.

In vectorized calculations, the choice between one algorithm and another is often determined by considerations of causality. Since block implicit and global implicit algorithms often entail recursion or tridiagonal systems of equations, the basic approach found most efficient on a scalar computer is often worst on a vector machine.

B. *Problems with Causality*

Physicists invoke causality to discuss events which occur simultaneously, or nearly simultaneously, at different locations. Causality problems occur in a computer when the operands of a series of operations are mutually dependent. If the programmer attempts to use parallelism, he finds that some results are needed in one place while they are simultaneously being calculated in another. The second of the following DO loops, a form of recursion relation, has problems with causality.

```
      DO 4 I = 1, 99
    4 A(I) = A(I) + A(I+1)
      DO 5 I = 2, 100
    5 A(I) = A(I) + A(I-1)
```

When an array addition is processed on a vector computer, either physically distinct hardware is performing successive additions; that is, one AU is doing the ith addition while another is doing the $(i + 1)$th, or the complete addition (address decoding, operand fetch, addition, store, etc.) is performed in steps, and the ith addition is in one step while the $(i + 1)$th is in the step behind it. In either case, the result of the ith addition is not available until after the $(i + 1)$th is already under way. Thus in the example given above statement number 5 has causality problems.

The compiler will vectorize statement 4, but not statement 5. (And the compiler must be very sophisticated to recognize the difference in general cases.) If a statement like 5 seldom occurs in a program, it is reasonable to let it be performed in scalar operations. However, if its execution dominates the computation time, it must be rewritten so that it can be vectorized. The easiest way to do this begins with an examination of the desired results. They are

```
B(2) = A(1) + A(2);
B(3) = A(1) + A(2) + A(3);
B(4) = A(1) + A(2) + A(3) + A(4),
```

etc. Here B is used to represent the array A after execution.

The most obvious method is to directly vectorize these cumulative sums. The operation in statement 5 may be vectorized as

```
      DO 6 I = 2, N
      B(I) = 0
      DO 6 J = 1, I
6     B(I) = B(I) + A(J)
      DO 7 I = 2, N
7     A(I) = B(I)
```

Statement 6 vectorizes on J, and statement 7 also vectorizes. Note, however, that the operation count has been increased from N to $N(N + 1)/2$. Statement 5 has been vectorized, but it will probably take longer to execute.

The most efficient method of vectorizing statement 5 lies between the original form and this one. Suppose we rewrite the operations on the As in terms of even Bs. That is,

```
B(2) = A(1) + A(2)
B(4) = A(3) + A(4) + B(2)
B(6) = A(5) + A(6) + B(4)
```

Now we must perform $2N$ additions, but we have a scalar operation on only $N/2$ Bs. This scalar operation can in turn be replaced by a similar process on the Bs, at the cost of $2(N/2)$ vector additions plus $N/4$ scalar operations. This is a recursive prescription for reducing the length of the scalar operation until its computational cost is insignificant.

This process is a form of cyclic reduction, closely related to that employed in some direct Poisson solvers (Buneman, 1969; Hockney, 1970; McDonald, 1980). For large N, the N scalar operations in statement 5 are replaced by $N \ln N$ vector operations, with a substantial increase in computational complexity but usually reduced computational cost. There are many other cases where operations which appear to be purely scalar may be vectorized.

Fortunately, in most of the important cases, a subroutine has been written to perform that operation. Much of the effort is then transferred from coding to library search for the appropriate subroutine. The two reports (Boris, 1975, 1976c) contain algorithms for recursion relations and vectorized tridiagonal solvers of several types along with documentation and timings.

For practical purposes, these causality problems occur only in one dimension. If the causality problem is in only one of several dimensions, then the vectorization can be performed on the dimension without the problem. For example, if the source code is

```
      DO 8 I = 1, N
      DO 8 J = 2, N
    8 A(I, J) = A(I, J - 1) + B(I, J)
```

then an interchange of the two "DO 8" statements is enough to allow the inner loop (now on I) to vectorize. Note that most vector computers will perform well in this case only if the vectorized index (here "I") is the first subscript. This is why memory layout is very important on a vector computer. For a more complex illustration, consider

```
      DO 9 I = 2, N
      DO 9 J = 2, N
      DO 9 K = 2, N
    9 A(K, J, I) = A(K - 1, J - 1, I - 1) + B(I, J, K)
```

where cyclic reduction is needed only for the innermost loop. When the loop on K has been vectorized by cyclic reduction, accesses to the "dependent" data in the $J - 1$ and $I - 1$ planes are far removed in time from the J and I accesses (during succeeding cyclic reductions), so no causality problem occurs. However, the ASC compiler requires special help to vectorize such a statement. The remainder of this section treats the problems of unacceptably short time scales in chemical kinetics and fluid dynamics. These are based on asymptotic splitting approaches to the short timestep problems.

C. Stiff Equations in Chemical Kinetics

The chemical kinetic rate equations for the species number densities can be written as a set of first-order, coupled, nonlinear, ordinary differential equations of the following form for the jth species:

$$\frac{dn_j}{dt} = P_j - \frac{n_j}{\tau_j}, \tag{7}$$

where n_j is the density, P_j the production rate, and τ_j the characteristic loss time. The "selected asymptotic method" (Young and Boris, 1973, 1977; Young, 1980) is often used at NRL to solve these equations. It first determines which equations satisfy an appropriately chosen stiffness criterion. The equations determined to be nonstiff are solved by a second-order Adams method while a very stable asymptotic method is applied to the stiff equations. This method, in which selected equations are treated asymptotically, is suitable for the situation near equilibrium states where the production and loss rates can be large but nearly cancel each other. Small numerically produced oscillations, which result from short time constants in the kinetics and would produce prohibitively short timesteps, are effectively removed. The timestep is carefully monitored to ensure the accuracy and convergence of the method. In coupled chemical and hydrodynamic models, the accuracy of the chemical integration can be relaxed somewhat, relative to pure kinetics calculations, due to the presence of unavoidable truncation errors in the fluid dynamics and the uncertainties in the rate constants.

The "selected asymptotic method" is very efficient for reactive flow problems since it has very low overhead and can be restarted inexpensively. Since the method is low order, only the initial values of the variables are needed at the beginning of a timestep. Values from several previous timesteps are not needed in CHEMEQ, in contrast to other high-order schemes. Furthermore, Jacobian linearization matrices are not needed. The simplicity of the algorithm also makes it easy to change the timestep in response to changes in the computed solution. Our major subprogram using selected asymptotics, CHEMEQ, has been programmed to make efficient use of the parallel processing capability of vector computers so the code runs at least an order of magnitude faster than Gear ODE solvers on reactive flow problems where many grid points are needed and where the rate equation integrations must be restarted hundreds of times—after each hydrodynamics timestep.

It is not clear a priori which equations are to be treated asymptotically and which are to be treated normally, so the two distinct solutions have to be computed at every grid point and timestep. A masking procedure (Boris, 1976b), developed for the Flux-Corrected Transport algorithm, is used here.

In general, four approaches for "data splitting" are available on the ASC computer; result merging with floating point masks (0.0 and 1.0), scalar code, vectorized "IF" statements, and use of special hardware instructions. The following four DO loops illustrate these approaches for the case where $C_i = \text{maximum}\,(A_i, B_i)$ for $i = 1, \ldots, N$:

```
      DO 10 I=1,N
10    C(I)=MASKA(I)*A(I) + MASKB(I)*B(I)      (resulting
                                                merging)

      DO 11 I=1,N
      C(I)=A(I)
      IF(A(I).LE.B(I))C(I)=B(I)               (scalar)
```

```
11  CONTINUE
    DO 12 I = 1, N                    (vectorized ''IF''
    C(I) = A(I)                              statement)
    IF (B(I).LE.A(I)) GO TO 12
    C(I) = B(I)
12  CONTINUE

    DO 13 I = 1, N                    (special ''hardware''
13  C(I) = AMAX1 (A(I), B(I))              instruction)
```

Even for this particularly simple example, only some computers have a hardware MAX, MIN capability. Then loop 13 is clearly the fastest implementation. In other situations, such as the FCT algorithm to be discussed, many computations are sufficiently complex that masking is most efficient. Even though masking involves more operations (3), some of which are not used, substantial speedups are possible because everything can be vectorized. When N is small the scalar loop is most efficient because using any of the vector techniques would incur an unacceptable premium for vector startup and overhead. The number N at which vectors become more efficient than scalars is typically 10–20 on the ASC, somewhat smaller on the Cray-1 computer, and somewhat larger on the CDC machines.

The ASC also has a feature in its optimizing FORTRAN which vectorizes the IF statement directly as used in loop 12. By branching to the end of the loop, the compiler can vectorize the test B(I).LE.A(I), constructing a table of indices I for which B(I).GT.A(I). These maverick cases are then handled at the end of the loop by a separate scalar loop. This approach executes faster than masking only when most B(I) are in fact less than A(I)s so only a few scalar cases have to be evaluated.

D. *Short Time Scales in Fluid Dynamics*

Earlier we described the "slow-flow" algorithm for dealing with very fast sound waves. It is an important part of constructing a complete flame propagation model (Jones and Boris, 1977; Boris, 1977). As in the case of the chemical kinetic rate equations, the need for an asymptotic treatment only arises when we are interested in simulating short time-scale "stiff" phenomena on a much longer time scale. When a "stiff" phenomenon is far from equilibrium, the governing profiles change rapidly and extensively on the short time scale, so the computation must be performed on this scale to acquire an accurate solution. Luckily this period of short timesteps and rapid readjustment does not last very long because the readjustment to equilibrium is rapid. As soon as the fast transients settle down, an asymptotic treatment of the fast scales yields a slowly evolving but stiff dynamic equilibrium.

In the case of flame propagation, the dynamic equilibrium is characterized by flow velocities small compared to the speed of sound and an essentially constant pressure field. These circumstances prohibit shocks and strong rarefactions, so Eq. (4), the energy conservation equation, can be written

$$\frac{dP}{dt} = -\gamma P \nabla \cdot \mathbf{v} + \nabla \cdot \lambda \nabla T + \left.\frac{\partial P}{\partial t}\right|_{\text{chem}} \tag{8}$$

without shock and viscous heating. When the spatially varying part of dP/dt can be ignored, Eq. (8) gives a simple algebraic solution for the divergence of the velocity field:

$$\nabla \cdot \mathbf{v} \approx \frac{1}{\gamma P}\left[\nabla \cdot \lambda \nabla T + \left.\frac{\partial P}{\partial t}\right|_{\text{chem}}\right]. \tag{9}$$

The curl of the velocity is still advanced by a convective partial differential equation,

$$\frac{\partial \xi}{\partial t} + \mathbf{v} \cdot \nabla \xi + \xi \nabla \cdot \mathbf{v} = \frac{\nabla \rho \times \nabla P}{\rho^2}, \tag{10}$$

where $\xi = \nabla \times \mathbf{v}$ (the vorticity) is a scalar in two dimensions. Compressional sound waves do not appear in the solutions of these equations. Strong compressions and rarefactions driven by chemistry and conduction evolve slowly compared to sound waves. There is an important advantage of this method; the way in which Eq. (10) is advanced has been left unspecified, so that the best finite difference techniques available can be applied. Both Eulerian techniques and Lagrangian techniques can be used. There are no restrictions which require the use of relatively poor linear-expansion finite-difference algorithms. This is in sharp contrast to the global implicit approach.

There are several complications which must be dealt with in any calculation using "slow flow." A net addition of heat to the system through chemistry, thermal conduction, or external sources changes the average pressure with time. In a closed container, this average heating $\langle dP/dt \rangle$ cannot lead to a velocity divergence and is therefore subtracted from Eq. (9) to get the correct value of $\nabla \cdot \mathbf{v}$. As a result, P is essentially constant in space but not in time. Furthermore, P appears in Eq. (10) as part of the vorticity source term, so its gradients must be handled with care. Even for buoyant flames, the vorticity source term can be handled adequately by using the average hydrostatic equilibrium pressure gradient, which changes slowly with the average temperature. In certain types of self-consistent fluid dynamic problems we must solve Eq. (9) to first order as well as zeroth order to reproduce accurately the kinetic vorticity generation. This can still be done without jeopardizing flexibility in the choice of convective transport algorithms.

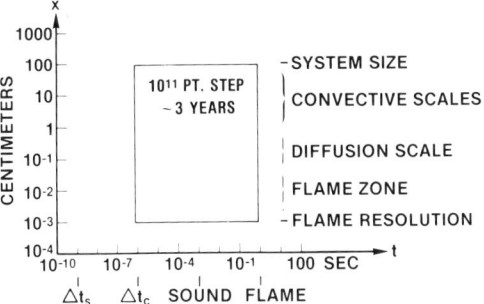

FIG. 5. Regions of the x, t diagram spanned by a solution of the *Gedanken* flame problem solved using asymptotic techniques to remove the fastest chemical and fluid dynamic time scales from the problem. A factor of 1000 speed improvement is achieved.

In closing this section on asymptotic treatment of fast time-scale phenomena, we wish to note that special care is required when two or more distinct phenomena are stiff at the same time. Since both chemistry and sound waves are usually stiff in the flame propagation problem, a difficulty arises in coupling the chemical heat release to the fluid dynamics. The overriding principle is always that the macroscopic variables should change by a controlled amount (10–20%) in each timestep. Thus the energy released in a flame goes primarily into fluid expansion and thus does not heat the fluid nearly as much as it would if fluid dynamic expansion were prohibited. This particular coupling difficulty is overcome by evaluating the change in pressure which would occur at constant pressure, but not allow this energy to heat the fluid during the kinetics part of the calculation. This pressure change is then used as an energy source term for the hydrodynamics partial timestep to get the correct fluid expansion. The correct temperature is obtained when the two independent calculations are complete because the density in the flame front has decreased at constant pressure.

Figure 5 shows the region of the x,t plane which defines our flame example after asymptotic techniques have been used to remove the timestep limitations incurred by stiff chemical kinetics and a very fast sound speed. The grid is still assumed to be uniform with a spacing of 10^{-3} cm, but the estimated computation time has been lowered to ~ 3 years.

IV. TECHNIQUES FOR MODELING SHORT SPACE SCALES

This section describes the modeling of physically important but very short space scales in a macroscopic system. In a previous paper (Boris, 1976a) on the numerical solution of continuity equations, this problem was addressed for convective transport of physical quantities. Here the focus is on techniques for dealing with convective flow in a chemically reactive environment. The methods we present here maximize the accuracy of representing continuous fluid dynamics using a minimum amount of stored data and computer time. We have found in general that both Eulerian and Lagrangian finite-difference techniques are needed for the highly nonlinear, transient phenomena which are simulated in detailed reactive flow modeling. Nevertheless, there are still a number of difficulties associated with their uses. In the case of Lagrangian methods, the major outstanding problems arise from gross distortions of the grid which quickly destroy the accuracy of physically realistic flow calculations. In the case of Eulerian methods, the major weakness in a very broad class of problems of real interest is the need for a large artificial damping (numerical diffusion) to fill in what would otherwise be pits of "negative density" in the calculated profiles. Since the "Eulerian" positivity problem may be encountered even in Lagrangian calculations, it demands constant attention.

A. *Flux-Corrected Transport Algorithms*

No algorithm has yet been found with which a numerical calculation is able to reproduce faithfully physical phenomena with characteristic scale lengths of variation shorter than a cell size. Therefore, to calculate realistic profiles of physical variables, the cell spacing must be made small enough to obtain a given accuracy. Choosing a basic finite difference method which maximizes accuracy with a minimum number of grid points is a major concern in detailed modeling. With this problem in mind, we use the Eulerian Flux-Corrected Transport (FCT) algorithm (Boris, 1976a,b; Boris and Book, 1976; Book *et al.*, 1980) to solve the convective parts of Eqs. (1)–(4). FCT reproduces steep gradients particularly well, and thus accurately treats a situation which occurs frequently in the fluid dynamics of shocks and in the chemical dynamics of combustion fronts and flames. Furthermore, because negative densities cannot occur using this method, many types of numerical instabilities which plague reactive flow calculations can be avoided without excessive nonphysical numerical diffusion.

The particular FCT algorithm suggested is a simple variant of the ETBFCT program (Boris, 1976b). Instead of following the cell center locations on the generalized one-dimensional grid and averaging these to define cell boundaries, the cell boundaries are followed and then averaged to define cell centers. In this way numerical diffusion can be minimized and the treatment of system boundaries is simplified.

An important feature of these FCT algorithms is the lack of an artificial viscosity needed to stabilize shocks. The flux-correction procedure itself ensures that the shocks are one or two zones wide and have maximal resolution. Therefore the detailed shock model developed at NRL uses FCT exclusively (Oran et al., 1978, 1980). Physically correct mass diffusion, viscosity, and thermal conduction terms are included in the model by separate diffusion subprograms which take advantage of timestep splitting. The fluid dynamics time integration is performed by an explicit two-step predictor–corrector technique. The usual Courant–Friedrich–Lewy stability conditions apply to the hydrodynamics timestep. The chemistry can proceed on a faster scale provided that the hydrodynamics does not respond to these chemical changes on this faster time scale.

The FCT algorithm solves a major part of the reactive flow problem and often consumes a significant fraction of the required computer time. The innermost kernels of the algorithm contain absolute values and inequalities as well as algebraic expressions. Thus it is an appropriate example for illustrating the principles of vector optimization. It has array, scalar, and conditional operations. Thus it is not obvious whether it should be efficient on a vector computer.

The ETBFCT algorithm itself (Boris, 1976b) has component parts. The first three of its stages (transport, diffusive fluxes, and antidiffusive fluxes) and the last (calculation of the flux-limited solution) are linear operations on arrays of fluid quantities. Thus they are easy to vectorize except for application of boundary conditions.

The remaining operations in ETBFCT associated with the flux limiting procedure are defined by partial operations in various special cases, determined by inequalities. We have seen that expressions involving IF tests do not fully vectorize. Thus vectorizing the flux limiter requires casting it in a form which does not explicitly test for these inequalities. In order to see how this is done, let us look at the actual limiting operations.

The purpose of the limiter is to ensure that no new nonphysical maxima or minima are formed in the physical quantity being convected. In principle, this means comparing each set of three adjacent values of variables, at successive timesteps, determining whether a new maximum or minimum has formed, and taking corrective action if necessary. The substantial number of comparisons involved can be reduced by focusing on the element of transport—the flux.

The flux $F_{i+1/2}$ transports density from the cell with density ρ_i to the cell with density ρ_{i+1}. This transport creates a new maximum or minimum only if it reverses the gradient before it $(\rho_i - \rho_{i-1})$ or after it $(\rho_{i+2} - \rho_{i+1})$. Examination of the flux and density gradients therefore appears to reduce the required number of tests to two for each cell. The next step is to eliminate the tests. This can be done if the unlimited flux and the result of flux limiting can together be expressed in terms of FORTRAN library functions.

The action of flux limitation consists of reducing the calculated flux to a value which is small enough that it does not reverse adjacent gradients. If $F_{i+1/2}$ is positive, the limited flux is (in Cartesian coordinates)

$$F_{i+1/2}^{c+} = \min(F_{i+1/2}, \tilde{\rho}_{i+2} - \tilde{\rho}_{i+1}, \tilde{\rho}_i - \tilde{\rho}_{i-1}), \tag{11}$$

since $F_{i+1/2}$ will subtract from ρ_n and add to ρ_{n+1}. If $F_{i+1/2}$ is negative, the gradients must be limited in the opposite direction:

$$F_{i+1/2}^{c-} = \min(F_{i+1/2}, \rho_{i+1} - \rho_{i+2}, \rho_{i-1} - \rho_i). \tag{12}$$

These two expressions can be combined with the absolute value and the signum function

$$\text{sgn}(x) = \begin{cases} 1, & x \geq 0, \\ -1, & x < 0, \end{cases} \tag{13}$$

in the form

$$F_{i+1/2}^{c\pm} = \sigma_{i+1/2} \min[\text{abs}(F_{i+1/2}), \sigma_{i+1/2}(\tilde{\rho}_{i+2} - \tilde{\rho}_{i+1}), \sigma_{i+1/2}(\tilde{\rho}_i - \tilde{\rho}_{i-1})], \tag{14}$$

where

$$\sigma_{i+1/2} = \text{sgn}(F_{i+1/2}). \tag{15}$$

Even this is not quite enough, since the limiting process should not be allowed to reverse the flux. Thus the final limited flux is determined by

$$F_{i+1/2}^c = \sigma_{i+1/2} \max\{0, \min[\text{abs}(F_{i+1/2}), \sigma_{i+1/2}(\tilde{\rho}_{i+2} - \tilde{\rho}_{i+1}), \sigma_{i+1/2}(\tilde{\rho}_i - \tilde{\rho}_{i-1})]\} \tag{16}$$

when the cell volumes are unity. This scalar formula for the limited flux $F_{i+1/2}$ can be expressed entirely in terms of standard FORTRAN library functions which compile vector code inline on the ASC.

Because the SIGN operation is not expanded in line on the ASC, the flux limiter Eq. (16) is evaluated by masking (Boris, 1976b). A single executable statement Eq. (16), which assigns the value of a complex arithmetic expression to the array containing the fluxes $F_{i+1/2}^c$, has been reexpressed using ten loops. The resulting code executes up to two orders of magnitude faster than the original scalar version, about 1.25 μsec per grid point per equation per timestep.

This completes the vectorization of the ETBFCT algorithm, but one more subject deserves explicit mention. When calculations on the interior of the grid are fully vectorized, a complicated boundary condition sometimes becomes the most time-consuming part of a calculation. The ETBFCT module, as implemented, incorporates the boundary condition in the vectorized subroutine, provided it is one of three general types. It treats the simplest, constant boundary values, by including the boundary values in the calculation, but only updating the interior values. Reflecting (or impermeable) boundary conditions are treated by placing a "guard" cell beyond the outermost cell and calculating the values there in such a way that the midpoint between the last cell and the guard cell is a reflection point. Finally, periodic boundary conditions are incorporated, again with a guard cell, by assigning to the guard cell the value at the other end of the grid. These three common boundary conditions are thus incorporated into the algorithm without degrading the vector performance of the algorithm.

B. *Adaptive Gridding and Rezoning*

In contrast to the shock model, the detailed NRL one-dimensional flame model uses a fully Lagrangian treatment which eliminates numerical diffusion (Boris, 1979; Oran and Boris, 1981). This is allowed because shocks cannot occur in the slow flow regime. In fact, it is questionable whether an ordinary Eulerian method could even deal with flame propagation because of the inherent numerical diffusion in these schemes. Fully Lagrangian treatments are difficult although some advances are discussed in Section V. The two-dimensional FLAME model is Eulerian and therefore we have used FCT and slow flow (Jones and Boris, 1977).

Another important feature of the FCT algorithms we have developed is their ability to divorce the grid motion from the fluid flow. We have used this freedom to include an efficient adaptive gridding procedure which follows regions such as a steep species gradient or a shock front where enhanced resolution is required.

Since the span of space scales is so large in many problems of interest, the requirement for good resolution in specific regions of a finite difference calculation is usually met by a local refinement of the discrete grid of computational cells. Refined localized gridding of boundary layers is a time-honored practice. Recent fluid dynamic calculations have even moved the locally adapted grid with the fluid regions requiring the fine resolution (Fry *et al.*, 1978).

In the NRL detailed shock model, the adaptive gridding moves with the shock. The region surrounding the shock front is gridded with finely spaced cells. The fine spacing transitions smoothly into the more coarsely resolved region. As the shock moves along the length of the tube, the finely spaced region moves with it and may be allowed to reflect off the boundary wall. We have found that the condition on the acceleration

$$\left|\frac{\Delta P}{\Delta r}\right| \frac{1}{\bar{\rho}} = \text{maximum} \tag{17}$$

is adequate for locating the shock front. Here P is the pressure, r is the generalized position coordinate, and $\bar{\rho}$ is the average mass density. In principle there can be any number of finely spaced regions, but more general procedures are clearly preferable in that case (Oran and Boris, 1981).

In the past the multiscale problem in flame propagation has been handled by artificially reducing the chemical reaction rates and increasing the diffusive transport proportionately. As a result the interaction region is nonphysically spread out to the point where numerical solution is economical. This is done in the same spirit as a large artificial viscosity is used in non-FCT algorithms to smear out shocks. Recent efforts at numerical modeling of unsteady flame propagation in an idealized problem have depended on an ad hoc criterion related to the local temperature gradients to determine where adaptive gridding refinement is required. Unfortunately there are many combustion problems where sharp species gradients precede the appearance of sharp temperature gradients.

Such problems are currently on the frontier of reactive flow modeling. There are no excellent techniques for adaptive gridding. Figure 6 shows the x,t region spanned in our Gedanken flame problem by an adaptively gridded calculation using temporal asymtotics. The saving over Fig. 5 is a factor of 500 when 100 cells of 1-cm length are used and a finely resolved region around the flame uses an additional 100 cells of 10^{-3}-cm length. The timestep is governed by the smallest cells, but now only 200 cells are needed rather than 10^5 because of the adaptive gridding.

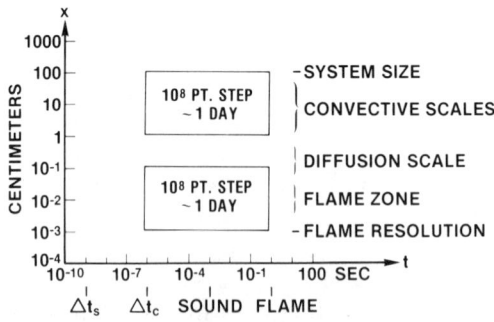

FIG. 6. Region of the x,t diagram spanned by a solution of the *Gedanken* flame problem once adaptive gridding has been included to give fine resolution only where it is needed.

The methodology of adaptive gridding in the Eulerian and the Lagrangian representations is intrinsically different. In Eulerian calculations, a general continuous sliding rezone method can be used. The Lagrangian calculations require discontinuous injection and removal of cell interfaces. While the Lagrangian framework provides some natural grid concentration in regions where the fluid is compressing, there are many exothermic situations where fine resolution is required but the fluid is expanding. Furthermore, in the case of a flame front, the region requiring the fine resolution actually propagates through the fluid rather than with it. In their present state, adaptive gridding methods are specialized in the sense that they are based upon a criterion using only a single dependent variable such as a pressure or temperature gradient. More generally, an adaptive grid technique should take into account all of the important gradients.

Figure 7 summarizes the computational expense of performing our flame propagation problem using a possible, but as yet unexploited technique, adaptive intermittent imbedding. Here a finely gridded region is imbedded into the calculation at intermittent timesteps, but often enough to update the properties of the finely spaced region. These are then used as interior boundary conditions for the coarsely spaced region. To investigate the possible savings, again assume that 100 cells are needed for the flame zone and 100 short timesteps are sufficient to resolve changes in the flame zone brought about by the relatively slowly changing outer boundary conditions. During this imbedded calculation, the flame front moves only 10 of the fine zones, but this should be enough to determine the flame speed and jump conditions to be used in the coarsely spaced regions. The imbedded calculation can be performed once in each large cell for a total of 100 sec + 100 × 10 sec. The total simulation cost is 1100 sec, including 100 sec for the large-scale macroscopic calculation.

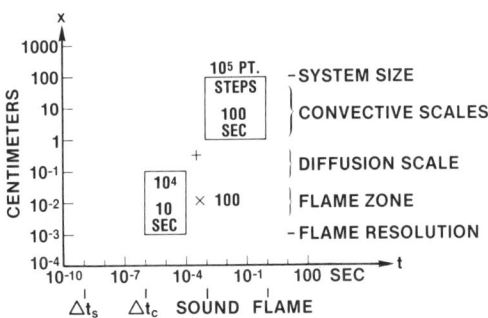

FIG. 7. Region of the x,t diagram spanned by a solution of the *Gedanken* flame problem using injected adaptive gridding to reduce the calculation to two asymptotically decoupled regions, each requiring a very short time to compute.

V. TECHNIQUES FOR DEALING WITH PHYSICAL AND GEOMETRIC COMPLEXITY

Disparate space and time scales are only part of the difficulty in constructing accurate detailed models. Usually, a large number of chemical species are involved in the reactions being studied. Generally the computer storage requirements and cost of a calculation limit the number of species and reaction rates. Such limits would not be too severe if the extent of a calculation merely scaled as the number of species M. Unfortunately, the operation count scales as M^3 whenever the physics or numerics demand that a matrix of size $M \times M$ be inverted. This section presents some methods for reducing the computational cost of modeling chemistry and phase boundaries or other geometric details.

A. Diffusion Fluxes in Multispecies Mixtures

When many chemical species are present, the evaluation of the molecular diffusion fluxes from Eqs. (5) and (6) requires computational expense but seldom involves any numerical difficulty. Because the strightforward inversion of the matrix Eq. (5) for the $\{\mathbf{V}_j\}$ requires $O(M^3)$ arithmetic operations, a more efficient method must be found. The most accurate and efficient algorithm we have found is an $O(M^2)$ iteration based on a special initial guess.

When both Eq. (5) and Eq. (6) are summed over all species, they give zero,

$$\sum_{j=1}^{M} S_j = 0, \tag{18}$$

where we have dropped the vector notation indicating spatial direction. Since $\{S_j\}$ sums to zero, the matrix W is singular as defined in Eq. (5) and the M different diffusion velocities cannot all be independent. The extra equation needed is the constraint

$$\sum_{j=1}^{M} \rho_j V_j = 0. \tag{19}$$

Once a particular solution $\{V_j^P\}$ is known for Eq. (5) [given $\{S_j\}$ from Eq. (6)], any constant velocity may be added, giving another equally correct solution $\{V_j^P + \delta V\}$ for $j = 1, \ldots, M$. We are free to use this unknown constant velocity to enforce Eq. (19).

Direct matrix methods applied to the system of Eqs. (5) and (19) are both practical and computationally economical for four or five species. For more complex systems, we recommend an iterative solution based on an expansion which is exact when there are two species and becomes asymptotically exact in a number of other important limits. Let the diffusion coefficient of species j through the background provided by the sum of all the other species be defined as $D_{j\Sigma}$, then higher-order terms arise as corrections to the lowest-order solution to the diffusion velocity:

$$V_j^0 \equiv -\frac{(\rho - \rho_j)}{\rho} \frac{N^2 D_{j\Sigma}}{(N - n_j)n_j} S_j. \tag{20}$$

If the correct diffusion velocity is $V_j \equiv V_j^0 + \delta V_j$, the equations for $\{\delta V_j\}$ are found by substituting $V_j^0 + \delta V_j$ into Eq. (5) using Eq. (20). The equation which results is

$$\delta S_j \equiv \sum_{k=1}^{M} A_{jk} S_k = \sum_{k=1}^{M} W_{jk}(\delta V_k - \delta V_j). \tag{21}$$

For each j, the value of $\{D_{j\Sigma}\}$ can be calculated directly from

$$\frac{D_{j\Sigma}}{N - n_j} \sum_{k \neq j} \frac{n_k}{D_{jk}} \equiv 1. \tag{22}$$

The matrix elements of A are then given by

$$A_{jk} \equiv \frac{\rho_j}{\rho} \delta_{jk} + \frac{n_j}{D_{jk}} \frac{(\rho - \rho_k)}{\rho} \frac{D_{k\Sigma}}{(N - n_k)} (1 - \delta_{jk}). \tag{23}$$

Equation (21) defines a linear system of equations which can be solved for the $\{\delta V_j\}$. The right-hand side of Eq. (21) vanishes when summed on j. Furthermore, it is easy to see that the choice of A_{jk} here is not unique, so it is treated in a manner analogous to Eq. (5). Since Eq. (18) constrains its solutions, each row of the matrix A can have an arbitrary constant added according to

$$\tilde{A}_{jk} \equiv A_{jk} - C_j \tag{24}$$

without changing Eq. (21). Such constants leave $\{\delta V_j\}$ unchanged. The general form of the complete solution is

$$V_j = \frac{-(\rho - \rho_j)}{\rho} \frac{N^2 D_{j\Sigma}}{(N - n_j)n_j} [\delta_{jk} + A_{jk} + A_{jl}A_{lk} + \cdots] S_k, \tag{25}$$

where the matrix in square brackets is just the formal expansion of $[1 - A]^{-1}$.

Numerical evaluation of Eq. (25) can take advantage of the fact that none of the indicated matrix multiplies actually has to be performed. Since $\{S_j\}$ is known, multiply from the right first. Each additional power of A is obtained by multiplying a vector, rather than a matrix, by A. This expansion gives an $O(M^2)$ algorithm. In practice we have found it convenient to take $C_j = 0$ in Eq. (24) and recommend truncating the expansion in Eq. (25) after the A^2 term. At least the first correction A_{jk} must be included to get the correct sign for all the diffusion fluxes. Experience has shown that the quadratic term adds significant extra accuracy and further iteration is unnecessary. The errors remaining are at most a few percent.

The fast convergence we have observed seems to result from the good initial approximation provided by the solution of Eqs. (20) and (27). The factor $(\rho - \rho_j)/\rho$ in Eq. (25) is crucial. When there are only two species, i.e., $N = n_1 + n_2$, this factor becomes ρ_2/ρ as required to give the two-species result. Note that terminating the expansion of Eq. (25) at the δ_{jk} term does not give Fickian diffusion. The effective diffusion coefficient differs from Fickian by the factor $(\rho - \rho_j)/\rho$. Furthermore, Fickian diffusion need not even give the correct sign for the diffusive flux.

B. *Equations-of-State Fits versus Table Lookup*

The treatment of equations of state and thermochemistry is another example of the physical complications which increase the expense of a simulation. The temperature, which is required for evaluation of chemical reaction rates and diffusion coefficients, is determined implicitly by an equation of the form

$$\varepsilon = h(T, \{n_j\}) - NkT, \qquad (26)$$

and is found by iteration. Here ε, the internal energy density, is in ergs per cubic centimeter, h is the enthalpy, $\{n_j\}$ and N are the species and total number density, respectively, and k is Boltzmann's constant. Thus during a chemical integration step, T is found only after the equations are solved for $\{n_j\}$. We assume that the internal energy remains constant for each chemical timestep.

The enthalpies for the species we monitor have been taken from the JANAF tables (Stull and Prophet, 1971) and from the sixth-order enthalpy coefficients of Gordon and McBride (Gordon and McBride, 1976). For each species the enthalpies are transformed into a seventh-order polynomial fit which goes from 0°K to the maximum temperature given in the JANAF tables. The enthalpy of a species j can then be written

$$h^i = \sum_{n=0}^{6} a_n^i \left[\frac{T}{1000}\right]^n, \qquad (27)$$

where h^i is in units of ergs per cubic centimeter and T is in degrees Kelvin. The motivation for using one fit over a wide temperature range instead of table lookups or different sets of coefficients for the temperature range below and above 1000°K lies in the speed and ease with which the evaluation of expressions such as Eq. (27) can be vectorized for efficient computation.

Recent improvements in vectorized table lookup routines for the ASC (Young, 1981) make Eq. (27) about the most complex formula which one would want to compute directly, however. A single, one-dimensional table lookup can be performed for about the same cost as a vectorized square root. Several table references with the same independent variable, e.g. temperature, can be performed even faster since vectorization across all the different tables being referenced at once can reduce the relative cost of the expensive part of the calculation, the random access memory references. Indeed, it now seems that substantial gains can also be achieved by looking up all the chemical reaction rates as a function of temperature as well. Factors of three to four will be realized since one lookup can replace several transcendental function evaluations needed to describe even a simple Arrhenius rate. Ganging all the rates together as a single vector lookup on temperature promises still further savings.

Such gains as are possible using table lookups do not come completely free. First, much additional memory is required to store all the tables. Second, accuracy greater than $\sim 0.1\%$ is difficult to obtain without very extensive tables. Speed is not sacrificed, but core requirements, particularly for multidimensional tables, can become excessive. Finally, the range of the independent variable generally must also be limited, to keep table size within acceptable limits.

C. *General Connectivity Lagrangian Meshes*

Geometric complexity is often another source of difficulty in detailed reactive flow modeling. Obviously, complicated boundaries and heterogeneous media cause problems. For example, suppose we wish to consider reactive flow in a region bounded by triangular walls with an interior circular boundary. Then even a Lagrangian grid of two-dimensional stretched rectangles is difficult to adapt to it. Furthermore, complicated flow fields invariably stretch such a Lagrangian grid unacceptably.

The distortion problem for Lagrangian grids can be handled adequately by using a finite-difference mesh of triangles rather than rectangles (tetrahedrons in three dimensions). Figure 8 shows the representational advantage of using triangles for a breaking gravity wave. This technique, which has been used in many finite element representations, was adapted to finite differences of incompressible flow (Crowley, 1971; Boris *et al.*, 1975; Fritts and Boris,

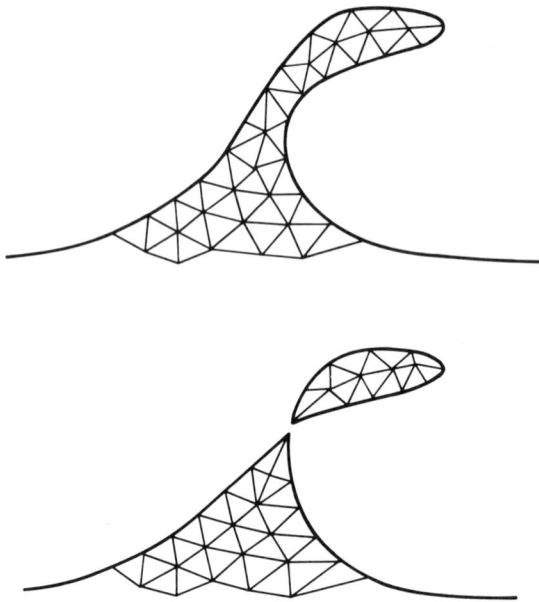

Fig. 8. Schematic diagram of reconnecting triangular grid being used to solve the separating flow problem of a breaking wave.

1979; Emery and Winsor, 1981). The noticeable improvements in performance arise for two reasons. First, the general connectivity permitted by triangles allows smooth representation of much more complicated shapes than can be treated smoothly by an equal number of rectangular cells. Second, since no specific symmetries are built into the grid, it can be varied during a calculation to prevent severe grid distortions.

This approach may not be the only method of dealing with complex flow thus avoiding severe grid distortions. However, it is the most physical and the least dissipative technique we have found so far. Fritts has recently employed Lagrangian triangular grid techniques in studying nonlinear aspects of free-surface waves including strongly sheared flows and flows over obstacles. Ten to fifteen full cycles of the waves can be integrated reversibly without significant deterioration of the solution and the reconnection procedure in the presence of strong shear is reversible (i.e., nondissipative). There is another advantage to using triangular cells. The nonlinear mesh separation instabilities, which plague low-dissipation rectangular cell techniques, seem to be absent or damped with triangular cells.

The triangular grid approach is burdened by the complexity of the programming required, usually involving linked lists. The arithmetic operations seem to require random access and sequential processing. Currently, vectorization procedures for these triangular grid techniques are being developed. As in the case of tridiagonal solvers and recursion relations, vectorization is possible for most of the computation provided the substantial fraction of scalar work is carefully organized. In table lookups this scalar segment is the single random access double fetch required for each of the N different values of the independent variable. In recursion relations and tridiagonal solvers, the folding logic and the necessity to close out the cyclic reduction procedure with a substantial scalar computation constitutes an appreciable scalar overhead.

In the case of reconnecting triangular grids, the scalar penalty is the necessity to sort triangle quantities onto vertex lists and vertex quantities onto triangle lists. Rule-of-thumb indications are that the vectorization gains over pure scalar code for typical triangular grid computations are similar to those for tridiagonal solvers on the ASC. Table IV supplements Table II by giving timing data on the tridiagonal solvers reported by Boris (1976c), and thus indicating the speed gains available in complex operations. In both single precision and double precision (64-bit), the asymptotic rate of the scalar algorithm is four times slower than the vectorized (but relatively inefficient) folding algorithm. For comparison recall that the most advantageous ratio depicted in Table II for the vector addition was 35:1. Typically the ratio is more like 16:1 when several scalar operations are executed in the same loop. Thus the following rule of thumb suggests itself: Algorithms which can be vectorized with difficulty, such as a single tridiagonal matrix system, a table lookup, or linked list searching as in the triangular grid algorithms, run at a speed which is the geometric mean between the asymptotic vector speed and the limiting scalar speed.

TABLE IV

Timing Formulas and Size of Tridiagonal Solvers

Solver	Single precision	Double precision
Scalar Short System	TRIDSS $\tau_{SS} \approx 100 + 19 N_c$ μsec ~ 310 words	TRIDDS $\tau_{DS} \approx 100 \times 20.5 N_c$ μsec ~ 310 words
Vector Long System	TRIDSV $\tau_{SV} < 460 + 4.5 N_c$ μsec ~ 1000 words	TRIDDV $\tau_{DV} < 540 + 5.3 N_c$ μsec ~ 1100 words
Multiple Tridiagonal Systems	TRIDSM $\tau_{SM} \approx 125 + 1.6 N_c$ μsec ~ 2000 words	TRIDDM $\tau_{DM} \approx 120 + 3.8 N_c$ μsec ~ 2000 words

Even though the Lagrangian triangular grid algorithms run about a factor of four or so slower than the full vector speed of which the ASC is capable, the basic algorithm is so much better than Eulerian methods for certain types of problems that the speed disadvantage is more than overcome by accuracy advantages. An early simulation using these techniques needed about 280 vertices to resolve and calculate a complex flow with material interfaces and free surfaces which had required 5000 vertices in an Eulerian calculation. The apparent factor-of-four disadvantage from speed became a factor-of-four advantage when accuracy was factored in as well.

Lagrangian calculations have virtually no numerical diffusion arising as a monotonicity or stability constraint, and thus they are ideally suited to reactive flow simulations in which real molecular diffusion and thermal conduction dominate the physics. They very effectively avoid the spurious diffusion arising from an imperfect convection algorithm. These Lagrangian techniques are sufficiently promising, in fact, that special-purpose computer architectures and hardware should be considered to implement them. Each fluid element could be represented as an independently programmed VLSI chip running out of synchronization with all the other fluid element chips. Local data sharing with all the triangle (or tetrahedra in three dimensions) nearest neighbors would ensure physical conservation conditions. If a networking type of data bus is used, the fluid element chips could communicate with their logically nearest neighbor chips as if they were independent computer terminals or computers on a network. The need for a supervising instruction processor would be replaced by redundant but independent programming for each fluid element. The advantage would be maximal parallelism in the form of multiple independent tasking. A computer with 10^5 fluid element chips might require as much as 10^{-2} sec per timestep per chip, but would run 100 times faster than a conventional or pipeline computer requiring 10 μsec per point step.

VI. PROGRAMMING GUIDELINES AND SUMMARY OF PARALLELISM PRINCIPLES

In Section II we considered several aspects of optimization and vectorization. Here we recapitulate the five optimization goals discussed earlier, specifically from the point of view of programming a complex computational model for a vector computer.

Good programming is an art, working in a difficult medium and requiring a special talent. Therefore it is fitting to provide general guidelines for good programming whereas rigidly enforced standards are appropriate for program documentation. The purpose of these guidelines is to make the resulting programs

1. clear,
2. simple,
3. fast,
4. accurate, and
5. robust.

A *clear* program can be understood and modified by any good programmer accurately and quickly. What is often even more important in a research environment, a clear program can be modified by the original programmer when he returns to the problem after a year or two.

A *simple* program is much less prone to error and is more easily modified than a complicated program. It is also much easier to satisfy the other four goals when all the interacting program modules are simple. Simplicity and clarity also lead to flexibility in the great majority of cases; but flexibility per se cannot be said to be a major programming goal. A program can still be good even if it does only one task well.

Making the program *fast* is clearly of economic benefit, and greater speed usually permits a trade-off in which some of the speed is traded for greater *accuracy*. The compromise between speed and accuracy has often posed the most troublesome and provocative challenge for the scientific programmer. Almost all state-of-the-art advances in numerical analysis have resulted from trying to reach these twin goals since algorithm changes generally yield much greater improvements in performance than minor program optimization.

The final goal is that the working program be *robust*. A robust code works adequately over a broad spectrum of input data and has few if any blind spots, i.e., special cases which it cannot handle. A robust code can be trusted to give correct answers reliably.

Much has been written about structured programming and programming style. Two particularly simple and clear discussions are found in *Programming Proverbs for Fortran Programmers* (Ledgard, 1975) and *The Elements of Programming Style* (Kernighan and Plauger, 1974). We take the *summary of rules* developed in the latter work as general coding guidelines. These "rules" are not intended to force rigid form on all programmers but rather to ensure a useful uniformity. As with their documentation, all subroutines and codes should be edited and tested before being entered into a user library. Therefore some help should be obtained in polishing the final version. Good programs usually arise iteratively.

A. General Programming Guidelines

Write clearly—do not try to be too clever. Say what you mean, simply and directly. Fancy tricks have a way of backfiring, and it seldom pays to sacrifice clarity for "efficiency."

Claritymeansinpartreadability.Readabilitycancomeinmanyformsandshapessothebestsingle guidelinewecancomeupwithistoforcethecodetolookmostlikewhatmostpeopleseemoreofthanany othertypeofvisualcommunication,Englishprose.Inthiswaytheeasyandunconciousassimilation ofthecomputerintelligibleinformationwillbemostdirectlyaidedandabetted.Thereareundoubtedly avastnumberofwaystoforcetheprogramstatementsintoamuchmoreclearandreadableform. Unfortunatelymanyselfproclaimedprogrammersofrenowntakeaboutasmuchadvantageofthe possibilitiesasthisrunonbutratherfortranlikeparagraphdoes.Thefirstoftheratherobvioustothe meanestintelligenceyetoftenneglectedbyreputedlygoodprogrammersmaximsforbetter legibilityisreallyonlyanextensionofthegoldenrule.

Put blanks in your FORTRAN statements. Put blanks after commas, between variables, after subroutine names, around + and − signs, and in the middle of "goto" (GOTO). In fact, use your terminal or keypunch as if it were a technical typewriter. Extra "blank comment" lines rarely hurt. Comments which are correctly spelled and make syntactic sense are part of legal FORTRAN. Do not break statements in strange places. Try to use a convenient operation at an outer level to begin each continuation card. Use short constructions if they suffice, and do not sacrifice readability to save cards, lines, or spaces. Just as in English, do not try to mix several ideas into a single line or a single statement of the program. You will seldom go wrong if your FORTRAN comes across like clear English.

Proper commenting is the most flexible tool in the programmer's bag of tricks to achieve clarity. You should make sure your comments and code agree but should not just echo the code with comments. Make every comment count and do not overcomment your program. Indent your code from your comments for readability, as in an outline, even to the extent of simulating the block structures allowed in ALGOL and PL/1. Alternatively, indent your comments to make them headings for short sections of code. Comments beginning in a column different from the code are easier to find.

The layout of the code can be almost as useful as comments in helping to explain what is going on. You should format and structure a program to help the reader understand it. He needs the most help in finding

    a.   the structure of program execution—DO loops, IFs, GO TOs,
    b.   the data structure—DIMENSION statements, DATA statements that initialize the variables, and assignment statements that change their values,
    c.   the other parts of compound structures—the ends of DO loops, the labels that are targets of GO TOs, and the FORMAT statements that go with READ and WRITE statements.

Different people have quite different concepts of good layout. Whatever practice you adopt, it should make the above structures easy to find and easy to scan for specific elements.

Document the data layouts and use variable names and grammatical constructions that mean something in English as well as in the programming language. Choose variable names that are difficult to confuse with each other and choose a data representation that makes the program simple. Modularity cuts down the scope of what must be comprehended simultaneously, so use subroutines and library functions.

There are some places where it pays to be wordy in the code. In expressions it never hurts to use parentheses to avoid ambiguity. The FLOAT and FIX functions seldom cost anything to use and they eliminate many mixed-mode ambiguities. If a logical expression in the code is hard to understand, try transforming it—even if the more legible form is somewhat longer.

There are also some places where it pays to be terse. You do not have to spell everything out to the computer. Let the machine do some of the dirty work; it will often do a better job than you could. For example, avoid temporary variables and unnecessary branches. Avoid the *FORTRAN* arithmetic IF which has three destinations rather than one or two. Use only logical IF statements, and put the controlled statement on a continuation card, where you usually see it. A logical expression will often substitute for a conditional branch, will usually compile at least as efficiently, and can often shorten the program by eliminating the need for an explanatory comment. Keep in mind, however, considerations which derive from the possibility of vectorizing certain types of IF statements.

Clarity is also greatly aided by adhering to a rigorous top-to-bottom FORTRAN logical path, and bugs will be easier to find and fix. "Top-to-bottom" means that branches and test of any sort should not refer, unless absolutely necessary, to previous statements. When a back reference is necessary, a comment explaining what and why is usually in order. Some authors suggest eliminating the GO TO statement altogether. While this is somewhat extreme, you certainly should avoid GO TO if you can keep the program readable and should use GO TO only to implement a fundamental structure. No amount of patching can fix badly structured code, so do not hesitate to rewrite sections of your program completely. The time lost in doing this is usually saved many times over by avoiding catastrophy further along.

One of the most tedious and hence poorly programmed aspects of many codes is I/O. Test input for plausibility and validity so that you can recover or crash gracefully if bad input is identified. In particular, if the algorithm has limitations, the input should be checked to make sure these limits are not violated. Make input easy to prepare and printed output self-explanatory. In this connection, uniform input formats make the data easy to proofread.

Many debugging techniques go hand in hand with good programming techniques. All variables should be initialized before being used. Constants should be initialized via DATA or BLOCK DATA statements while variables should be initialized by input, executable code, or both. When free-form input, i.e., NAMELIST, is used to overwrite default values, echo both the current value and the default in the output. Test programs at their boundary values and keep a sharp lookout for "off-by-one" errors. Take care to branch the correct way on equality and be careful when a loop exists to the same statement from both the bottom and an internal GO TO. Do not compare floating point numbers solely for equality since 10.0 times 0.1 is usually not exactly 1.0 in a computer. Check some answers by hand; in particular, make sure your code "does nothing" gracefully.

Only after ensuring that the code is properly understood by both the computer and the programmer should attempts to speed up the code be undertaken. In particular, you should convert and test the routines before undertaking any optimization. Once the subroutine and test programs are running satisfactorily, techniques for scalar and vector optimization may be investigated. Let the compiler do the simple optimizations. The ASC NX compiler, for example, will probably do a better job than you would. For more sweeping types of optimization do not be afraid to reorganize. The laziness that keeps the programmer from redoing blocks of code is "penny wise and pound foolish." Finally, do not diddle with the code to speed it up—find a better algorithm.

The above suggestions indicate some of the ways programs can be made easier to read, debug, and modify. There are a number of easy ways, within these general programming guidelines, to ensure that the resultant code vectorizes efficiently. This chapter ends with a summary of these parallelism principles.

B. *Summary of Parallelism Principles*

There are a number of parallelism principles that apply to most fast computers. These have applicability to general vectorization efforts both in and out of reactive flow problems. The methods are straightforward, but must be applied many times, and often with considerable ingenuity, to completely vectorize a code. Programs which are coded following these principles will generally run faster than programs which are not, provided that an optimizing compiler is used. The factor may be 2 or more on machines such as the IBM 360/91 or the CDC-7600, or it may be as much as a factor of 10 or 100 for a vectorizable code on a vector computer. The corresponding cost is usually an increase in the computer memory required for data (and arrays of intermediate results in vector computations).

Vectorized Computation of Reactive Flow

First, certain principles apply to any fast computer. They are

1. Plan programs and subroutines to operate on arrays of data rather than on individual elements.
2. Store the arrays in memory so that the elements are usually read from or stored into consecutive memory locations.
3. The DO loop index (the innermost one, if there is more than one) should be the *first* subscript of an array wherever possible.
4. If a scalar appears on the left-hand side of an assignment in a DO loop, replace it by an array.
5. Remove IF statements from DO loops wherever possible.
6. Use optimized or vectorized subroutines to eliminate IF statements or replace code which the compiler does not optimize.

Note that principle 2 usually means maximizing the application of principle 3. Principle 6 applies to IF statements which are performing a function such as MIN, MAX, or a cyclic reduction or merge.

Second, there are principles which relate to computers which have more than one arithmetic unit. The application of these principles is fairly machine dependent. For pipeline machines with more than one pipeline (TI ASC, CYBER 203) operations on independent data streams should be made adjacent:

7. Write short FORTRAN lines.
8. Code independent pairs of array operations in neighboring lines.

Principle 7 also applies to loop-mode machines like the IBM 360/91, where DO loops must be short enough for the required instructions to fit into the "loop stack." For machines which have independent functional units (adder, multiplier, etc.) such as the CDC 7600 and Cray-1, the principles are

9. Write long FORTRAN lines.
10. Code expressions so that independent operations alternate on dependent data.

Beyond these ten principles, it is necessary to consider the detailed architecture of the target machine. However, codes which follow these principles should require nearly the minimum amount of restructuring in order to take advantage of the detailed hardware features.

A number of specialized numerical techniques have been presented which are aimed specifically at solving multiple time scale, multiple space scale and real system complexities. Our goal has been to develop flexible, economic, and reliable detailed reactive flow models which can be used to test theory and in conjunction with experiments. You will find numerous examples of the detailed application of these principles in the references.

Acknowledgments

The authors would like to acknowledge the profound and valuable contributions of our colleagues in the Laboratory for Computational Physics, the Plasma Physics Division, and the Chemistry Division of NRL and in particular the innovative techniques and ideas of Elaine Oran, David Book, Martin Fritts, and Theodore Young which we have been privileged to report here. We would also like to acknowledge the extraordinary efforts of Francine Rosenberg, Nancy Ciatti, and Louise McDonald in the preparation of this manuscript and the support of the Naval Research Laboratory, the Office of Naval Research, the Defense Nuclear Agency, the Department of Energy, and the National Aeronautics and Space Administration during the years of development and application efforts leading up to this paper.

References

Book, D., Boris, J., Kuhl, A., Oran, E., Picone, M., and Zalesak, S. (1980). Simulation of complex shock reflections from wedges in inert and reactive gaseous mixtures. *Proc. Internat. Conf. Numer. Methods Fluid Dynamics, Stanford University, Palo Alto, California, June*, p. 84. Springer-Verlag, Berlin and New York, 1981.

Boris, J. P. (1975). Vectorization of recursion relations, Memo. Rep. 3144, Naval Research Laboratory, Washington, D.C.

Boris, J. P. (1976a). *Comput. Phys. Comm.* **12**, 67; see also Memo. Rep. 3327, July, Naval Research Laboratory, Washington, D.C.

Boris, J. P. (1976b). Flux-corrected transport modules for solving generalized continuity equations. Memo. Rep. 3237, Naval Research Laboratory, Washington, D.C.

Boris, J. P. (1976c). Vectorized tridiagonal solvers. Memo. Rep. 3408, Naval Research Laboratory, Washington, D.C.

Boris, J. P. (1977). Requirements for a slow-flow algorithms to simulate chemically reactive flows. *J. Phys. Chem.* **81**, 2525.

Boris, J. P. (1979). ADINC: An implicit Lagrangian hydrodynamic code. Memo. Rep. 4022, Naval Research Laboratory, Washington, D.C.

Boris, J. P., and Book, D. L. (1976). Solution of continuity equations, by the method of flux-corrected transport. *In* "Computer Applications to Controlled Fusion Research" (J. Killeen, ed.), Methods in Computational Physics, Vol. 16, p. 85. Academic Press, New York.

Boris, J. P., Fritts, J. M., and Hain, K. (1975). Free surface hydrodynamics using a Lagrangian triangular mesh. *Proc. Internat. Conf. Numer. Ship Hydrodynamics, 1st, NBS, Gaithersburg, Maryland, October*, p. 683. Document 215-572, U.S. Govt. Printing Office, Washington, D.C.

Buneman, O. (1969). A compact non-iterative Poisson solver. SUIPR Rep. No. 294, Institute for Plasma Research, Stanford University, Palo Alto, California.

Crowley, W. P. (1971). *Proc. Internat. Conf. Numer. Methods Fluid Dynamics, 2nd*, p. 34, Springer-Verlag, Berlin and New York, 1972.

Emery, M. H., and Winsor, N. K. (1981). A fully two-dimensional equilibrium and transport model of the poloidal director. Memo. Rep. 4498, Naval Research Laboratory, Washington, D.C.

Fritts, M. J., and Boris, J. P. (1979). The Lagrangian solution of transient problems in hydrodynamics using a triangular mesh. *J. Comput. Phys.* **31**, 2.

Fry, M. A., Ganong, G. P., and Aubrey, J. W. (1978). Effects of nonideal surfaces on air blast from one megaton yields. Rep. AFWL-TR-77-179, February, Air Force Weapons Laboratory, Albuquerque, New Mexico.

Gear, C. W. (1971). "Numerical Initial Value Problems in Ordinary Differential Equations." Prentice-Hall, Englewood Cliffs, New Jersey.

Gordon, S., and McBride, B. J. (1976). Computer program for calculation of complex chemical equilibrium compositions, rocket performance, incident and reflected shocks, and Chapman-Jouguet detonations. NASA SP-273, National Aeronautics and Space Administration, Washington, D.C.

Hockney, R. W. (1970). The potential calculation and some applications. In "Plasma Physics" (B. Alder, S. Fernbach, and M. Rotenberg, eds.), Methods in Computational Physics, Vol. 9, pp. 136–210. Academic Press, New York.

Jones, W. W., and Boris, J. P. (1977). Flame and reactive jet studies using a self-consistent two-dimensional hydrocode. *J. Chem. Phys.* **81**, 2532.

Kernighan, B. W., and Plauger, P. J. (1974). "The Elements of Programming Style." McGraw-Hill, New York.

Ledgard, H. F. (1975). "Programming Proverbs for Fortran Programmers." Hayden, Rochelle Park, New Jersey.

McDonald, B. E. (1980). The Chebychev method for solving non-self-adjoint elliptic equations on a vector computer. *J. Comput. Phys.* **35**, 147.

Oran, E. S., and Boris, J. P. (1981). Detailed modelling of combustion systems. *Progr. Energy Combust. Sci.* **7**, 1–72.

Oran, E. S., Young, T. R., and Boris, J. P. (1978). Application of time-dependent numerical methods to the description of reactive shocks. *Proc. Internat. Symp. Combust., 17th, Leeds, England*, p. 43. Combustion Institute, Pittsburgh, Pennsylvania.

Oran, E. S., Boris, J. P., Young, T., Flanigan, M., Burks, T., and Picone, M. (1980). Detonations in hydrogen-air and methane-air mixtures. *Proc. Internat. Symp. Combust. 18th, Waterloo, Canada*, p. 1641. Combustion Institute, Pittsburgh, Pennsylvania, 1981.

Stull, D. R., and Prophet, H. (1971). "JANAF Thermodynamical Tables," 2nd ed. No. 37, June, National Standards Reference Data Service, U.S. National Bureau of Standards, Washington, D.C.

Williams, F. A. (1965). "Combustion Theory." Addison-Wesley, Reading, Massachusetts.

Young, T. R. (1980). CHEMEQ—A subroutine for solving stiff ordinary differential equations. Memo. Rep. 4091, Naval Research Laboratory, Washington, D.C.

Young, T. R. (1981). Vectorized table lookup routines. Personal communication.

Young, T. R., and Boris, J. P. (1973). *DNA Atmospheric Effects Symp.* Vol. 4, DNA Rep. No. 3131, p. 9 (VI 75-442), p. 571; see also Memo. Rep. 2611, Naval Research Laboratory, Washington, D.C.

Young, T. R., and Boris, J. P. (1977). A numerical technique for solving stiff ordinary differential equations associated with the chemical kinetics of reactive-flow problems. *J. Phys. Chem.* **81**, 2424.

A Fully Implicit, Factored Code for Computing Three-Dimensional Flows on the ILLIAC IV

Harvard Lomax
Thomas H. Pulliam

Ames Research Center
National Aeronautics and Space Administration
Moffett Field, California

I. INTRODUCTION

Although many important fluid-flow phenomena are approximately one- or two-dimensional in their behavior and can therefore be analyzed by equations formulated in these dimensions, flows in the real world are three-dimensional and are often quite unsteady. Many, if not most, two-dimensional fluid flows of interest to aerodynamicists can be resolved to a fairly high degree of accuracy (within engineering standards) and with reasonable run times by the standard scientific computers of the 1970s. This is not at all true for unsteady three-dimensional flows. Even the class VI computers, defined as computers capable of computing over 20 million floating-point operations per second (megaflops), such as the Cray-1 and the CDC 205, will give only marginal resolution to most types of three-dimensional flows. In particular, this is true of flows governed by the unsteady, Reynolds-averaged, Navier–Stokes equations. Nevertheless, it is appropriate to begin the investigation of these flows on the class VI computers, even if the codes produced are only prototypes for still more advanced machines.

In this chapter we identify some of the problems that arise in constructing efficient programs to be used for large three-dimensional flow simulations. We then show how these problems can be dealt with on one particular computer—the ILLIAC IV. It should be pointed out that the ILLIAC qualifies as a class VI computer. For example, in 32-bit arithmetic, it can compute a fast Fourier transform at the rate of 70 megaflops. However, like all top speeds quoted for these machines, this is quite unrealistic as a measure of speed for complete programs. This is especially true for codes operating on large data bases, which we define to be data bases of more than two million words (still small for three-dimensional Euler or Navier–Stokes problems). Nevertheless, it should be noted that the ILLIAC has computed the three-dimensional flow fields reported in Section VI at a rate of over 20 megaflops, based on complete runs with full input/output and wall-clock time.

In our experience with designing efficient codes for three-dimensional simulations that have large data bases, there are three areas of concern. One relates to the architecture of the computer, one to the algorithm chosen for the solution, and one to the geometry and boundary conditions of the flow field. In the first place, in the case of the ILLIAC, a convenient vector of length 64 has to be identified, and the very small high-speed in-core memory has to be fed by disks in such a way that the computing time covers up the latency as much as possible. In the second place, it is well known that algorithms used to solve the Reynolds-averaged, Navier–Stokes equations can have a severe numerical stability limitation. One way to overcome the latter is to use implicit methods. That is the choice made here. However, this leads to the problem of providing enough temporary storage space necessary to solve the simultaneous equations brought about by the implicit algorithm used at every step. Finally, although the geometry for most realistic three-dimensional problems is not simple, we must provide ways to solve for a wide variety of flows without having to recode for every problem. Furthermore, application of the boundary conditions should not have a significant effect on the efficiency of the vectorization.

The following material is a discussion of how these difficulties were treated by one particular code, ARC3, which is outlined in Section V. We should mention some aspects that did *not* cause any particular trouble. For any flow field that can be mapped into a topological box, the geometry and the boundary conditions caused no problem. Finding a suitable vector length and constructing efficient second- and fourth-order implicit algorithms using this vector length caused no problem. Working with full efficiency in all three of the space directions caused no problem. (The data base could be efficiently permuted.) By far the most important problems that did occur were (1) providing an efficient way to carry out the necessary disk-to-core data transfers and (2) providing enough temporary storage space for the efficient use of the tridiagonal solver used in the implicit algorithm. Both of these problems are discussed in detail in Section V.

II. BASIC EQUATIONS

A. Coordinate Systems

The unsteady, Reynolds-averaged, Navier–Stokes equations in inertial Cartesian coordinates (x, y, z, t) are taken as the basic set of equations. The Cartesian space represents the physical domain of consideration. If the Cartesian velocity components are retained as the dependent variables, the three-dimensional unsteady, Navier–Stokes equations can be transformed to an arbitrary curvilinear coordinate space (ξ, η, ζ, τ), shown in Fig. 1, while retaining the strong conservation-law form (Peyret and Viviand, 1975; Lapidus, 1967; Viviand, 1974; Vinokur, 1974). The curvilinear space is referred to as the computational domain. The curvilinear coordinate transformations allow for the use of a wide class of geometries and grid systems in one basic set of equations.

B. Transformed Navier–Stokes Equations

The transformed, unsteady, three-dimensional Navier–Stokes equations in dimensionless form are given by

$$\partial_\tau Q + \partial_\xi(E - E_v) + \partial_\eta(F - F_v) + \partial_\zeta(G - G_v) = 0, \tag{1}$$

where

$$Q = J^{-1}\begin{bmatrix} \rho \\ \rho u \\ \rho v \\ \rho w \\ e \end{bmatrix}, \quad E = J^{-1}\begin{bmatrix} \rho U \\ \rho u U + \xi_x p \\ \rho v U + \xi_y p \\ \rho w U + \xi_z p \\ (e+p)U - \xi_t p \end{bmatrix},$$

$$F = J^{-1}\begin{bmatrix} \rho V \\ \rho u V + \eta_x p \\ \rho v V + \eta_y p \\ \rho w V + \eta_z p \\ (e+p)V - \eta_t p \end{bmatrix}, \quad G = J^{-1}\begin{bmatrix} \rho W \\ \rho u W + \zeta_x p \\ \rho v W + \zeta_y p \\ \rho w W + \zeta_z p \\ (e+p)W - \zeta_t p \end{bmatrix}, \tag{2}$$

and

$$E_v = (J\text{Re})^{-1} \begin{bmatrix} 0 \\ \xi_x\tau_{xx} + \xi_y\tau_{xy} + \xi_z\tau_{xz} \\ \xi_x\tau_{yx} + \xi_y\tau_{yy} + \xi_z\tau_{yz} \\ \xi_x\tau_{zx} + \xi_y\tau_{zy} + \xi_z\tau_{zz} \\ \xi_x\beta_x + \xi_y\beta_y + \xi_z\beta_z \end{bmatrix},$$

$$F_v = (J\text{Re})^{-1} \begin{bmatrix} 0 \\ \eta_x\tau_{xx} + \eta_y\tau_{xy} + \eta_z\tau_{xz} \\ \eta_x\tau_{yx} + \eta_y\tau_{yy} + \eta_z\tau_{yz} \\ \eta_x\tau_{zx} + \eta_y\tau_{zy} + \eta_z\tau_{zz} \\ \eta_x\beta_x + \eta_y\beta_y + \eta_z\beta_z \end{bmatrix},$$

$$G_v = (J\text{Re})^{-1} \begin{bmatrix} 0 \\ \zeta_x\tau_{xx} + \zeta_y\tau_{xy} + \zeta_z\tau_{xz} \\ \zeta_x\tau_{yx} + \zeta_y\tau_{yy} + \zeta_z\tau_{yz} \\ \zeta_x\tau_{zx} + \zeta_y\tau_{zy} + \zeta_z\tau_{zz} \\ \zeta_x\beta_x + \zeta_y\beta_y + \zeta_z\beta_z \end{bmatrix},$$

$$\begin{aligned}
\tau_{xx} &= \lambda(u_x + v_y + w_z) + 2\mu u_x, \\
\tau_{xy} &= \tau_{yx} = \mu(u_y + v_x), \\
\tau_{xz} &= \tau_{xz} = \mu(u_z + w_x), \\
\tau_{yy} &= \lambda(u_x + v_y + w_z) + 2\mu v_y, \\
\tau_{yz} &= \tau_{zy} = \mu(v_z + w_y), \\
\tau_{zz} &= \lambda(u_x + v_y + w_z) + 2\mu w_z, \\
\beta_x &= \gamma\kappa\text{Pr}^{-1}\partial_x e_I + u\tau_{xx} + v\tau_{xy} + w\tau_{xz}, \\
\beta_y &= \gamma\kappa\text{Pr}^{-1}\partial_y e_I + u\tau_{yx} + v\tau_{yy} + w\tau_{yz}, \\
\beta_z &= \gamma\kappa\text{Pr}^{-1}\partial_z e_I + u\tau_{zx} + v\tau_{zy} + w\tau_{zz}, \\
e_I &= e\rho^{-1} - 0.5(u^2 + v^2 + w^2),
\end{aligned} \quad (3)$$

where

ρ is the density, made dimensionless by ρ_∞,
u, v, w are Cartesian velocity components made dimensionless by a_∞,
e is the total energy, made dimensionless by $\rho_\infty a_\infty^2$,
p is the pressure $(=(\gamma - 1)[e - 0.5\rho(u^2 + v^2 + w^2)])$,
γ is the ratio of specific heats,
μ is the dynamic viscosity,
λ is from Stokes's hypothesis, $= -\tfrac{2}{3}\mu$,
κ is the coefficient of thermal conductivity,
Re is the Reynolds number, and
Pr is the Prandtl number.

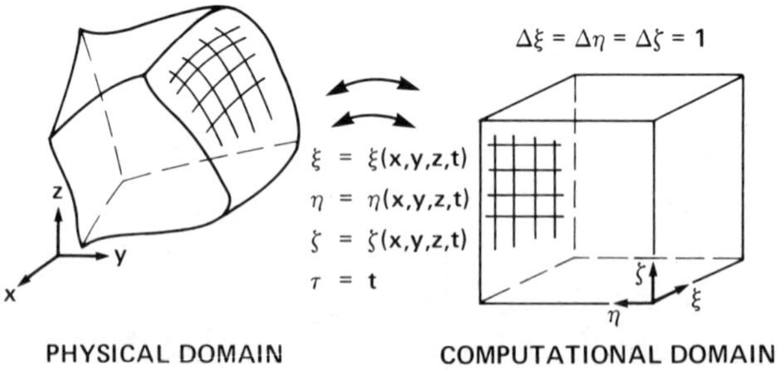

(4)

FIG. 1. Transformation from physical space to computational space.

The (ξ, η, ζ, τ) transformations applied to the Cartesian derivatives in the Navier–Stokes equations produce the metric terms

$$
\begin{aligned}
\xi_x &= J(y_\eta z_\zeta - y_\zeta z_\eta), & \eta_x &= J(z_\xi y_\zeta - y_\xi z_\zeta), \\
\xi_y &= J(z_\eta x_\zeta - x_\eta z_\zeta), & \eta_y &= J(x_\xi z_\zeta - x_\zeta z_\xi), \\
\xi_z &= J(x_\eta y_\zeta - y_\eta x_\zeta), & \eta_z &= J(y_\xi x_\zeta - x_\xi y_\zeta), \\
\zeta_x &= J(y_\xi z_\eta - z_\xi y_\eta), & \xi_t &= -x_\tau \xi_x - y_\tau \xi_y - z_\tau \xi_z, \\
\zeta_y &= J(x_\eta z_\xi - x_\xi z_\eta), & \eta_t &= -x_\tau \eta_x - y_\tau \eta_y - z_\tau \eta_z, \\
\zeta_z &= J(x_\xi y_\eta - y_\xi x_\eta), & \zeta_t &= -x_\tau \zeta_x - y_\tau \zeta_y - z_\tau \zeta_z,
\end{aligned}
\tag{5}
$$

and

$$J^{-1} = x_\xi y_\eta z_\zeta + x_\zeta y_\xi z_\eta + x_\eta y_\zeta z_\xi - x_\xi y_\zeta z_\eta - x_\eta y_\xi z_\zeta - x_\zeta y_\eta z_\xi$$

and the contravariant velocity components

$$
\begin{aligned}
U &= \xi_t + \xi_x u + \xi_y v + \xi_z w, \\
V &= \eta_t + \eta_x u + \eta_y v + \eta_z w, \\
W &= \zeta_t + \zeta_x u + \zeta_y v + \zeta_z w.
\end{aligned}
\tag{6}
$$

In terms of a physical representation, the contravariant velocity components are the decomposition of a velocity vector into components along each of the curvilinear coordinates. The use of the contravariant velocities produces the compact form of the equations and simplifies the application of boundary conditions, to be discussed.

C. Grid System

A grid system is defined in the physical domain where each node point is assigned an x, y, z location, indexed by J, K, and L, respectively. The grid is almost always nonuniform in the physical domain, where body-conforming mesh lines and grid clustering are used for better resolution. The type and amount of clustering is very problem dependent. The formation of the grid does not have to be done on the main computer. It can be constructed preliminary to the calculations on a separate computer and transferred by tape to the main frame for the final simulation. This was the strategy carried out for the ILLIAC code. However, the metric evaluations given by Eq. (5) were computed numerically on the ILLIAC.

The (ξ, η, ζ, τ) transformation is constrained so that the grid spacing in the computational domain is uniform and of unit length. In our applications this computational space is a topological box, and, since the grid is uniform, all difference schemes can be based on equispaced finite-difference formulas. All

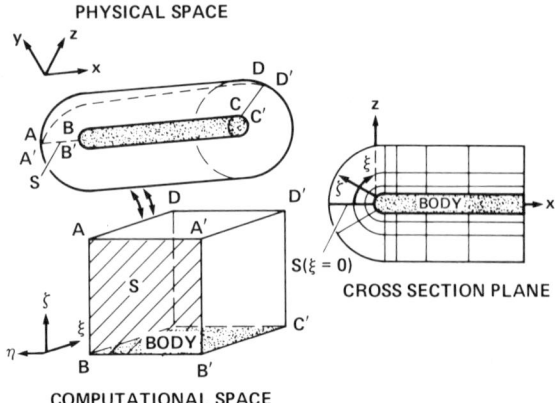

FIG. 2. Warped spherical system for hemisphere–cylinder body.

physical boundaries are mapped onto coordinate surfaces in the computational domain. With such a relation between the physical and computational domains, problems associated with nonuniform differencing and complicated boundary conditions are avoided.

A good example of a grid system that meets the above requirements is the warped spherical system for pointed and blunt semi-infinite bodies (Fig. 2). A two-dimensional array of body-conforming lines is generated and then rotated about the axis of symmetry. The singularity produced at the nose of the body is unavoidable when mapping a closed three-dimensional body into a coordinate system (for a discussion, see Section II.F).

D. *Viscous Approximations*

1. *Near-Rigid Surfaces*

A fundamental objective of the code we are describing is that it be capable of approximating high-Reynolds-number flows with regions of moderate separation. This means that near a body surface the code must be capable of producing turbulent boundary-layer profiles that exhibit regions of reversed flow; this, in turn, means that the flow very near a body surface must be highly resolved, at least in the direction normal to the surface. In order to provide the necessary resolution, grid lines near the body are exponentially clustered. Along the body surface, the spacing is much coarser. A typical case is shown in Fig. 3. Ratios of the spacing along the body to the spacing normal to the body of 1000 are not unusual. The reasons why the flow along the surface is not more highly resolved are twofold, one practical and the

A Three-Dimensional Implicit Code for the ILLIAC IV

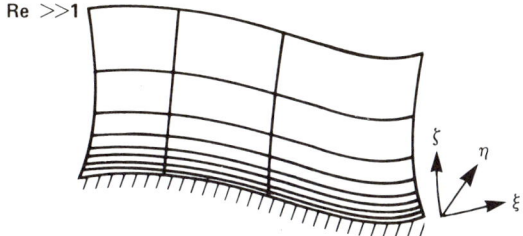

FIG. 3. Clustered grid near body surface, showing disparity between grid spacing normal to surface and along surface.

other physical. The practical reason is the limitation of computer memory and speed; the physical one is the lack of need for extremely high resolution in the tangential directions. Thus, just as in classical compressible boundary-layer theory, we make use of the assumption that it is only the viscous terms in the surface normal direction that need be resolved, and the viscous terms in the other two directions can be neglected. However, unlike classical boundary-layer theory, we do not assume constant pressure in any region of the flow, and we do not neglect any of the inertial terms in the momentum equations.

This "thin-layer" approximation is used throughout the code. If ζ is the coordinate containing the body surface, all viscous derivatives in the ξ and η directions are neglected and terms in the ζ direction are retained. This requires that any body surfaces be mapped onto $\zeta =$ constant surfaces in the computational domain. Applying the thin-layer approximation to Eqs. (1)–(3) produces the following thin-layer, Navier–Stokes equations:

$$\partial_\tau Q + \partial_\xi E + \partial_\eta F + \partial_\zeta G = \mathrm{Re}^{-1} \partial_\zeta S, \tag{7}$$

where

$$Q = J^{-1} \begin{bmatrix} \rho \\ \rho u \\ \rho v \\ \rho w \\ e \end{bmatrix}, \quad E = J^{-1} \begin{bmatrix} \rho U \\ \rho u U + \xi_x p \\ \rho v U + \xi_y p \\ \rho w U + \xi_z p \\ (e+p)U - \xi_t p \end{bmatrix},$$

$$F = J^{-1} \begin{bmatrix} \rho V \\ \rho u V + \eta_x p \\ \rho v V + \eta_y p \\ \rho w V + \eta_z p \\ (e+p)V - \eta_t p \end{bmatrix}, \quad G = J^{-1} \begin{bmatrix} \rho W \\ \rho u W + \zeta_x p \\ \rho v W + \zeta_y p \\ \rho w W + \zeta_z p \\ (e+p)W - \zeta_t p \end{bmatrix}, \tag{8}$$

and

$$S = J^{-1} \begin{bmatrix} 0 \\ \mu(\zeta_x^2 + \zeta_y^2 + \zeta_z^2)u_\zeta + (\mu/3)(\zeta_x u_\zeta + \zeta_y v_\zeta + \zeta_z w_\zeta)\zeta_x \\ \mu(\zeta_x^2 + \zeta_y^2 + \zeta_z^2)v_\zeta + (\mu/3)(\zeta_x u_\zeta + \zeta_y v_\zeta + \zeta_z w_\zeta)\zeta_y \\ \mu(\zeta_x^2 + \zeta_y^2 + \zeta_z^2)w_\zeta + (\mu/3)(\zeta_x u_\zeta + \zeta_y v_\zeta + \zeta_z w_\zeta)\zeta_z \\ \{(\zeta_x^2 + \zeta_y^2 + \zeta_z^2)[0.5\mu(u^2 + v^2 + w^2)_\zeta + \kappa\text{Pr}^{-1}(\gamma - 1)^{-1}(a^2)_\zeta] \\ + (\mu/3)(\zeta_x u + \zeta_y v + \zeta_z w)(\zeta_x u_\zeta + \zeta_y v_\zeta + \zeta_z w_\zeta)\} \end{bmatrix}$$

(9)

2. Turbulence Modeling

An algebraic mixing length model is included to approximate the effect of turbulence. It is a conventional two-layer model. The inner layer is governed by the Prandtl mixing length with Van Driest damping, and the outer layer follows the Clauser approximation. Computer vorticity is used in defining the reference mixing length required for the outer layer. The turbulence model, detailed by Baldwin and Lomax (1978), was designed for use with the thin-layer approximation. The model is most appropriate to attached and moderately separated boundary layers. No attempt is made to model wake regions and massively separated flows. The turbulence model has been "successfully" applied to two- and three-dimensional separated flows for many practical cases (Steger, 1978; Kutler *et al.*, 1978; Pulliam and Steger, 1980; Steger and Bailey, 1980).

E. Numerical Algorithm

1. Space Differencing

Consider now the thin-layer form of the Reynolds-averaged, Navier–Stokes equations expressed in conservation-law form as given by Eq. (7) and rewritten here as

$$Q_\tau = -(\partial_\xi E + \partial_\eta F + \partial_\zeta G - R_e^{-1}\partial_\zeta S) = R. \tag{10}$$

First, we shall discuss the space differencing of the right-hand, or explicit, side. Generally, the finite difference δ that replaces the spatial derivatives ∂ are not unique. For example, they can be forward-, backward-, or central-difference schemes of any order of accuracy. In our case, every finite-difference point operator used to approximate a space derivative is based on a centered difference. Furthermore, the order of the schemes can vary. We use a combination of second- and fourth-order methods. However, order is not so easy to define in computational space where $\Delta\xi$, $\Delta\eta$, and $\Delta\zeta$ are all equal to 1. The actual accuracy depends on the metrics and the grid spacing relative to the local variation of the functions being approximated. However, conventional fourth-order methods can be expected to give better accuracy than conventional second-order methods.

The scheme used to approximate the convective derivatives on the right-hand side of Eq. (10) is the five-point, central-difference, fourth-order-accurate operator

$$(\delta_\xi u)_j = (-u_{j+2} + 8u_{j+1} - 8u_{j-1} + u_{j-2})/(12\,\Delta\xi). \tag{11}$$

Defining the scheme used to approximate the viscous flux is more complicated because the viscous flux term itself contains spatial derivatives. Writing the viscous term in the general form

$$\delta_\zeta(\alpha\,\delta_\zeta\beta), \tag{12}$$

we can express our approximation by the generalized, three-point, central-difference scheme for a second derivative given by

$$\delta_\zeta(\alpha_l\,\delta_\zeta\beta_l) = [(\alpha_{l+1} + \alpha_l)\beta_{l+1} - (\alpha_{l+1} + 2\alpha_l + \alpha_{l-1})\beta_l + (\alpha_l + \alpha_{l-1})\beta_{l-1}]/(2\,\Delta\zeta^2). \tag{13}$$

The above operator is second-order accurate; this is adequate for the viscous terms because grids are clustered in regions where the viscous terms are important.

2. Time Differencing

The numerical method chosen for the time march is the implicit θ method described by Richtmyer and Morton (1967). Using n for the time index and h for the time step, we can apply this method to Eq. (10) and arrive at

$$Q_{n+1} = Q_n + h[(1-\theta)R_{n+1} + \theta R_n]. \tag{14}$$

This formula represents the fully implicit Euler method, which is first-order accurate in time, when $\theta = 0$, and the trapezoidal rule (Crank–Nicolson), which is second-order accurate in time, when $\theta = \frac{1}{2}$.

In order to carry forward the implicit operation, the nonlinear term R_{n+1} must be locally linearized in terms of Q_n to form a set of coupled simultaneous equations that can be solved for each time step. This is done by making a Taylor series expansion of each term. For example,

$$E_{n+1} = E_n + \left(\frac{\partial E}{\partial Q}\right)_n (Q_{n+1} - Q_n) + O(h^2), \tag{15}$$

and so forth for F, G, and S. This leads to the formulation of the flux Jacobian matrices

$$A_n = \left(\frac{\partial E}{\partial Q}\right)_n, \quad B_n = \left(\frac{\partial F}{\partial Q}\right)_n, \quad C_n = \left(\frac{\partial G}{\partial Q}\right)_n, \quad M_n = \left(\frac{\partial S}{\partial Q}\right)_n, \quad (16)$$

where, for example, the elements are defined by

$$A_n = [a_{ij}] = \partial E_i/\partial Q_j. \quad (17)$$

For the sake of completeness, and to express the full nature of the code, we present the flux Jacobians for each of the four terms in Eq. (16) (see also Pulliam and Steger, 1980). The inviscid flux Jacobians are

$$A, B, \text{ or } C = \begin{bmatrix} K_0 & K_1 & K_2 & K_3 & 0 \\ K_1\phi^2 - u\theta & K_0 + \theta - K_1(\gamma - 2)u & K_2u - (\gamma - 1)K_1v & K_3u - (\gamma - 1)K_1w & K_1(\gamma - 1) \\ K_2\phi^2 - v\theta & K_1v - K_2(\gamma - 1)u & K_0 + \theta - K_2(\gamma - 2)v & K_3v - (\gamma - 1)K_2w & K_2(\gamma - 1) \\ K_3\phi^2 - w\theta & K_1w - K_3(\gamma - 1)u & K_2w - K_3(\gamma - 1)v & K_0 + \theta - K_3(\gamma - 2)w & K_3(\gamma - 1) \\ \theta[2\phi^2 - \gamma(e/\rho)] & \{K_1[\gamma(e/\rho) - \phi^2] - (\gamma - 1)u\theta\} & \{K_2[\gamma(e/\rho) - \phi^2] - (\gamma - 1)v\theta\} & \{K_3[\gamma(e/\rho) - \phi^2] - (\gamma - 1)w\theta\} & K_0 + \gamma\theta \end{bmatrix}$$

$$(18)$$

where

$$\phi^2 = 0.5(\gamma - 1)(u^2 + v^2 + w^2),$$
$$\theta = K_1 u + K_2 v + K_3 w, \quad (19)$$

and, for example, to obtain A

$$K_0 = \xi_t, \quad K_1 = \xi_x, \quad K_2 = \xi_y, \quad K_3 = \xi_z. \quad (20)$$

Finally, the viscous flux Jacobian is

$$M = J^{-1} \begin{bmatrix} 0 & 0 & 0 & 0 & 0 \\ m_{21} & \alpha_1 \delta_\zeta(\rho^{-1}) & \alpha_2 \delta_\zeta(\rho^{-1}) & \alpha_3 \delta_\zeta(\rho^{-1}) & 0 \\ m_{31} & \alpha_2 \delta_\zeta(\rho^{-1}) & \alpha_4 \delta_\zeta(\rho^{-1}) & \alpha_5 \delta_\zeta(\rho^{-1}) & 0 \\ m_{41} & \alpha_3 \delta_\zeta(\rho^{-1}) & \alpha_5 \delta_\zeta(\rho^{-1}) & \alpha_6 \delta_\zeta(\rho^{-1}) & 0 \\ m_{51} & m_{52} & m_{53} & m_{54} & \alpha_0 \delta_\zeta(\rho^{-1}) \end{bmatrix} J, \quad (21)$$

with

$$m_{21} = \alpha_1 \delta_\zeta(-u/\rho) + \alpha_2 \delta_\zeta(-v/\rho) + \alpha_3 \delta_\zeta(-w/\rho),$$
$$m_{31} = \alpha_2 \delta_\zeta(-u/\rho) + \alpha_4 \delta_\zeta(-v/\rho) + \alpha_5 \delta_\zeta(-w/\rho),$$
$$m_{41} = \alpha_3 \delta_\zeta(-u/\rho) + \alpha_5 \delta_\zeta(-v/\rho) + \alpha_6 \delta_\zeta(-w/\rho),$$
$$m_{51} = \alpha_1 \delta_\zeta(-u^2/\rho) + \alpha_2 \delta_\zeta(-2uv/\rho) + \alpha_3 \delta_\zeta(-2uw/\rho)$$
$$\quad + \alpha_4 \delta_\zeta(-v^2/\rho) + \alpha_6 \delta_\zeta(-w^2/\rho) + \alpha_5 \delta_\zeta(-2vw/\rho)$$
$$\quad + \alpha_0 \delta_\zeta(-e/\rho^2) + \alpha_0 \delta_\zeta[(u^2 + v^2 + w^2)/\rho],$$
$$m_{52} = -m_{21} - \alpha_0 \delta_\zeta(u/\rho), \tag{22}$$
$$m_{53} = -m_{31} - \alpha_0 \delta_\zeta(v/\rho),$$
$$m_{54} = -m_{41} - \alpha_0 \delta_\zeta(w/\rho)$$

$$\alpha_0 = \gamma\kappa \Pr^{-1}(\zeta_x^2 + \zeta_y^2 + \zeta_z^2), \qquad \alpha_1 = \mu[(\tfrac{4}{3})\zeta_x^2 + \zeta_y^2 + \zeta_z^2],$$
$$\alpha_2 = (\mu/3)\zeta_x\zeta_y, \qquad \alpha_3 = (\mu/3)\zeta_x\zeta_z,$$
$$\alpha_4 = \mu[\zeta_x^2 + (\tfrac{4}{3})\zeta_y^2 + \zeta_z^2], \qquad \alpha_5 = (\mu/3)\zeta_y\zeta_z,$$
$$\alpha_6 = \mu[\zeta_x^2 + \zeta_y^2 + (\tfrac{4}{3})\zeta_z^2].$$

By means of the local linearizations given above, the implicit unfactored form of Eq. (14) can be expressed as

$$[I + (1 - \theta)h(\delta_\xi A_n + \delta_\eta B_n + \delta_\zeta C_n - R_e^{-1}\delta_\zeta M_n)]\Delta Q_n = hR_n, \tag{23}$$

which is said to be in the "delta form" because the left-hand, or implicit, side contains the factor $Q_{n+1} - Q_n$, which defines ΔQ_n in conventional differencing notation. The spatial differencing of the terms composing R_n on the right-hand side of Eq. (23) has been described in Section II.E.1. The spatial differences involving the flux Jacobians on the left-hand side of Eq. (23) are all three-point, second-order-accurate central differences.

The left-hand side of Eq. (23) can be factored to form three one-dimensional operators, without changing the order of accuracy (Beam and Warming, 1976). This results in the factored delta form of Eq. (23):

$$[I + (1 - \theta)h\,\delta_\xi A_n][I + (1 - \theta)h\,\delta_\eta B_n]$$
$$\times [I + (1 - \theta)h(\delta_\zeta C_n - R_e^{-1}\delta_\zeta M_n)]\Delta Q_n = hR_n. \tag{24}$$

Each of the operators in brackets represents a block-tridiagonal matrix based on the mesh size in its coordinate direction. The implications of using both the unfactored and factored forms are discussed in Section IV.B.

3. Subgrid Truncation

Whenever discrete methods are used to "capture" shocks (as opposed to fitting them), or to compute high-Reynolds-number viscous behavior, scales of motion appear that cannot be resolved by the numerics. These are brought about by the nonlinear interactions in the convection terms of the momentum equations. If scale is represented by wavelength or frequency, it can be easily shown that two waves interacting as products (e.g., $u\partial_x u$) form a wave of higher frequency (the sum of the original two) and one of lower frequency (the difference). The lower frequencies do not cause a problem, but the continual cascading into higher and higher frequencies does. It is accounted for physically by shock formation (the harmonic analysis of a discontinuity contains all frequencies of motion) or by viscous dissipation of the very high wave numbers. In numerical computations it cannot be ignored and must be accounted for somehow in the algorithm constructed. In any finite, discrete mesh, the cascading frequencies can eventually exceed the capacity of the mesh [even with spectral methods, an equispaced mesh with M points cannot support a frequency $\sin(kx)$ with k greater than $M/2$], at which point they can either (1) alias back into the lower frequencies or (2) if alias-free methods are used, pile up at the higher frequency side. In either case, if uncontrolled, these terms can lead to serious inaccuracies and possible numerical instability.

The most common way of coping with the high-frequency cascade is to add to the complete algorithm some form of numerical dissipation with an error level that does not interfere with the accuracy of any physical viscous effects. This can be done in a variety of ways. For example, it is accomplished by using upwind differencing, by using numerical schemes (e.g., Lax–Wendroff forms) that have hidden higher-order dissipation terms, which appear in their modified equation representation, or by simply overtly inserting dissipation terms into the solution process. The latter course is chosen for the code being discussed. However, the crucial point being made is that, without some form of nonlinear cascade control, a code simulating a flow in which the cascade is important would be incorrect.

Fourth-order dissipation terms are added to the right-hand (explicit) side and second-order dissipation terms to the left-hand (implicit) side of Eq. (24). The total effect is to modify Eq. (24) to the form

$$[I + (1 - \theta)h \, \delta_\zeta A_n - \varepsilon_I J^{-1}(\nabla_\xi \Delta_\xi)J]$$
$$\times [I + (1 - \theta)h \, \delta_\eta B_n - \varepsilon_I J^{-1}(\nabla_\eta \Delta_\eta)J]$$
$$\times [I + (1 - \theta)h(\delta_\zeta C_n - R_e^{-1} \delta_\zeta M_n) - \varepsilon_I J^{-1}(\nabla_\zeta \Delta_\zeta)J] \Delta Q_n$$
$$= hR_n - \varepsilon_E J^{-1}[(\nabla_\xi \Delta_\xi)^2 + (\nabla_\eta \Delta_\eta)^2 + (\nabla_\zeta \Delta_\zeta)^2]JQ_n, \quad (25)$$

where the coefficients ε_I and ε_E are based on linear stability theory. They are chosen as

$$\varepsilon_E = h, \qquad \varepsilon_I = 3\varepsilon_E. \quad (26)$$

The numerical operators ∇ and Δ are defined as

$$(\nabla_\xi u)_j = (u_j - u_{j-1})/\Delta\xi, \qquad (\Delta_\xi u)_j = (u_{j+1} - u_j)/\Delta\xi. \quad (27)$$

F. Boundary Conditions

We include the discussion of boundary conditions in this section because it is only in the actual numerical formulation that the boundary conditions take their final form. Since all the boundary surfaces have been mapped to coordinate surfaces in the computational domain, the boundary condition application is unique to each coordinate direction. A sketch of the particular geometry and boundary surfaces for a hemisphere–cylinder body is shown in Fig. 4; it will be used to illustrate the application of boundary conditions. In the subsequent discussions, take $JMAX$, $KMAX$, and $LMAX$ as the maximum number of nodes in the coordinate directions ξ, η, and ζ, respectively.

At the free-stream surface, $L = LMAX$, the variables are fixed at their free-stream values. Along the vertical plane indexed by K, bilateral symmetry is imposed. Here the variables ρ, u, w, and e are reflected positively (e.g., u at $K = 1$ equals u at $K = 3$) and velocity v is reflected negatively (i.e., v at $K = 1$ equals $-v$ at $K = 3$). Note that this requires that $K = 2$ and $K = KMAX - 1$ lie on the symmetry plane and that nodes at $K = 1$ and $K = KMAX$ are only necessary for boundary-condition application. At the outflow boundary $J = JMAX$, two situations can exist. For supersonic flow, all the variables are extrapolated to zeroth order (e.g., ρ at $JMAX = \rho$ at $JMAX - 1$). In transonic cases, all the variables except e are extrapolated to zeroth order. Then e is found by using these extrapolated values and a fixed value of pressure.

A coordinate singularity occurs at the $J = 1$ coordinate surface. However, the ξ metrics of Eq. (8) are zero at the singularity, and therefore the transformed flux E vanishes there. This means that the finite-difference equations do not see the variables along the singularity line. An exception is made for the subgrid truncation terms, which require a smooth value of the variables.

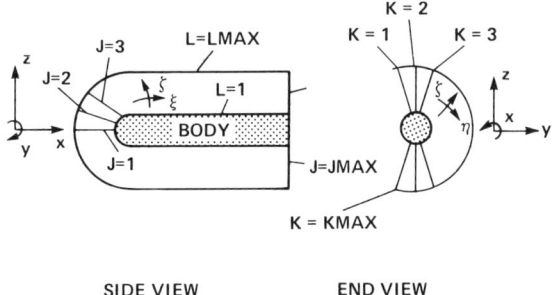

FIG. 4. Sketch of boundary surfaces for hemisphere–cylinder geometry.

For these terms, values of the variables at $J = 1$ are found from linear extrapolation of the variables at $J = 2$ and $J = 3$.

There remains the treatment of the variables at the surface $L = 1$. Since we are interested here only in high-Reynolds-number viscous flows, the surface conditions are relatively simple. All of the contravariant velocities are set to zero at $L = 1$, and the pressure and density (adiabatic wall condition) are given by

$$\partial p/\partial \zeta = \partial \rho/\partial \zeta = 0, \qquad (28)$$

which is approximated by setting p and ρ at $L = 1$ to p and ρ at $L = 2$. Remember that the mesh spacing in the $\zeta(L)$ direction is extremely fine near the surface.

At present, all boundary conditions are treated explicitly. Our investigations at this time (carried out on several computers) have shown no advantage in incorporating implicit conditions. However, this is still an open question. There is no reason why implicit boundary conditions could not be directly incorporated into the present algorithm and coding.

III. ILLIAC ARCHITECTURE

A. Parallelism

The ILLIAC IV computer is a parallel processor. It is composed of 64 processing elements (PEs) that work strictly in a lock-step mode. Each processor has the equivalent speed of a CDC 6600 and a local memory (PEM) of 2048 floating-point, 64-bit words. The PEs are given instructions by a control unit (CU) that issues an identical set of instructions (e.g. fetch, store, add, multiply, and divide) to each of the 64 PEs. A masking capability is provided that can cripple the store operation of a PE. When the mask is on, a PE is said to be "turned off." Typically, though, all the PEs are on during a calculation. A FORTRAN-like language called CFD (Ames Research Center, NASA, 1974) is provided that can be used to instruct the processors to carry out arithmetic and logical manipulations on the data base.

B. Memory

The total processor memory (PEM) is composed of 64×2048 words, about $64 \times 2040 = 130{,}560$ of which are available to the user. This memory is backed up by a synchronized disk (a logical drum) holding about 7 million usable 64-bit words. Data are stored on the disk in bands. Each band holds 300 pages, and each page contains 16 "rows" of 64-bit floating-point words. The total data base is mapped onto the disk memory in such a way that predetermined groups of data can be moved in blocks of pages to and from each processor memory in parallel transmissions. No operating system or virtual memory is provided. All computing is in batch mode, and each user has the total machine during the course of his calculations.

C. Bit Accuracy

The ILLIAC IV can operate in 64-bit or 32-bit floating-point accuracy. The total memory of both PEM and disk are double or, equivalently, the same problem size will run about twice as fast in 32-bit mode as it does in 64-bit mode. The code described here has been run in both 32- and 64-bit modes. In all cases studied to date, the 32-bit word size has provided adequate accuracy. The special coding required to perform calculations in the 32-bit mode is peculiar to the ILLIAC, and so it will not be described in detail here.

IV. DATA-BASE CONSIDERATIONS

A. Data-Base System Definition

Let us describe what is meant by a "data-base system" used to compute a fluid-flow simulation. First of all, we assume that memory requirements for the dependent variables are such that at least two levels of storage are necessary. The arithmetic units are connected to one of these levels. This level is called high-speed store. The other level(s), called low-speed store, can communicate to the arithmetic units only by first transferring to the high-speed store. We are interested only in codes and calculations that require the storage of masses of floating-point words (at least 2 million), and a large percentage of these are constantly being streamed through the arithmetic units. We take considerable care to design our codes so that the variables required during the computation can be moved to and from the storage levels as seldom and as efficiently as possible. Any system of data storage and management that relieves the user of having to worry about data transfer efficiencies in this environment is referred to as a "data-base system."

Consider an example. Probably the most efficient possible arrangement occurs when the problem can be cast in a form such that (1) the computational space is a topological box and (2) the boundary conditions are applied only along the sides of this box. Any physical space, and boundaries that can be mapped into and onto this box by appropriate coordinate transformations, can be used for a numerical simulation without regard for the data management—if one stays within the constraints of some data-base system constructed specifically for this arrangement. If properly designed, the system will handle all transfers of the required data in an optimal manner while the computations are being carried out.

To design such a system, one must be aware of the particular machine architecture to be used and of the type of algorithm to be processed. Consider the constraints imposed by these considerations for the ILLIAC computer and our particular code.

B. Constraints of Architecture

1. *Memory Hierarchies*

Generally, modern scientific computers have several levels of memories with greatly varying capacities, latencies, and interconnecting transfer rates. In particular, the ILLIAC has three: a disk, the memory in the processing elements (PEM), and five vector registers in the arithmetic units. For our purposes, we can consider this memory to be stored by vectors in the manner illustrated in Fig. 5. Note that the disk storage is arranged so that 16 vectors, or one page, are moved to and from the next higher level at one time.

2. *Speed of Data Flow and Computation*

Data flow from one memory level to the next is in minimum chunks of 16 vectors for disk-to-PEM transfers, and 1 vector for PEM-to-arithmetic-unit transfers. First, consider the transfer from the disk to the processing

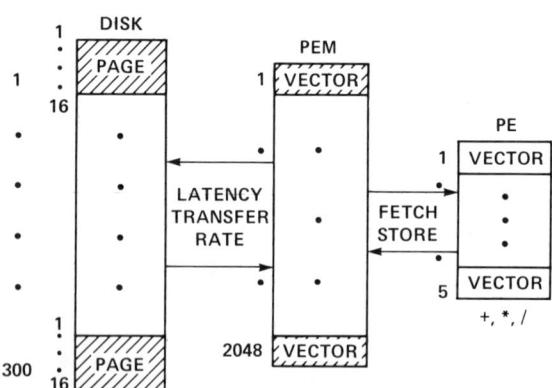

FIG. 5. ILLIAC memory hierarchy.

element memory. The potentially slowest part of this transfer is the latency time required to access the starting location of the memory location on the disk. In computer terminology, the disk is a logical drum and the maximum wait is the length of time it requires the disk to make one revolution, which is 4×10^7 nsec. The basic time unit of a computer is often measured in "clocks." The clock time varies greatly from computer to computer, even at times from month to month in a given computer. The standard clock time for the ILLIAC is 80 nsec, which means that the maximum latency time is 500,000 clocks. Since there are 300 pages of 16 vectors each stored on the disk, one page can be transferred in 1666 clocks. This amounts to just over 100 clocks per vector.

Next, consider briefly the transfer rate between the memory of the processing elements and the arithmetic unit. These fetches and stores occur vector by vector and require 9 and 5 clocks per transfer, respectively. Both stores and fetches can be partially covered up by arithmetic.

Finally, the arithmetic itself is timed as follows: add, 9 clocks; multiply, 9 clocks; and divide, 56 clocks.

3. *Input/Output Capabilities*

The disk cannot be reading to or writing from the processor memories while computations are being carried out, but the latency period can be covered up by computation. On the ILLIAC, the user is expected to optimize this operation since there is no operating system or virtual memory.

4. *Vector Length*

The ILLIAC vector length in 64-bit precision is 64 words long; in 32-bit precision it is 128 words long.

5. *Objective*

The maximum use of the machine occurs when all 64 of the processing units are active all of the time. This condition can only be approached when the entire data base is carried in the core of the processing elements. When part of the data is carried on the disk, it must be moved to and from the processing element memories in order to proceed with the calculation. The arithmetic units must remain idle while this transfer is carried out. From the statistics given above, we see that the transfer of one vector is equivalent to about 11 multiplies. This means that an important part of designing an ILLIAC code, which is using massive amounts of information stored on the disk, is to minimize the number of DISK/PEM data transfers.

C. Constraints of Algorithm

1. Basic Implicit Formulation

The complete expression for the form of the Reynolds-averaged, Navier–Stokes equations we are considering is given in Section II.B. Our discussion here concerns the manner in which implicit solutions of these equations put constraints on the data organization in the computer. To simplify the discussion, we use x, y, z to represent the coordinate system, but these are intended to represent any generalized set with the appropriate metrics in the flux terms. Let us express the equations in Section II using A_x, A_y, and A_z to represent the flux Jacobians. The two-level, time-march, theta method given in Section II can be used to write the unfactored, implicit, delta form as

$$\{I + (1 - \theta)h(\delta_x A_x + \delta_y A_y + \delta_z A_z)\}(Q_{n+1} - Q_n) = hR_n. \qquad (29)$$

This represents a coupled set of difference equations that can be expressed by

$$[L_{x+y+z}]_n(Q_{n+1} - Q_n) = hR_n, \qquad (30)$$

where the vector R and the matrix L are known and we must solve for the vector $Q_{n+1} - Q_n = \Delta Q_n$. The form of the matrix operator is illustrated in Fig. 6. Only the structure matters, and a $3 \times 3 \times 3 = 27$-point mesh is sufficient to bring out the general form. In constructing the matrix we have assumed that it multiplies a Q vector, for which the x data are closely packed;

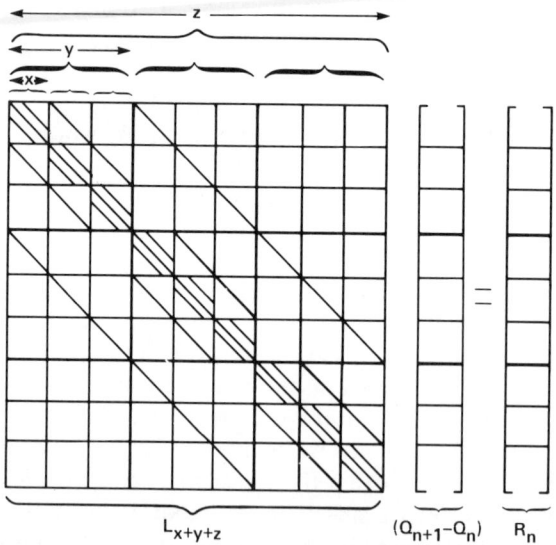

FIG. 6. Structure of unfactored matrix used in Eq. (30).

the y data skip across blocks of x; and the z data skip blocks of y. The solution is advanced one time step, with first- or second-order accuracy in the time advance, by solving the set of coupled simultaneous equations illustrated in Fig. 6. For each advancing time step, the simultaneous equations are solved once without iteration. Then the nonlinear terms in the 5×5 A matrices are reevaluated in the next time advance. If the size of the mesh is Mx, My, and Mz in the x, y, and z directions, respectively, the Q and R vectors have $(5)(Mx)(My)(Mz)$ elements, and the order of L is $[(5)(Mx)(My)(Mz)]^2$. The factor of 5 enters because each "element" in the matrix structure is a 5×5 block matrix stemming from the flux Jacobian. The form of the matrix is clearly block-banded and, in the x, y, z data sequencing discussed above, the bandwidth is $(2(My) + 1)(Mx)$, 5×5 blocks.

2. Factored Delta Form

As discussed in Section II.C, we can, without changing the order of accuracy in time or space, factor the operator given inside { } in Eq. (29) to form the sequence:

$$\{I + (1 - \theta)h\,\delta_x A_x\}\{I + (1 - \theta)h\,\delta_y A_y\}\{I + (1 - \theta)h\,\delta_z A_z\}, \quad (31)$$

which leads to the definition of the one-dimensional operator:

$$[L_p]_n \equiv I + (1 - \theta)h\,\delta_p A_p. \quad (32)$$

These one-dimensional operators can be used to form the predictor–corrector sequence:

$$\begin{aligned}
[L_x]_n Q^* &= hR_n, \\
[L_y]_n Q^{**} &= Q^*, \\
[L_z]_n Q^{***} &= Q^{**}, \\
Q_{n+1} &= Q^{***} + Q_n.
\end{aligned} \quad (33)$$

The sequence into which the L_p operators are factored is arbitrary. The manner in which they were actually coded is given in Section V.

The structure of the matrix L_p is shown in Fig. 7. In order that the figure be representative of all three space directions, the data in Q and R must be properly permuted before each operator is applied. Two properties of the matrix are especially important. First, each diagonal block is completely uncoupled. This means that a tridiagonal solver can be applied to each block independently. Furthermore, the order in which each block is solved is arbitrary. Second, the bandwidth of the matrix is independent of the mesh size, being in every case three 5×5 blocks.

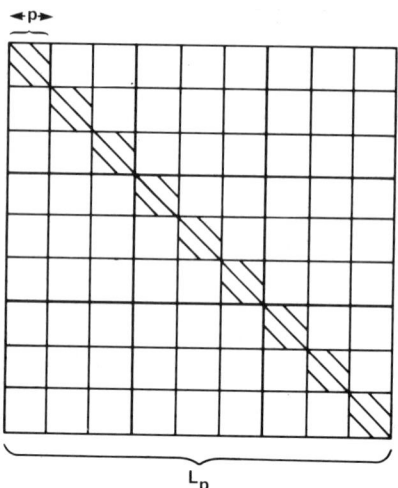

FIG. 7. Structure of factored matrix after appropriate data permutation; p represents x, y, or z of Eq. (31).

3. Tridiagonal Solvers

The actual application of the sequence given by Eq. (33) is carried out by using a block-tridiagonal solver for each line in the mesh in each of the three space orientations. For the data base used in Fig. 7 this solution process is carried out by the application of nine 5×5 block-tridiagonal solvers for each of the three space directions, the data representing the vectors Q and R being properly permuted before each space direction is processed. In general, $(Mx)(My) + (Mx)(Mz) + (My)(Mz)$ block-tridiagonal solvers would be used to advance the data base one time step. Each solver amounts to

1. Converting (by Gaussian elimination using LUD in each of the individual blocks) the basic tridiagonal to a normalized upper-diagonal, shown as the X vector in Fig. 8—referred to as the forward sweep;

2. Constructing the final solution using the modified right-hand side and the X vector saved from the forward sweep—referred to as the backward sweep.

The details of this algorithm (often called the Thomas algorithm) are described in many texts on numerical analysis.

4. Nonlinear Effects

The forward sweep is carried out by computing the X matrices in Fig. 8 one by one, proceeding from the top down. For each calculation of X, the

A Three-Dimensional Implicit Code for the ILLIAC IV

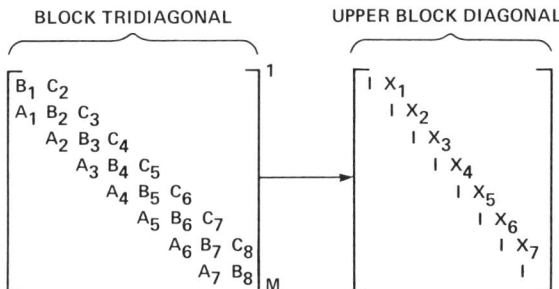

FIG. 8. Block tridiagonal structure after forward sweep; A, B, and C can be computed and discarded term by term, but X must be available for backward sweep.

necessary A, B, and C must be computed, using their corresponding Q at the time level n. This means that Q must be available during the calculation of the forward sweep.

D. *Planar Data Bases*

A popular way to treat massive data transfers is plane by plane. In this case, the dependent variables for the computational box are stored in such a way that an entire plane of data is brought into high-speed core at each data request. This system is used extensively in the study of homogeneous turbulence. In a typical case, a ξ plane (see Fig. 9a) is sent from low-speed to high-speed core. While resident there, fast Fourier transforms are applied to it, proceeding from boundary to boundary in the two directions of the plane, η and ζ in Fig. 9a. The plane is square and the identifiable vector length is determined by the number of points on one side of the plane. The planes are permuted while in core to permit a vectorized solution in both directions. The third direction is resolved by bringing the data into high-speed core in a different orientation, for example, in the form of η planes, as shown in Fig. 9b.

Efficient planar data-base systems can be designed for many architectures. In particular, one has been written for the ILLIAC that can manage the DISK/PEM transfers for a 128-cube mesh and for a 128 × 64 × 64 mesh (Kim and Moin, 1979).

E. *Pencil Data Bases*

Another way to arrange for systematic massive data transfer is to group the data by pencils. This mode of operation allows for much more flexible grid dimensioning and can be adapted to long or short vector lengths. The

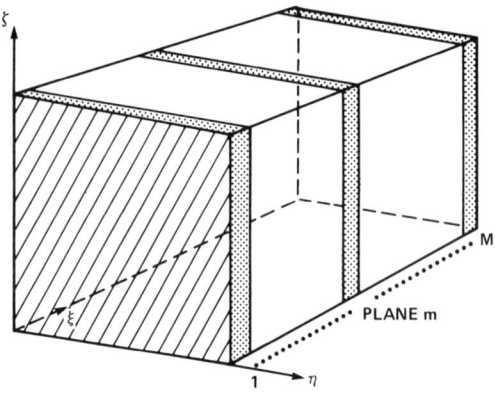

(a) ξ PLANES

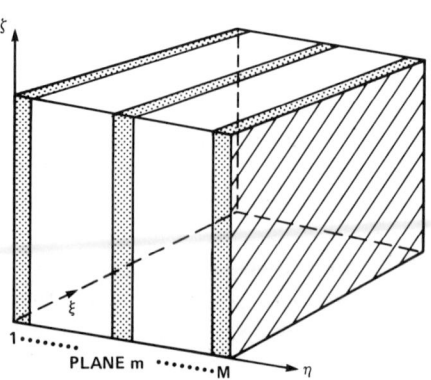

(b) η PLANES

Fig. 9. Planar data-base concept: (a) ζ plane; (b) η plane.

computational space, which can be rectangular (see Fig. 10a), is subdivided into cubes of mesh locations. A string of cubes parallel to a coordinate axis and extending from boundary to boundary is referred to as a pencil. The pencils are numbered as shown in Fig. 10b, in which there are 12 ξ pencils, 15 η pencils, and 20 ζ pencils.

To use the pencil system, a user sets a direction and asks for a pencil number. The desired pencil is entered into a computation buffer area so that its cross section is indexed by the first two arguments of a FORTRAN statement. Thus the pressure in ξ pencil 5 (see Fig. 10c) could be designated

A Three-Dimensional Implicit Code for the ILLIAC IV

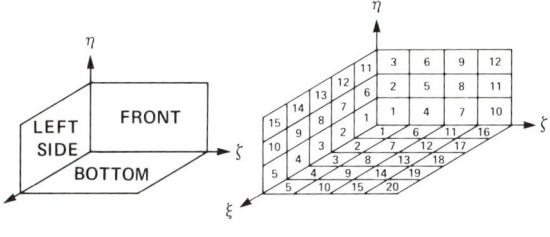

(a) COMPUTATIONAL SPACE (b) PENCIL NUMBERING

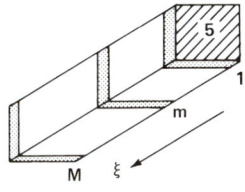

(c) ξ PENCIL #5

FIG. 10. Data arrangement of computational space for pencil data-base system: (a) computational space; (b) pencil numbering; (c) ζ pencil.

in FORTRAN by $P(L, K, J)$. The dimensions for L and K are the same and equal to the square root of the vector length chosen for the computer. The dimension for J depends on the number of blocks in the pencil. It is convenient to designate the vector by an asterisk (∗). Thus, for the example, if we (1) flag the ξ direction, (2) ask for pencil 5, and (3) write the code $S(*) = P(*, 21)$, vector $S(*)$ is filled with the values of the pressure in plane 21 of ξ pencil 5.

It is important to mention that the pencils are mapped by the data-base system in such a way that they can be called into core in *any order*, and that this order can be set by the user. Thus, the 12 ξ pencils shown in Fig. 10b can be processed in the order 1, 7, 2, 8, 3, 9, 4, 10, 5, 11, 6, 12, if the user so specifies. This can be very useful in tuning the latency to the arithmetic computational load.

V. THE ILLIAC CODE ARC3

The following discussion outlines a computational procedure designed for making time-accurate simulations of rather general compressible fluid flows. It is designed specifically for the ILLIAC, but the concepts are useful for other computers with similar architectures. The governing equations are the unsteady, Reynolds-averaged, Navier–Stokes equations presented in Section II, with the thin-layer approximation and an algebraic, eddy-viscosity turbulence model. The numerical algorithm is implicit, at least second-order accurate in both space and time, and adaptable to any three-dimensional flow field that can be mapped onto a computational box. The constraints imposed by the algorithm choice are described in Section IV.C. The computer to be used is the ILLIAC IV described briefly in Section III. The constraints imposed by the choice of this computer are described in Section IV.B. The data-base system chosen to help cope with these constraints is the pencil data-base system described in Section IV.E. We emphasize that the use of this system assumes that an entire pencil will be solved when all of the data necessary for its solution are resident in the PEMs (i.e., "in core").

First of all, we note that the numerical algorithm is required to be implicit. This necessitates the solution of a large number of simultaneous equations for each time step. Consider first the possibility of using the unfactored implicit method given by Eq. (29). A solution algorithm for the coupled set of equations with the bandwidth shown in Fig. 6 requires a very large amount of temporary storage capacity, even if the number of grid points is only moderate. For example, a mesh of $20*20*20$ carrying five dependent variables at each point would produce a matrix having a half bandwidth of $20*20$ dense block matrices, each block having $5*5$ elements. The total amount of storage required for some form of Gaussian elimination would be about $20*25^5 = 8 \times 10^7$ words. For this reason, use of the unfactored implicit algorithm was abandoned. Consider next the possibility of using the factored implicit method given by Eq. (31). Block-tridiagonal solvers require 25 words of temporary store for each A, B, C, and X in Fig. 8. The A, B, and C can be discarded as the solution proceeds, but all of the X matrices must be available for the backward sweep. As a consequence, these one-dimensional tridiagonal sweeps require $25*M$ words of temporary storage, where M is the largest number of points in any one of the three directions. For example, a $40*60*80$ mesh would require a temporary storage of 2000 words for the block-tridiagonal sweep in the 80-point direction. Use of this kind of algorithm, therefore, is at least feasible on our chosen computer.

Next consider the architectural constraints imposed by the parallelism (64 locked-step processors) and the limited PE memory (2048 "vectors" of length 64). For maximum efficiency, we must be constantly performing all arithmetic operations on strings of 64 words, that is, on a complete "vector." The pencil data base identifies a vector of length 64 with the $8*8$ cross-sectional plane of the pencil. This immediately removes the necessity that the user identify a vector in the algorithm. It also means, however, that the block-tridiagonal algorithm chosen in the preceding paragraph requires 2000 *vectors* of temporary storage for one forward and backward step. But only 2048 vectors can be stored in the entire PE memory, and the pencil system is designed only for those algorithms that can process a complete pencil while it resides in core.

The above problem is compounded by further algorithm constraints. Recall the factored tridiagonal sequence shown in Eq. (33). In order to compute this algorithm, we must reserve in the PEM 5 vectors of Q for each plane in the pencil (see Section IV.C.4), 8 vectors of R (3 of these are for the eddy viscosity model), 4 vectors for the metrics in any one direction (the time metric is included), and 1 vector for the metric Jacobian J. This means that 18 vectors per pencil plane must be set aside in PEM before we can even begin to allocate temporary storage.

There are three known ways to cope with this situation. The most obvious one is simply to limit the length of the pencil. About 40 planes are maximum for this route; this severely limits the resolution for many problems. The second way is to reduce the algorithm to first-order accuracy, diagonalize the 5×5 blocks, and replace the block-tridiagonal solver with five scalar solvers. This reduces the required temporary storage area by a factor of 25! In this way, an 80-plane pencil can easily be processed in core. This method, which has been used on the ILLIAC, is reported by Pulliam and Chaussee (1981). The third way, and the one actually used most often on the ILLIAC, is to discard some of the X matrices on the forward sweep and recompute them on the backward sweep. This clever strategy is referred to as the Marshal plan (Merriam, 1980). If used in moderation, it does not drastically increase the computation time; for example, it about doubles the computing time for an 80-plane pencil in which only 300 vectors (instead of the 2000 required by the standard processes) are used for temporary storage.

Finally, we consider the most severe programming constraint in the ILLIAC system. This is the large latency time (500,000 clocks or 40 msec), which represents the time required for the disk to make one revolution. It is crucial that the data be mapped onto the disks in such a way that the largest amount of data possible be interchanged in the course of one revolution. It is also crucial that these interchanges be made as seldom as possible (or, at least, in an optimal manner). In the pencil data-base system this is accomplished by storing the individual blocks, illustrated in Fig. 8b, in a skewed fashion across the disk bands.

Of course, one of the principal arguments for using a data-base system is that the user does not have to bother with the details of disk mapping. However, just to illustrate the concept, we consider the simplified arrangement shown in Fig. 11. Arcs $\Delta\theta$ of nine disk bands are shown. Each arc has enough capacity to store seven blocks of a pencil. If the blocks are mapped out in the manner shown, any pencil in any orientation is a $3 * 3 * 3$ cube can be brought into core in the time required to revolve $\frac{3}{7}$ of the arc length, and in no more

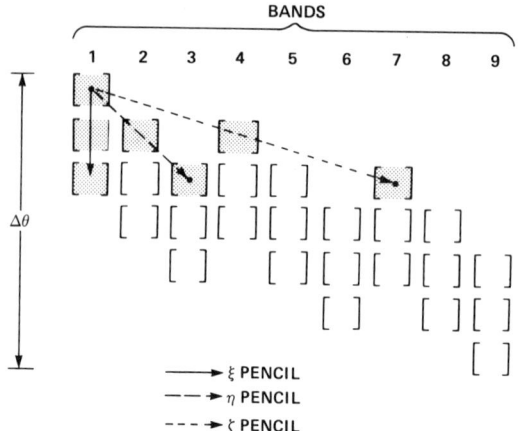

FIG. 11. A form of disk map for a pencil data base.

than one revolution. Sophisticated applications of this concept are provided the ILLIAC user who accepts the pencil data-base system (by means of the subroutine SWEEP written especially for the ILLIAC). For the most part we work around the latency and disk-transfer constraints by the following techniques: (1) using a space-factored algorithm; (2) solving all block-tridiagonal systems in core, using the Marshal plan (the disk is never used for temporary storage); (3) processing the data, pencil by pencil, from disks that have been previously mapped for pencil access in some near-optimal fashion; and (4) using the sequence of data flow to be outlined.

We close this section with an outline of the predictor–corrector sequence given by Eq. (33) as it is actually coded. This outline shows how a cycle is carried out after the process has been started. That is, the first step assumes a previous Q has already been computed. Special starting logic has to be provided.

In studying this outline, notice that the data permutation, which allows us to compute all three directions without going across the processors during the computation, is accomplished in two steps—first by bringing the appropriate blocks in from the disk so that all of the information for the desired pencil is in core and, second, by permuting the data internal to each block in core so that the vector is always a plane of the pencil. Only in the latter permutations are the data moved across processors. It is estimated that these in-core permutations take less than 2% of the computation time.

1. Solve for $Q_n = Q_{n-1} + [L_\xi]_{n-1}^{-1} Q^{**}$, pencil by pencil.
2. While each pencil in step 1 is in core, compute $[\partial E/\partial \xi]_n$, label it R_n, and store Q_n and R_n on disk.

A Three-Dimensional Implicit Code for the ILLIAC IV

3. Fetch Q_n and R_n as η pencils and permute them in core.
4. Compute $[\partial F/\partial \eta]_n$, add it to R_n, permute R_n, and store it on disk, overstoring previous R_n.
5. Fetch Q_n and R_n as ζ pencils and permute them in core.
6. Compute $[\partial G/\partial \zeta]_n$, add it to R_n, solve $Q^* = [L_\zeta]_n^{-1} R_n$, permute Q^*, and store it on disk, overstoring previous R_n.
7. Fetch Q_n and Q^* as η pencils and permute them in core.
8. Solve $Q^{**} = [L_\eta]_n^{-1} Q^*$, permute Q^{**}, and store in on disk, overstoring previous Q^* and R_n.
9. Fetch Q_n and Q^{**} as ξ pencils.
10. Solve for $Q_{n+1} = Q_n + [L_\xi]_n^{-1} Q^{**}$; cycle is completed.

In the actual code R, Q^*, and Q^{**} are all labeled R and stored in the same area on both disk and PEM.

Note that the calculation on the right-hand side of

$$[L_\xi]_n [L_\eta]_n [L_\zeta]_n \Delta Q_n = -h[\delta_\xi E_n + \delta_\eta F_n + \delta_\zeta G_n] \tag{34}$$

is coupled as much as possible with the left-hand-side tridiagonal solvers so that, whenever possible, both are carried out while a pencil is still in core. This minimizes the number of fetches and stores needed to carry out one complete time step. By studying the above outline we see that in one cycle:

1. Q has been stored once, pencil by pencil,
2. Q has been fetched four times, pencil by pencil,
3. R has been stored four times, pencil by pencil,
4. R has been fetched four times, pencil by pencil.

It is estimated that latency and disk input/output operations account for between 15% and 30% of the total running time of the ILLIAC codes, using data-base systems.

VI. RESULTS

A. *Introduction*

The computational results presented here are for demonstration purposes only. The details of the calculations can be found in Pulliam and Lomax (1979) and Deiwert (1980). The code described above has been applied to several types of viscous flow-field calculations. The two described here are for semiinfinite axisymmetric bodies at high angles of attack and at transonic to supersonic Mach numbers. The first application is a hemisphere–cylinder body on which we concentrate on the forebody region. The second application is to the boattail region of a cylindrical afterbody. This problem arises in the study of exhaust plumes. In both cases, the geometry is the warped spherical system described in Section II.C. Applications to three-dimensional wings spanning a wind-tunnel wall are currently under way, but they are not reported here.

B. *Supersonic Flow over Hemisphere–Cylinder: Laminar*

Laminar results are presented for a free-stream Mach number $M_\infty = 1.2$, a Reynolds number based on diameter Re = 450,000, and an angle of attack $\alpha = 19°$ in Figs. 12–14. This case was originally reported by Pulliam

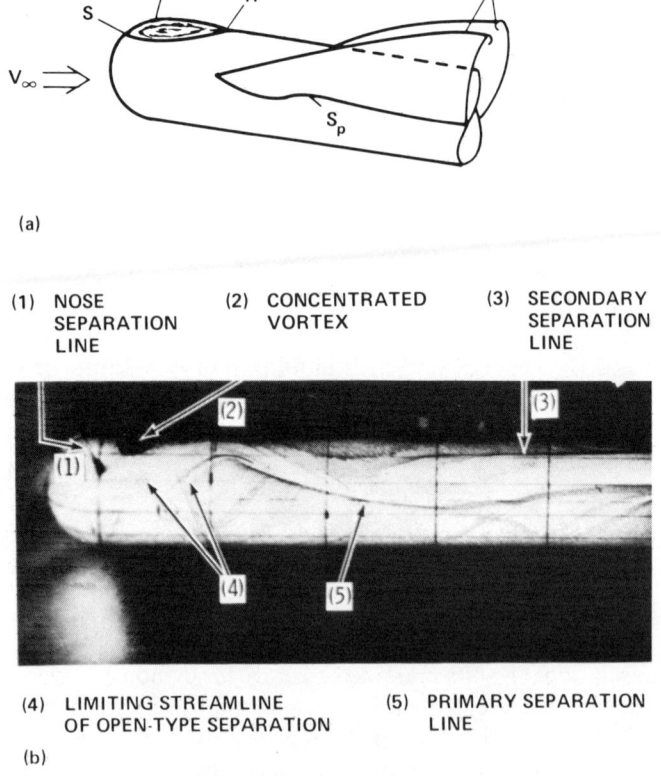

FIG. 12. Characteristic features of hemisphere–cylinder flow field: (a) hemisphere–cylinder at incidence; (b) different forms of separation [from Hseih (1976)].

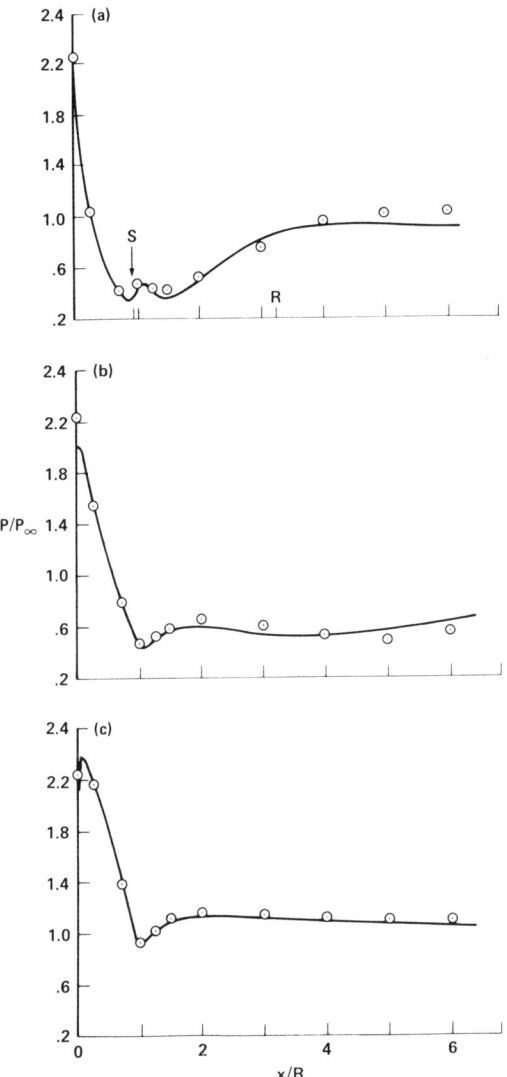

Fig. 13. Laminar flow results for hemisphere–cylinder. $M_\infty = 1.2$; $\alpha = 19°$; $Re_D = 450,000$; ⊙, experimental results; —, calculated results. (a) $\phi = 0°$ leeward, (b) $\phi = 90°$, and (c) $\phi = 180°$ windward.

and Steger (1980); they made the calculations on a CDC 7600 computer, using a code without vector structure. These results were used for comparison in designing the vectorized ILLIAC code. The results presented in this Chapter were all obtained on the ILLIAC computer (Pulliam and Lomax, 1979). The mesh size is 30 × 30 × 21, where 30 points are used in the axial direction, 30 in the radial direction, and 21 circumferentially. The mesh is clustered near the body with a minimum spacing of 0.0002 diameter. The axial spacings are of the order of 0.05 diameter. Figure 12a is a sketch of the body and some of the physical characteristics of the flow. A separation bubble is created at the nose of the body and leeward vortex sheets are shed from the crossflow due to the high angle of attack. An oil-flow picture, Fig. 12b, taken by Hsieh (1976), shows similar features. Computed pressures along axial rays are compared with the experimental data in Fig. 13. Leeward refers to the upper surface and windward refers to the lower surface. The comparison is quite good, even through the streamwise nose separation region. Crossflow velocity vectors are shown in Fig. 14 for an axial station of $x = 3.9$ diameters downstream of the nose. At this high angle of attack, the recirculation region is well developed with a large computed vortex clearly shown. In Pulliam and Lomax (1979), comparisons made of the crossflow separations angles show good agreement for the primary and secondary separations.

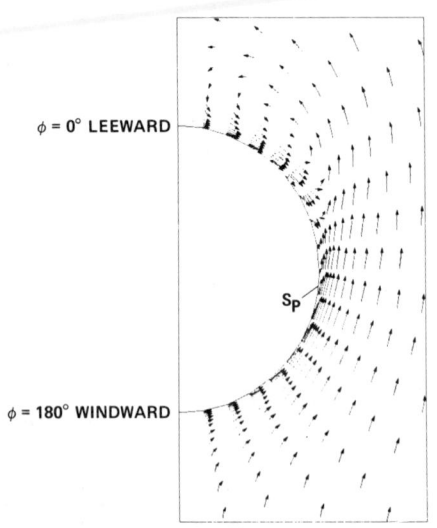

FIG. 14. Crossflow velocity vectors for hemisphere–cylinder flow field. $M_\infty = 1.2; \alpha = 19°$; $Re_D = 450{,}000; x = 3.9$ diam; S_P is the primary separation point.

C. Transonic Flow over Boattail: Turbulent

Transonic flow results, originally reported by Deiwert (1980), are presented for a blunt cone–cylinder–boattail body at a free-stream Mach number $M_\infty = 0.9$, Reynolds number based on the diameter $Re = 2.9 \times 10^6$, and an angle of attack $\alpha = 6°$. The calculations were performed on the ILLIAC and compared with experimental data taken by Shrewsbury (1968). The body geometry is sketched in Fig. 15. The experiment had a solid cylindrical plume simulator; it was simulated in the calculations. The mesh size was $80 \times 48 \times 39$, where 80 points were used in the axial direction, 48 radially, and 39 circumferentially. The clustered grid in the boattail region is shown in Fig. 15b, where the minimum radial spacing is 0.00008 diameter and the minimum axial spacing is 0.0041 diameter, which occurs at the boattail

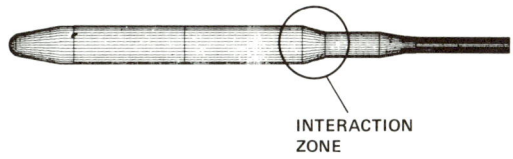

(a) BOATTAIL BODY GEOMETRY

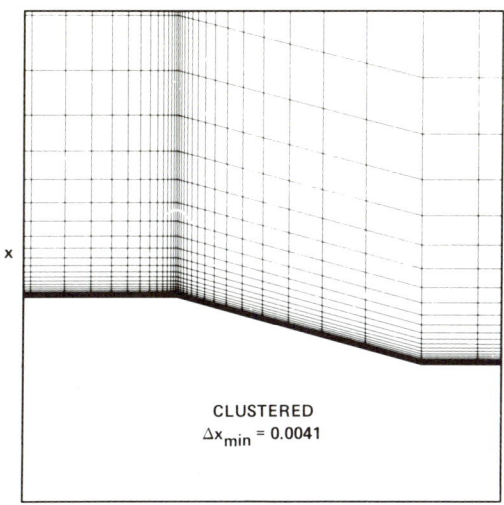

(b) BOATTAIL GRID

FIG. 15. Shrewsbury boattail geometry and grid system: (a) boattail body geometry; (b) boattail grid.

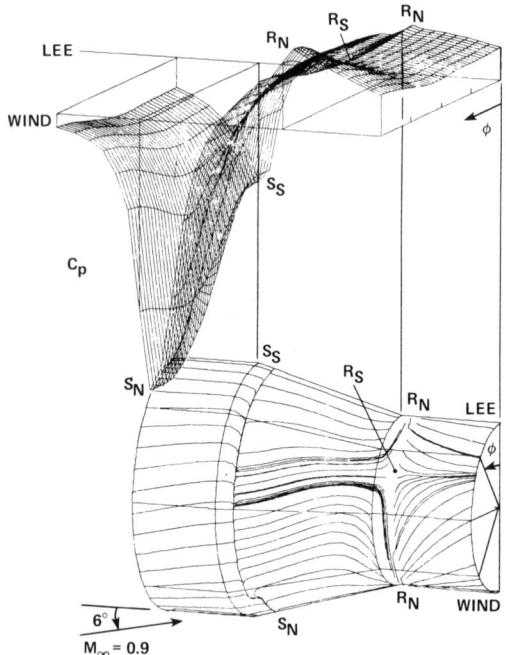

FIG. 16. Surface flow pattern and pressure map for Shrewsbury afterbody: $M_\infty = 0.9$, $Re_{dm} = 1.4 \times 10^6$, $\alpha = 6°$.

corner. A perspective view of the entire flow field in the vicinity of the boattail is shown in Fig. 16. The bottom part of the figure shows the computed surface flow pattern, which corresponds to an experiment oil-flow pattern. The upper part is a three-dimensional pressure surface map corresponding to the surface pressure. A ring of separation (right after the boattail juncture, indicated by S) and reattachment (on the plume simulator, indicated by R) occurs with a recirculation region between. There is a shock/boundary-layer interaction occurring right after the separation line. The nature of the interaction is indicated by the large pressure gradient in the pressure map. Details of this and other calculations and comparisons with experiment are given by Deiwert (1980).

VII. CONCLUDING REMARKS

The solution of three-dimensional, time-accurate problems based on the Reynolds-averaged, Navier–Stokes equations for compressible fluid flows can easily require several million words of storage. In this report we have concentrated on finite-difference mesh geometries that can be mapped into a box in the computational plane so that all boundary conditions can be imposed on the faces of the box. Such problems are ideally suited for fully implicit, factored algorithms with certain forms of turbulence modeling. It is demonstrated that, on the ILLIAC IV, codes based on a pencil data-base system can, when used to solve problems of this type, sustain, for the complete solution process, over 20 million floating-point operations per second on data bases having up to 4 million floating-point words. Similar strategies should be useful for other machines with parallel architectures.

References

Ames Research Center, NASA (1974). CFD: A FORTRAN-based language for ILLIAC IV. Computational Fluid Dynamics Branch, Ames Research Center, National Aeronautics and Space Administration, Moffett Field, California.

Baldwin, B. S., and Lomax, H. (1978). Thin layer approximation and algebraic model for separated turbulent flows. *AIAA Pap.* 78-257.

Beam, R., and Warming, R. F. (1976). An implicit finite-difference algorithm for hyperbolic systems in conservation-law form. *J. Comput. Phys.* **22**, 87–110.

Deiwert, G. S. (1980). Numerical simulation of three-dimensional boattail afterbody flow field. *AIAA Pap.* 80-1347.

Hsieh, T. (1976). An investigation of separated flow about a hemisphere-cylinder at 0- to 90-deg incidence in the mach number range from 0.6 to 1.5. AEDC TR 76-112. Arnold Engineering Development Center, Arnold Air Force Station, Tennessee.

Kim, J., and Moin, P. (1979). Large eddy simulation of turbulent channel flow-ILLIAC IV calculation. *AGARD Conf. Turbulent Boundary Layers, Experiments, Theory Modelling, The Hague, September*, No. 271.

Kutler, P., Chakravarthy, S. R., and Lombard, C. K. (1978). Supersonic flow over ablated nosetips using an unsteady implicit numerical procedure. *AIAA Pap.* 78-213.

Lapidus, A. (1967). A detached shock calculation by second-order finite differences. *J. Comput. Phys.* **2**, 154–177.

Merriam, M. L. (1982). On the inversion of block-tridiagonals without storage constraints. *NASA Tech. Memo* 84228.

Peyret, R., and Viviand, H. (1975). Computation of viscous compressible flows based on the Navier–Stokes equations. *AGARDograph* AGARD-AG-212.

Pulliam, T. H., and Chaussee, D. S. (1981). A diagonal form of an approximate-factorization algorithm. *J. Comput. Phys.* **39**, 347–363.

Pulliam, T. H., and Lomax, H. (1979). Simulation of three-dimensional compressible viscous flow on the ILLIAC IV computer. *AIAA Pap.* 79-206.

Pulliam, T. H., and Steger, J. L. (1980). On implicit finite-difference simulations of three-dimensional flow. *AIAA J.* **18**, 159–167.

Richtmyer, R. D., and Morton, K. W. (1967). "Difference Methods for Initial-Value Problems." Wiley, New York.

Shrewsbury, G. D. (1968). Effect of boattail juncture shape on pressure drag coefficients of isolated afterbodies. *NASA Tech. Memo* TM-X-1517.

Steger, J. L. (1978). Implicit finite-difference simulation of flow about two-dimensional geometries. *AIAA J.* **16**, 679–686.
Steger, J. L., and Bailey, H. E. (1980). Calculation of transonic aileron buzz. *AIAA J.* **18**, 249–255.
Vinokur, M. (1974). Conservation equations of gas dynamics in curvilinear coordinate systems. *J. Comput. Phys.* **14**, 105–125.
Viviand, H. (1974). Conservative forms of gas dynamic equations. *Rech. Aérospat.* No. 1, January–February, 65–68.

A Time-Split Difference Scheme for the Compressible Navier–Stokes Equations with Applications to Flows in Slotted Nozzles

John C. Strikwerda

Computer Sciences Department
University of Wisconsin-Madison
Madison, Wisconsin

I. INTRODUCTION

This paper presents a time-split difference scheme for solving the compressible Navier–Stokes equations and describes the implementation of that scheme on the Control Data Corporation STAR-100 vector processor. Results are presented of computations which use that scheme to compute the laminar flow in converging-diverging nozzles with suction slots. The scheme is highly vectorizable and suitable for pipeline processors. An interesting feature of the application is that the computational grid is not rectangular but rather two rectangular regions joined along part of one side.

The scheme has been used to compute flows in both two-dimensional and axisymmetric nozzles. The nozzles were either conventional converging-diverging Laval nozzles or nozzles with suction slots located ahead of the throat. The purpose of these computations was to aid in the design of the quiet wind tunnel being developed at NASA Langley Research Center (see, e.g., Beckwith, 1975; Anders *et al.*, 1977).

This work appears to be the first that uses the full Navier–Stokes equations to solve for the flow in a nozzle at high Reynolds number. In related work, Cline (1976) has used the full Navier–Stokes equations for nozzles at lower Reynolds numbers. Various researchers have employed the slender-channel approximation for nozzles (see, e.g., Rae, 1971; Mitra and Fiebig, 1975); however, the nozzles treated in this paper are not slender enough to be amenable to slender channel approximations. Thomas (1979) uses the Navier–Stokes equations with the parabolic approximation for three-dimensional flows in nonaxisymmetric nozzles.

A modification of the method described in this paper has also been used to compute the flow at a wing–elevon junction. This work will be described in a forthcoming report by Walsh and Strikwerda (1982).

Time-split difference schemes have been presented by several authors. Strang (1968) and Gottlieb (1972) have discussed time-split difference schemes for general hyperbolic and parabolic systems and MacCormack (1971) and Abarbanel and Gottlieb (1981) presented time-split schemes for the compressible Navier–Stokes equations. The scheme presented here differs from that of MacCormack in that the viscous terms are split from the inviscid terms. The viscous terms are not split further as advocated by Abarbanel and Gottlieb (1981).

II. THE DIFFERENCE SCHEME

The difference scheme for the Navier–Stokes equations is a time-split scheme with three splittings. One splitting encompasses the parabolic or viscous terms and two splittings are for the hyperbolic or inertial terms, one for each direction. (For the three-dimensional equations a fourth splitting would be added for the inertial terms in the extra dimension.) Each of the corresponding operators is of the predictor–corrector type. The total scheme can be described as a set of three operators applied in sequence so as to be consistent and second-order accurate.

Before describing the difference scheme in detail, we discuss time splitting in a more general setting (see also Gottlieb, 1972; Strang, 1968). Consider an evolution equation

$$u_t = Au + Bu + Cu, \qquad (1)$$

where A, B, and C are linear operators of some type. Then, given u at time t_0, to compute u at $t_0 + \Delta t$ we may approximate the above equation by

$$u_t = \begin{cases} 3Au & \text{for } t_0 \le t < t_0 + \tfrac{1}{3}\Delta t \\ 3Bu & \text{for } t_0 + \tfrac{1}{3}\Delta t \le t < t_0 + \tfrac{2}{3}\Delta t \\ 3Cu & \text{for } t_0 + \tfrac{2}{3}\Delta t \le t \le t_0 + \Delta t. \end{cases}$$

A Time-Split Difference Scheme

In this way, the exact solution $u(t_0 + \Delta t)$, which may be written as

$$e^{(A+B+C)\Delta t}u(t_0), \qquad (2)$$

is approximated by

$$e^{C\Delta t}e^{B\Delta t}e^{A\Delta t}u(t_0). \qquad (3)$$

Unless the operators A, B, and C commute the expression (3) is not equal to the exact solution (2), but will be an approximation to within $O(\Delta t^2)$. Strang (1968) has shown that by reversing the order of the splitting in the next time step $u(t_0 + 2\Delta t)$ can be approximated to within $O(\Delta t^3)$, and thus the overall method is second-order accurate in time. When the operators A, B, and C are differential operators and are approximated by difference operators, then the above approach gives rise to a time-split finite difference scheme.

The Navier–Stokes equations for two dimensions may be written as

$$U_t + F_x + G_y - V = 0,$$

where

$$U = \begin{bmatrix} \rho \\ \rho u \\ \rho v \\ E \end{bmatrix}, \quad F = \begin{bmatrix} \rho u \\ \rho u^2 + p \\ \rho uv \\ (E+p)u \end{bmatrix}, \quad G = \begin{bmatrix} \rho v \\ \rho uv \\ \rho v^2 + p \\ (E+p)v \end{bmatrix},$$

and

$$V = \begin{bmatrix} 0 \\ \dfrac{\partial}{\partial x}\tau_{xx} + \dfrac{\partial}{\partial y}\tau_{xy} \\ \dfrac{\partial}{\partial x}\tau_{xy} + \dfrac{\partial}{\partial y}\tau_{yy} \\ \dfrac{\partial}{\partial x}(u\tau_{xx} + v\tau_{xy}) + \dfrac{\partial}{\partial y}(u\tau_{xy} + v\tau_{yy}) + \dfrac{c_p}{\sigma}\left(\dfrac{\partial}{\partial x}\left(\mu\dfrac{\partial T}{\partial x}\right) + \dfrac{\partial}{\partial y}\left(\mu\dfrac{\partial T}{\partial y}\right)\right) \end{bmatrix}$$

where

$$\tau_{xx} = \frac{4}{3}\mu\frac{\partial u}{\partial x} - \frac{2}{3}\mu\frac{\partial v}{\partial y},$$

$$\tau_{xy} = \mu\frac{\partial u}{\partial y} + \mu\frac{\partial v}{\partial x},$$

$$\tau_{yy} = \frac{4}{3}\mu\frac{\partial v}{\partial y} - \frac{2}{3}\mu\frac{\partial u}{\partial x}.$$

The equation of state is

$$p = \rho T c_v (\gamma - 1),$$

and the viscosity is given by Sutherland's law.

The energy E is defined by

$$E = \rho(c_v T + \tfrac{1}{2}(u^2 + v^2)).$$

The gas constant γ is 1.4; the Prandtl number σ is 0.72; c_v is the coefficient of specific volume. The vectors F and G represent the inertial effects of the flow and V represents the viscous forces. The three splittings used in the scheme correspond to the terms F_x, G_y, and V.

Considering first the viscous terms, the system to be integrated for a step of size $\tfrac{1}{3}\Delta t$ can be written

$$\tfrac{1}{3}\rho_t = 0,$$

$$\tfrac{1}{3}\rho u_t = \tfrac{4}{3}(\mu u_x)_x + (\mu u_y)_y - \tfrac{2}{3}(\mu v_y)_x + (\mu v_x)_y,$$

$$\tfrac{1}{3}\rho v_t = (\mu u_y)_x - \tfrac{2}{3}(\mu u_x)_y + (\mu v_x)_x + \tfrac{4}{3}(\mu v_y)_y, \quad (4)$$

$$\tfrac{1}{3}\rho c_v T_t = (kT_x)_x + (kT_y)_y + \Phi,$$

where

$$\Phi = 2\mu\{u_x^2 + v_y^2 - \tfrac{1}{3}(u_x + v_y)^2\} + \mu(u_y + v_x)^2, \qquad k = c_p \mu/\sigma.$$

The subscripts t, x, and y denote differentiation.

We now rewrite the equations in terms of the independent variables (ξ, η). The mapping from the physical (x, y) coordinates to the computational (ξ, η) coordinates is described in more detail in the appendix, but for now it is sufficient to write it as

$$\xi = \xi(x), \qquad \eta = \eta(x, y).$$

In the transformed coordinates the equation for u is

$$\tfrac{1}{3}\rho u_t = \tfrac{4}{3}\xi_x(\mu\xi_x u_\xi)_\xi + \tfrac{4}{3}\eta_x(\mu\xi_x u_\xi)_\eta$$
$$+ \tfrac{4}{3}\xi_x(\mu\eta_x u_\eta)_\xi + \tfrac{4}{3}\eta_x(\mu\eta_x u_\eta)_\eta + \eta_y(\mu\eta_y u_\eta)_\eta$$
$$- \tfrac{2}{3}\xi_x(\mu\eta_y v_\eta)_\xi - \tfrac{2}{3}\eta_x(\mu\eta_y v_\eta)_\eta + \eta_y(\mu\xi_x v_\xi)_\eta + \eta_y(\mu\eta_x v_\eta)_\eta. \quad (5)$$

The equations for v and T are also transformed. The differencing of the above equations will be illustrated using only the first two terms on the right-hand side of the above equation since the other terms are differenced in a similar manner. The first term is differenced as

$$\tfrac{4}{3}(\xi_x)_i \left\{ (\mu\xi_x)_{i+1/2, j}\left(\frac{u_{i+1,j} - u_{i,j}}{\Delta\xi}\right) - (\mu\xi_x)_{i-1/2, j}\left(\frac{u_{i,j} - u_{i-1,j}}{\Delta\xi}\right) \right\} \bigg/ \Delta\xi, \quad (6)$$

where

$$(\mu\xi_x)_{i+1/2,j} = \tfrac{1}{2}(\mu_{i,j}(\xi_x)_i + \mu_{i+1,j}(\xi_x)_{i+1})$$

and

$$(\xi_x)_i = \xi_x(\xi_i), \qquad \mu_{i,j} = \mu(T_{i,j}).$$

The second term on the right-hand side of Eq. (5) is differenced as

$$\tfrac{4}{3}(\eta_x)_{i,j}\left\{(\mu\xi_x)_{i,j+1}\left(\frac{u_{i+1,j+1} - u_{i-1,j+1}}{2\,\Delta\xi}\right)\right.$$
$$\left. - (\mu\xi_x)_{i,j-1}\left(\frac{u_{i+1,j-1} - u_{i-1,j-1}}{2\,\Delta\xi}\right)\right\}\frac{1}{2\,\Delta\eta}. \qquad (7)$$

Denote the difference operator for the right-hand side of equation (4) by V; then the algorithm for the viscous terms can be written as

$$\begin{aligned}\overline{W}_{i,j} &= W_{i,j} + \Delta t\, V_{i,j}(W), \\ W^*_{i,j} &= \tfrac{1}{2}\{W_{i,j} + \overline{W}_{i,j} + \Delta t\, V_{i,j}(\overline{W})\},\end{aligned} \qquad (8)$$

where

$$W = \begin{bmatrix} \rho \\ u \\ v \\ T \end{bmatrix}.$$

Note that the factor of $\tfrac{1}{3}$ on the left-hand side of Eq. (4) cancels with the $\tfrac{1}{3}$ factor from the time step of $\tfrac{1}{3}\Delta t$. This defines the viscous splitting operator $S_V(\Delta t)$, i.e.,

$$W^* = S_V(\Delta t)W. \qquad (9)$$

The hyperbolic portion of the splitting for the two-dimensional equations can be written

$$\tfrac{2}{3}U_t + F_x + G_y = 0, \qquad (10)$$

where

$$U = \begin{bmatrix} \rho \\ \rho u \\ \rho v \\ E \end{bmatrix}, \qquad F = \begin{bmatrix} \rho u \\ \rho u^2 + p \\ \rho uv \\ (E+p)u \end{bmatrix}, \qquad G = \begin{bmatrix} \rho v \\ \rho uv \\ \rho v^2 + p \\ (E+p)v \end{bmatrix}.$$

By the coordinate transformation the equations in the (ξ, η) coordinates become

$$\tfrac{2}{3} U_t + \xi_x F_\xi + \eta_x F_\eta + \eta_y G_\eta = 0. \tag{11}$$

Equation (11) is split into the two one-dimensional systems

$$\tfrac{1}{3} U_t + \xi_x F_\xi = 0 \tag{12}$$

and

$$\tfrac{1}{3} U_t + \eta_x F_\eta + \eta_y G_\eta = 0. \tag{13}$$

The difference scheme for Eq. (12) is

$$\begin{aligned}
\bar{U}_{i,j} &= U_{i,j} - \frac{\Delta t}{\Delta \xi} (\xi_x)_i (F_{i+1,j} - F_{i,j}), \\
U^*_{i,j} &= \frac{1}{2} \left\{ U_{i,j} + \bar{U}_{i,j} - \frac{\Delta t}{\Delta \xi} (\xi_x)_i (\bar{F}_{i,j} - \bar{F}_{i-1,j}) \right\},
\end{aligned} \tag{14}$$

where $\bar{F}_{i,j} = F(\bar{U}_{i,j})$. The difference scheme for Eq. (13) is

$$\begin{aligned}
\bar{U}_{i,j} &= U_{i,j} - \frac{2 \Delta t\, a^-_{i,j}}{y_{i,j+1} - y_{i,j-1}} (b_{i,j}(F_{i,j} - F_{i,j-1}) + G_{i,j} - G_{i,j-1})), \\
U^*_{i,j} &= \frac{1}{2} \bigg(U_{i,j} + \bar{U}_{i,j} - \frac{2 \Delta t\, a^+_{i,j}}{y_{i,j+1} - y_{i,j-1}} \\
&\quad \times (b_{i,j}(\bar{F}_{i,j+1} - \bar{F}_{i,j}) + (\bar{G}_{i,j+1} - \bar{G}_{i,j})) \bigg),
\end{aligned} \tag{15}$$

where

$$\bar{F}_{i,j} = F(\bar{U}_{i,j}), \qquad \bar{G}_{i,j} = G(\bar{U}_{i,j}),$$

and

$$b_{i,j} = \left.\frac{\eta_x}{\eta_y}\right|_{i,j}, \qquad a^+_{i,j} = 1/a^-_{i,j} = (y_{i,j} - y_{i,j-1})/(y_{i,j+1} - y_{i,j}).$$

Note that the scheme (15) is second-order accurate in time and space.

Equations (14) and (15) define the splitting operators $S_\xi(\Delta t)$ and $S_\eta(\Delta t)$, respectively. The total scheme is then defined by

$$\begin{aligned}
U^{2n+1}_{i,j} &= T^{-1} S_V(\Delta t) T S_\eta(\Delta t) S_\xi(\Delta t) U^{2n}_{i,j}, \\
U^{2n+2}_{i,j} &= S_\xi(\Delta t) S_\eta(\Delta t) T^{-1} S_V(\Delta t) T U^{2n+1}_{i,j}
\end{aligned} \tag{16}$$

where T is the transformation from the conserved variables $U = (\rho, \rho u, \rho v, E)'$ to the physical variables $W = (\rho, u, v, T)'$, i.e., $W = T(U)$. Note that the operator sequences in (16) are reversed in successive time steps; this is done to maintain the overall second-order accuracy in time. The stability for each scheme of the splitting requires that the time step for the scheme be no larger than the largest allowable time step for each splitting. Let

$$\Delta t_\xi = \min_{i,j} \frac{\Delta \xi}{(\xi_x)_i(|u_{i,j}| + c_{i,j})},$$

$$\Delta t_\eta = \min_{i,j} \frac{(y_{i,j+1} - y_{i,j-1})/2}{|v_{i,j}| + |\eta_x/\eta_y|_{i,j}|u_{i,j}| + c_{i,j}\sqrt{1 + (\eta_x/\eta_y)^2_{i,j}}},$$

$$\Delta t_V = \min_{i,j} \frac{\sigma \rho_{i,j}}{2\gamma \mu_{i,j}} \left(\left(\frac{\xi_{x,i}}{\Delta \xi}\right)^2 + \left(\frac{\eta_{y,i,j}}{\Delta \eta}\right)^2 + \left(\frac{\eta_{x,i,j}}{\Delta \eta}\right)^2\right)^{-1},$$

where $c_{i,j} = \sqrt{\gamma p_{i,j}/\rho_{i,j}}$ is the local sound speed. The time step Δt was chosen as

$$\Delta t = 0.9 \min(\Delta t_\xi, \Delta t_\eta, \Delta t_V).$$

Because of the presence of the transformation T in the scheme (16), the stability of the overall scheme does not follow immediately from the stability of each of the splitting schemes. However, Abarbanel and Gottlieb (1981) show that the linearized Navier–Stokes equations can be transformed to a system which is symmetric. In this symmetric form it is easily seen that the linearization of the time-split scheme (16) is stable since each of the splittings is stable (Abarbanel and Gottlieb, 1981). Further splitting of the parabolic operator as advocated by Abarbanel and Gottlieb would also be efficient on a vector processor.

III. THE APPLICATION

A sketch of the slotted nozzle for which computations were made is given in Fig. 1. In the converging portion of the nozzle the flow is subsonic and accelerating. The flow becomes supersonic in the throat region of the nozzle and continues accelerating in the diverging portion. It has been shown experimentally by Anders *et al.* (1977) that the turbulent boundary layer on the tunnel wall propagates disturbances which interfere with measurements

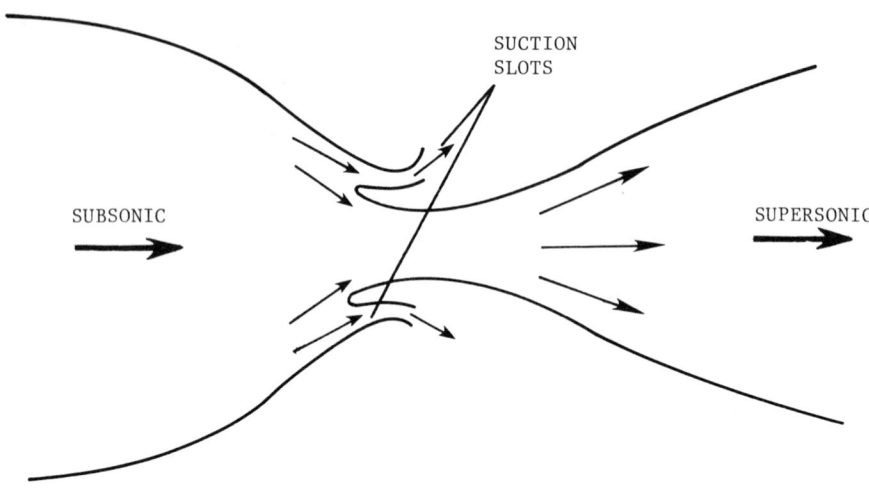

Fig. 1. A diagram of a two-dimensional nozzle with suction slots.

in the test section. The suction slot removes the upstream boundary layer so that a new boundary layer which remains laminar to a much higher Reynolds number begins at the slot lip. The disturbance level of the flow in the test section is then considerably reduced.

The slot is so designed that the flow within it quickly becomes supersonic, and for the computations the outflow boundary in the slot was chosen so that the flow there would be supersonic.

The computational grid for an inviscid flow is shown in Fig. 2. A finer grid spacing was used near the walls to resolve the boundary layers for viscous flow computations. The computational domain consisted of only the area above the centerline since the flow was assumed to be symmetric about the centerline.

The boundary conditions for the differential equations are as follows. At inflow, the density, temperature, and flow angle were specified, i.e.,

$$\rho = \rho_0, \qquad T = T_0, \qquad \text{and} \qquad u = v \tan \theta(y), \tag{17}$$

where $\theta(y)$ is a specified function. Three boundary conditions are appropriate at a subsonic inflow boundary (Oliger and Sundstrom, 1978). At outflow the flow is supersonic and no quantities are specified. Along the wall of the tunnel the no-slip and adiabatic conditions were used, i.e.,

$$u = v = 0 \qquad \text{and} \qquad \partial T/\partial n = 0.$$

Fig. 2. Grid lines for the slotted nozzle calculation. Insert shows slot region enlarged.

The above inflow boundary conditions were chosen because from a one-dimensional analysis of steady nozzle flow (Courant and Friedrichs, 1948, pp. 377–387) the Mach number at inflow, which is proportional to $u/\sqrt{T}$, and the mass flux at inflow, which is the integral of ρu, are determined by the conditions at the nozzle throat and the cross-sectional area ratios. Thus, if the inflow velocity component u were prescribed along with either T or ρ, then a steady flow may not have developed. Also, physically the steady flow is determined by the temperature and pressure, and hence density, of the fluid far upstream of the throat. Therefore, among all choices of three physical variables to specify at the inflow that given by (17) is most natural.

For the difference approximation, all the variables at the outflow boundary were determined by extrapolation from the interior. At the inflow boundary, the velocity component u was determined by extrapolation and then the component v was computed by Eq. (17). Along the wall of the tunnel the density was determined by extrapolation.

IV. THE IMPLEMENTATION

To utilize the capabilities of the STAR to their best advantage, the problem was organized so that as much of the computation as possible was done through long vector operations. The scheme presented in Section II when applied to the problem of Section III is quite suitable for a vector processor.

The problem had the advantage of being able to fit entirely in core memory on the STAR and thus avoided the difficulties of paging data from the virtual memory. The language used to encode the program was SL/1, which was developed at NASA Langley Research Center by Knight (1979).

To implement the difference scheme (16) on the STAR the values of each variable, ρ, ρu, ρv, E, etc., were assigned to a vector of length L equal to the total number of grid points. Thus, referring to Fig. 3, which illustrates the computational domain, $\rho(1), \rho(2), \ldots, \rho(IA), \rho(IA + 1), \ldots$ are the values of the density at the grid points $a_1, a_2, \ldots, a_{IA}, a_{IA+1}, \ldots$. There are $LA = IA \times JA$ points in the lower region, region A, and $LB = IB \times JB$ points in the upper region, region B. The points $b_1, \ldots, b_{IC}$ represent the same physical points as the points $a_k, \ldots, a_{k+IC-1}$, but are distinct in the computational domain and correspond to distinct vector locations. This line of grid points which is common to both regions and is doubly represented is called the common line. Note that there is no special differencing used at these points and the grid coordinate mapping is smooth across this line.

A vector then consists of $L = LA + LB$ locations, one location for each point in region A and region B.

In addition to vectors whose elements take on numerical values, the STAR also elmploys bit vectors. Bit vectors are vectors whose elements take on logical values and are used to mask out operations. Thus if the vectors A and B are to be added under the control of the bit vector b to give the result C, then when the kth element of b is FALSE the sum of the kth elements of A and B is computed but is not stored in the kth location of C. If the kth element of b is TRUE, then the sum is computed and stored in the kth location of the vector C.

Consider now the simplest splitting (14). Each component of the flux F is represented by a vector and is computed by vector instructions. Then the forward difference is taken for all points from index 1 to $L - 1$ and the result is multiplied by $\xi_x \Delta t / \Delta \xi$. Each of these are vector operations of length $L - 1$. However, for those indices k for which the $(k + 1)$th grid point is not a neighbor of the kth grid point, the "forward" difference taken

A Time-Split Difference Scheme 261

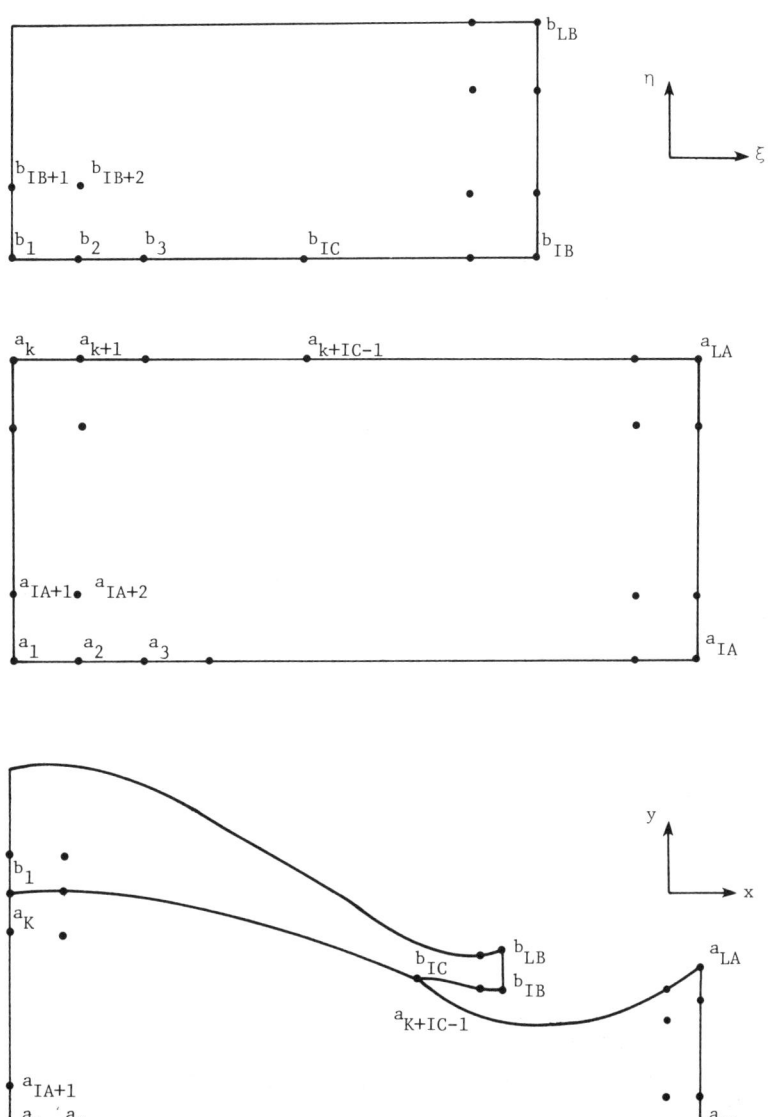

FIG. 3. A schematic diagram of the computational grid and mapping discussed in Section IV and the Appendix.

between the fluxes at the indices k and $k + 1$ is a physically meaningless quantity. Thus when the results of the forward difference operation are added to the value of the components of U, a bit vector is employed to mask out those additions involving physically meaningless quantities. By this means the predicted quantities $\bar{U}$ are computed by Eq. (14). Those elements of $\bar{U}$ for which the results of (14) where masked by the bit vector are computed by different means such as extrapolation or leaving the value unchanged if it is a specified boundary value. For example, if zeroth-order extrapolation is used, then the components of $\bar{U}$ at the kth location are set equal to the values of $\bar{U}$ at the $(k - 1)$th location under the control of the bit vector which is the negation of that mentioned earlier. The corrector portion of Eq. (14) is done similarly.

A computation such as the extrapolation in the example just mentioned in which all but a small percentage of the points are masked may be considered wasteful. However, on the STAR-100A the alternative is to employ scalar operations, which is quite slow. Thus these "wasteful" operations may be the fastest. On the CYBER 203, which has a much faster speed for scalar arithmetic, it is almost certainly more economical to handle the boundary conditions with scalar operations.

The splitting (15) is implemented in a similar fashion, but with the difference that the forward and backward differences in the η variable require three vector operations due to the geometry of the computational domain. For a forward difference, the first operation is for the lower region A and involves points whose indices in the vector differ by IA. This is because grid points in region A which are neighbors in the η direction correspond to vector locations a distance of IA apart. This operation is valid for all points of region A except the top row. The second vector operation is to do the same thing for region B, but here the neighbors occupy locations which are IB apart in the vector.

The third operation is to set the values for the points of the common line which are in part A of the vector, and not computed in the first operation, equal to the corresponding points in part B of the vector which were computed in the second operation. These vector operations are of lengths $LA - IA$, $LB - IB$, and IC, respectively. The remaining points are again treated by an extrapolation.

To implement the viscous splitting operator required a large number of temporary vectors. Therefore care was taken so that few, if any, temporary vectors were generated by the compiler to evaluate complicated expressions. Thus all computations to compute $V(W)$ in Eq. (8) were written to involve only two operands. Also the terms such as those in Eq. (6) were computed in such an order so as to minimize as much as possible the number of temporary vectors that had to be carried along in the computation. For example, the values of $(\mu\xi_x)_{i+1/2, j}$ were computed and stored in a temporary vector and all those divided differences requiring $(\mu\xi_x)_{i+1/2, j}$ were computed and stored in a temporary vector and all those divided differences requiring $(\mu\eta_x)_{i+1/2, j}$ were computed and added to W. Next $(\mu\eta_x)_{i, j+1/2}$ was computed and stored in the same temporary vector which had stored $(\mu\eta_x)_{i+1/2, j}$. After all divided differences requiring $(\mu\eta_x)_{i, j+1/2}$ were computed and added to W, the temporary vector was used for another quantity. Through the use of the LITERALLY statement in SL/1 this same temporary vector could be given a different appropriate name each time its use changed. In this way the viscous splitting (8) could be implemented without using an excess number of temporary vectors which may have caused the storage to exceed the core storage.

As with the η splitting most of the differences in the viscous splitting involved three vector operations, one for region A, one for region B, and one for the common line. The program had to be well documented with comments to make it understandable.

The calculations were performed using 32-bit arithmetic. The standard word size on the STAR is 64 bits, but through the use of SL/1 the half-word arithmetic can be used giving a substantial increase of speed without a serious loss of accuracy.

V. RESULTS

The numerical method and computer code described here have been used in design studies for several nozzles, both two-dimensional and axisymmetric. We present here only the results for a particular two-dimensional slotted nozzle.

The nozzle contour with the numerically generated grid is shown in Fig. 2. The total number of grid points was 12,006. The stagnation pressure and temperature were 3600 psf and 520°R, respectively, and the Reynolds number was 3.2×10^5, being evaluated at the sonic line on the center line using the throat radius as the reference length.

The computations were started with a flow field derived from an inviscid one-dimensional analysis (see, e.g., Courant and Friedrichs, 1948, pp. 377–387) with a correction for a boundary layer. The computations were stopped when the maximum relative change in the density between two successive time steps was sufficiently small or when the solution appeared to be nearly converged in the main portion of the nozzle. For the results to be discussed, this was achieved in 40,000 time steps in several stages of 2000 or 8000 steps each. The cpu time per time step per grid point was about 1.1×10^{-5} sec on the CYBER-203.

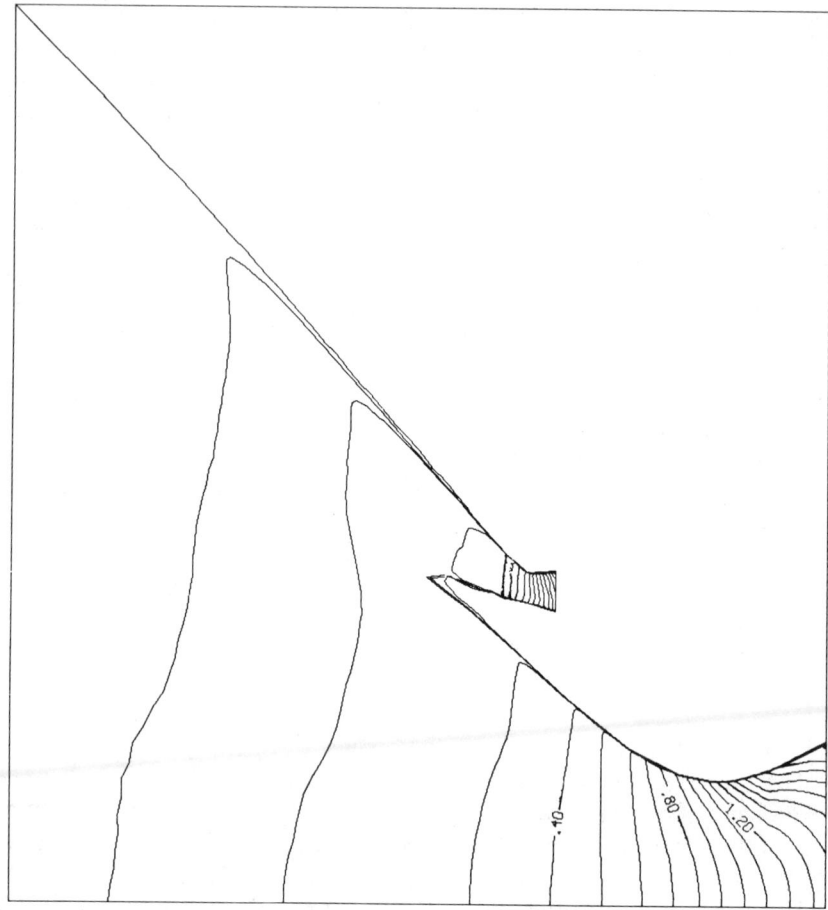

Fig. 4. Contour plot of Mach number.

The Mach number contours for this solution are shown in Fig. 4. The Mach number varied from about 0.01 at the inflow to 2.0 at the main outflow and 1.7 at the slot outflow. Figure 5 shows the streamfunction computed for this run. The streamfunction was computed as an approximation to

$$\psi = \int_0^y \rho u \, dy.$$

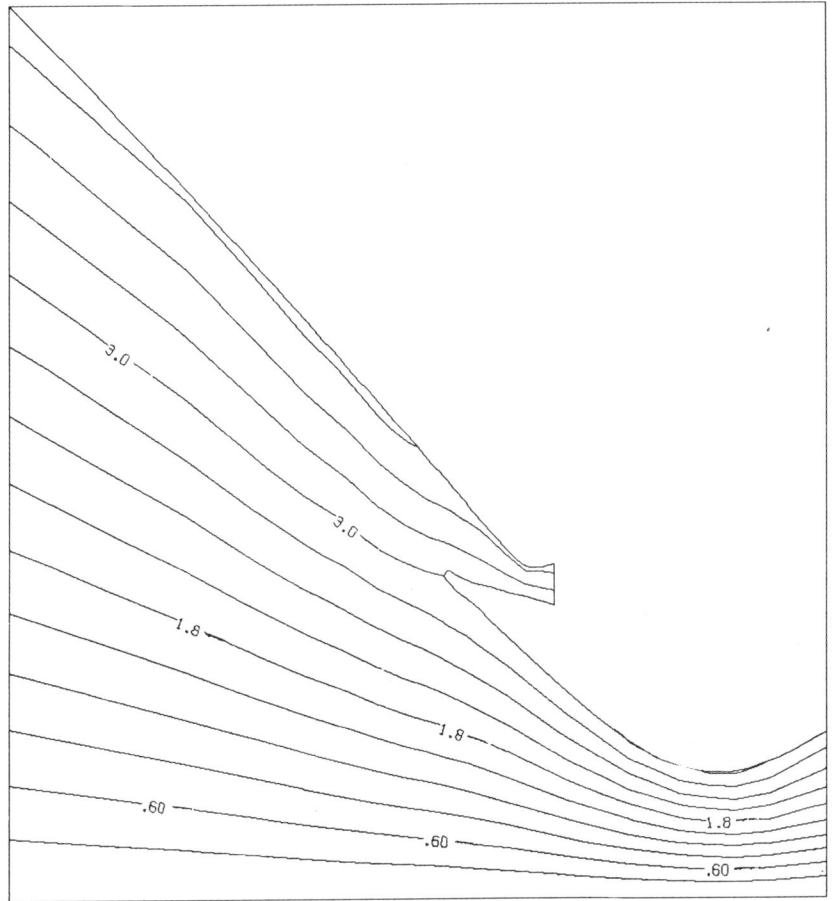

Fig. 5. Contour plot of the streamfunction.

APPENDIX. NUMERICAL GRID GENERATION

The procedure for numerically computing the grid for the slotted nozzle calculations did not use the vector processor, but for completeness it will be briefly described.

The numerical method for the grid generation is essentially that of Thompson *et al.* (1974). The basic idea is to map the computational domain consisting of rectangles joined along a portion of one side onto the physical domain (see Fig. 3). The mapping function $x(\xi)$ was specified as a piecewise polynomial defined so that $x'(\xi)$ was small where clustering of grid lines was desired and $x''(\xi)$ was continuous. The function $y(\xi, \eta)$ was determined by requiring that the inverse function $\eta(x, y)$ satisfy

$$\frac{\partial^2 \eta}{\partial x^2} + \frac{\partial^2 \eta}{\partial y^2} = f(\xi, \eta) \tag{A.1}$$

on the region surrounded by the contour with η specified on the boundaries. The function $f(\xi, \eta)$ will be described later.

If the Jacobian J of the mapping is nonvanishing, then Eq. (A.1) can be transformed to

$$\alpha y_{\xi\xi} - 2\beta y_{\xi\eta} + \gamma y_{\eta\eta} + J^2(\xi_{xx} y_\xi + f(\xi, \eta) y_\eta) = 0 \tag{A.2}$$

on the computational domain illustrated in Fig. 3. The coefficients in Eq. (A.2) are given by

$$\alpha = y_\eta^2, \quad \beta = y_\xi y_\eta, \quad \gamma = x_\xi^2 + y_\xi^2, \quad \text{and} \quad J = x_\xi y_\eta,$$

and note that $\xi_{xx} = -x_{\xi\xi}/(x_\xi)^3$.

The values of y were specified on the boundary, and thus Eq. (A.2) is a nonlinear elliptic boundary value problem with Dirichlet boundary data.

Equation (A.2) was solved by an iterative procedure similar to SOR but with a variable iteration parameter. The variable parameter was required because of the high degree of stretching in the grid transformation. The procedure is discussed in Strikwerda (1980).

The function $f(\xi, \eta)$ was determined as follows. On the boundaries on which x is constant, $\xi = \xi_0$ and f was specified by

$$f(\xi_0, \eta) = \eta_{yy} = -y_{\eta\eta}/(y_\eta)^3. \tag{A.3}$$

This formula is obtained by observing that at these boundaries it is desirable to have the coordinate lines to be nearly straight, i.e., $\eta_{xx} = 0$. If $\eta_{xx} = 0$, then from (A.1) we obtain (A.3). Along the boundaries $\eta = 0$ and $\eta = 1$, the value of f was set to zero. To cluster the grid points near the lip of the slot a large positive value was assigned to f at the grid point at the lip. The function $f(\xi, \eta)$ was then determined by a polynomial interpolation from these boundary points.

By adjusting the value of f at the tip, the polynomial interpolation, and the relative number of grid points in regions A and B, satisfactory grids were obtained quite easily.

Acknowledgments

The author would like to acknowledge the programming assistance of Geoffrey Tennille and the technical assistance of Ivan Beckwith. Work was supported under NASA Contract No. NAS1-15810 while the author was in residence at ICASE, NASA Langley Research Center, Hampton, Virginia 23665.

References

Abarbanel, S., and Gottlieb, D. (1981). Optimal time splitting for two and three dimensional Navier–Stokes equations with mixed derivatives. *J. Comput. Phys.* **41**, 1–33.

Anders, J. B., Stainback, P. C., Keefe, L. R., and Beckwith, I. E. (1977). Fluctuating disturbances in a Mach 5 wind tunnel. *AIAA J.* **15**, 1123–1129.

Beckwith, I. E. (1975). Development of a high Reynolds number quiet tunnel for transition research. *AIAA J.* **13**, 300–306.

Cline, M. C. (1976). Computation of two-dimensional, visocus nozzle flow. *AIAA J.* **14**, 295–296.

Courant, R., and Friedrichs, K. O. (1948). "Supersonic Flow and Shock Waves." Wiley (Interscience), New York.

Gottlieb, D. (1972). Strang type difference schemes for multidimensional problems. *SIAM J. Numer. Anal.* **9**, 650–661.

Knight, J. C. (1979). SL/1 Reference Manual. Analysis and Computation Division, NASA Langley Research Center, Hampton, Virginia.

MacCormack, R. W. (1971). Numerical solution of the interaction of a shock wave with a laminar boundary layer. *Lect. Notes Phys.* **8**, 151.

Mitra, N. K., and Fiebig, M. (1975). Low Reynolds number hypersonic nozzle flow. *Z. Flugwiss.* **23**, 39–45.

Oliger, J., and Sundstrom, A. (1978). Theoretical and practical aspects of some initial boundary value problems in fluid dynamics. *SIAM J. Appl. Math.* **35**, 419–446.

Rae, W. J. (1971). Some numerical results on viscous low-density nozzle flows in the slender-channel approximation. *AIAA J.* **9**, 811–820.

Strang, G. (1968). On the construction and comparison of difference schemes. *SIAM J. Numer. Anal.* **5**, 506–517.

Strikwerda, J. (1980). Iterative methods for the numerical solution of second order elliptic equations with large first order terms. *SIAM J. Sci. Statist. Comput.* **1**, 119–130.

Thomas, P. D. (1979). Numerical method for predicting flow characteristics and performance of nonaxisymmetric nozzles theory. *NASA [Contract. Rep.] CR* **NASA-CR-3147**.

Thompson, J. F., Thames, F. C., and Mastin, C. W. (1974). Automatic numerical generation of body-fitted curvilinear coordinate system for fields containing any number of arbitrary two-dimensional bodies. *J. Comput. Phys.* **15**, 299–319.

Walsh, J., and Strikerda, J. (1982). Numerical solutions of compressible Navier–Stokes equations for two-dimensional, laminar, hypersonic flow at a wing-elevon junction. To appear.

Geophysical Fluid Simulation on a Parallel Computer

James G. Welsh

The Geophysical Fluid Dynamics Laboratory
The National Oceanic and Atmospheric Administration
Princeton University
Princeton, New Jersey

I. INTRODUCTION

In 1972 the Geophysical Fluid Dynamics Laboratory of the National Oceanic and Atmospheric Administration signed a contract with Texas Instruments, Inc., for delivery of a multipipe Advanced Scientific Computer (ASC). Although the delivery of the computer was delayed approximately one year, the ASC became the sole computing resource for the Laboratory in the summer of 1975, and has remained so ever since. Recognizing that traditional FORTRAN programs are not especially written in a form that states the natural parallelism of the problem, the contract for the ASC also included 25 man-years of program conversion effort to convert the existing Laboratory codes to a more efficient, highly vectorizable form of ASC FORTRAN. This chapter relates the experience of converting this large body of programs to a parallel computer and assesses the resulting vectorization that has been achieved.

II. THE SALIENT CHARACTERISTICS OF THE ASC

The ASC is a multipipe, vector processor. Our ASC has four pipes, the maximum. Each pipe executes a separate vector or scalar instruction. A vector instruction passes many consecutive sets of operands from memory to a pipe and returns a corresponding consecutive set of results to memory. A scalar instruction, on the other hand, passes only one pair of operands, one from a register and the other from memory, through a pipe and usually returns the result to the same register. The difference in processing speeds between a vector instruction and a comparable sequence of scalar instructions is about tenfold. With four pipes, all processing vectors, computation can proceed up to 40 times faster than the same computation can proceed with scalar instructions. Thus there is considerable incentive to vectorize a program.

Each pipe can be configured for any of the operations in the pipe repertoire, but multiple vector instructions must actually be issued to keep all pipes busy. Since each pipe requires a separate vector instruction, and since the Instruction Processing Unit issuing these instructions can issue only one vector instruction at a time, it takes four consecutive vector setup times to get all four pipes busy.

The vector operands on the ASC are not restricted to one-dimensional vectors that occupy ascending storage locations. The operands can be two- or three-dimensional. Only the first, innermost dimension of such arrays must correspond either to consecutive storage locations (forward or backward) or to the same (nonvarying) location. If a vector operand cannot occupy such consecutive or nonvarying locations, its first dimension is one, and it may therefore contain only one or two further dimensions. In this case the fetching of operand data from memory cannot proceed at the maximum bandwidth of the memory. The practical effect of multidimensional vector operands is that the effective length of a vector instruction is determined by the product of the dimensions, and not solely by the innermost dimension.

III. THE FORTRAN COMPILER ON THE ASC

The FORTRAN compiler on the ASC analyzes standard DO loop structures to use parallel, vector instructions, instead of scalar instructions, as much as possible to carry out the calculation. In a multiple-pipe ASC, the compiler must also try to keep vectors busy in all pipes concurrently. To that end, when only one vector instruction can be found, the compiler will attempt to partition it into sufficient vector pieces to fill all pipes. A broad constraint on vectorizing DO loops is that calling, branching, or IF statements not be present. Although there are hardware vector instructions that implement selective storing, and therefore, a certain large class of conditions within DO loops, these instructions are not generated by the compiler from conventional FORTRAN syntax. Further constraints on vectorizing statements within DO loops are that

(a) the result of an assignment be a subscripted array, and
(b) all subscripts in the statement be nonrecursive, linear functions of the DO variable.

A common practice on conventional machines is to compute intermediate results as local, undimensioned variables in a DO loop; such a practice will usually thwart vectorization.

The FORTRAN compiler on the ASC includes language extensions for arrays. Array assignment statements are allowed: the result of an assignment may be either a nonsubscripted array name or an array cross section, which is an array with a subscript of one or more *s. The expression on the right side of an array assignment can also contain array cross sections. It is not possible to define an ARRAY FUNCTION capability, but the compiler recognizes certain intrinsic functions that accept array cross sections as arguments and may produce an array cross section as a result. These functions correspond to specific hardware instructions. Typical of such functions are

SUM, which computes the sum of an array cross section,
DOTPRD, which computes the inner sum of two array cross sections,
MAXVAL, which returns the index of the maximum value of an array cross section,
MINVAL, which returns the index of the minimum value of an array cross section.

These extensions provide considerable convenience to programmers, but their use (except for MAXVAL and MINVAL) is not obligatory for achieving vectorization. In fact, the compiler generally converts these extensions into standard FORTRAN structure internally, which is then scanned for vectorizability. Using the extensions merely ensures that the vectorizing algorithm will be successful with such a statement. In the following sections I shall describe how successfully a geophysical model can be brought into the ASC FORTRAN syntax that allows vectorization.

IV. THE PHYSICAL PROCESSES OF A MODEL

The scientific endeavor of the Laboratory is basic research in the numerical modeling of meteorological and oceanic dynamic processes, with particular emphasis on global, long-term phenomena. In a general sense, the real-life physical processes that are modeled do proceed in parallel in the geophysical fluids, so we expect to state the programs that simulate these processes in a parallel form of suitable FORTRAN. When these physical processes are cast into numerical form, subject to the practical constraints of existing computing machines, however, there are a number of processes that seem to resist computer parallelism.

Geophysical fluid models have traditionally used an Eulerian formulation of the Navier–Stokes equations because of its simplicity over Lagrangian methods. The particle-following Lagrangian methods and their intrinsic nonparallelism have not been generally used in the atmosphere and oceans. Thus the fluid state is straightforwardly advanced by computing local tendency fields for wind, temperature, humidity, and pressure, and then by applying a simple time integration scheme, usually termed "leap frog,"

$$f(t + \Delta t) = f(t - \Delta t) + 2 \Delta t \cdot f'(t). \tag{1}$$

It is in the tendency fields f' that the great bulk of computation takes place and the question of parallelism arises.

Both the atmosphere and the ocean are essentially two-dimensional fluids. Their depths are roughly three orders of magnitude less than their horizontal extents. In models this means that the number of vertical levels is usually on the order of 10, while the resolution in a horizontal direction is on the order of 100. Therefore, vectorization must occur with respect to at least one horizontal direction of the model to be efficient. There is a natural division of the physical processes into horizontal and vertical components. In the horizontal, the principal physical processes are the advective transport of momentum, heat, water, and mass, and the diffusion of momentum, heat, and water. The horizontal processes are almost completely parallel, whether described in finite-difference form or in spectral form, which has become popular in the past decade.

Normally the data for a model exceed memory capacity, so the data must be retained on disk and cycled through memory as needed. The data on disk represent three-dimensional states of the atmosphere and/or ocean. They are often processed in a series of two-dimensional, vertical slabs, with three consecutive slabs in central memory at any time to allow horizontal differencing. All such differencing is fully vectorized over the two-dimensional slab, so vectors of length order 10 × 100 are possible. Processes that do not vectorize with respect to the vertical direction will still be of length order 100.

In the vertical, there are additional processes, such as long- and short-wave radiation, condensation of water vapor, the drag of the earth's surface, and the computation of hydrostatic pressure. It is in these additional vertical processes that we encounter the most problems realizing full parallelism. The basic difficulties may be divided into four categories:

(a) empirical, tabulated functions, such as the saturation vapor pressure of water,
(b) processes that turn on and off locally, such as condensation,
(c) processes that iterate locally to a value, such as convection, and
(d) recursive or integrating processes, such as solving Poisson's equation by relaxation.

Consider examples of each of these categories.

To determine relative humidity, the local saturation vapor pressure of water e_s must be determined from an empirical function of local temperature:

$$v_{ik} = e_s(T_{ik}), \quad i = 1, \ldots, N, \quad k = 1, \ldots, L. \tag{2}$$

In ASC FORTRAN this may look like

```
      DO  1 K = 1, L
      DO  1 I = 1, N
   1  V(I,K)  =  ES(IT(I,K))
```
(3)

While there is no parallel vector instruction on the ASC to implement this construct, a highly optimized nonvector subroutine has been written to carry it out, and the compiler produces a call to such a routine when the construct is encountered. The same approach has been used on other systems, and still others implement this construct in hardware as a pseudovector, although the required random accessing of memory thwarts any full parallel implementation. The problem for the ASC programmer is to isolate this construct syntactically so that the compiler can recognize it. This approach on the ASC achieves a speedup two to five times over the compiled scalar code.

Wherever the relative humidity exceeds 100%, moisture must be condensed and latent heat released into the atmosphere. In a scalar machine this means bypassing the condensation wherever the humidity is less than 100%. On the ASC, with vectors 10–40 times faster than scalars, it is more efficient to do the process everywhere and then set its result to zero wherever it does not apply, provided the frequency of occurrence is at least about 0.05. Ideally one would like to write (in FORTRAN)

```
      CNDNSE(I,K) = HUMID(I,K) .GE. 100.0,
```
(4)

where CNDNSE would have either the value 1.0 or 0.0. FORTRAN (including ASC FORTRAN), unfortunately, does not provide conversion between logical and numeric data types. Therefore, the programmer must resort to an unnatural, machine-specific construct which involves shifting the sign bit over so it becomes an integer 1 or 0. Equation (4) is thus transformed to

```
CNDNSE(I,K) = 1 - LSHF(HUMID(I,K) - 100.0, -31).        (5)
```

The construct LSHF(, -31) shifts the sign of the argument expression to the low-order bit of an ASC word.

When the latent heat is liberated by condensation, it tends to destabilize the atmosphere with respect to the layers above and below, which can lead to a convective mixing. Convection is turned on and off locally in the same way that condensation is, but because the convective adjustment can propagate to both the layer above and the layer below and can also lead to additional condensation, the convective process must be iterated over an entire vertical column. This is equivalent to maintaining an array CNVECT(I,K) with 1s, where convection is taking place (and 0s elsewhere), and then turning off all 1s in a K-column when it has been stabilized. Computing convection everywhere with vectors is probably effective for the first iteration, but the number of 1s in the matrix falls off quickly with further iterations. An acceptable solution is to iterate everywhere a fixed number of times (say, two or three). That means that some small percentage of the columns may not be completely stablized, but they will be treated again at the next time step.

A recursive or integrating process in an ocean model occurs when Poisson's equation is solved to determine the flow field. The traditional scheme has been overrelaxation using a mixture of updated and nonupdated points. The equation being solved for G in terms of the forcing function F is

$$\nabla^2 G = F, \tag{6}$$

or, in finite difference form,

$$G_{i+1,j} + G_{i-1,j} + G_{i,j-1} + G_{i,j+1} - 4 \cdot G_{ij} = F_{ij}. \tag{7}$$

This equation is solved by computing first a residual R_{ij}, where superscripts m have been introduced to denote iteration number:

$$R_{ij} = G^m_{i+1,j} + G^{m+1}_{i-1,j} + G^{m+1}_{i,j-1} + G^m_{i,j+1} - 4 \cdot G^m_{ij} - F_{ij}. \tag{8}$$

Note that the already updated values for $G_{i-1,j}$ and $G_{i,j-1}$ are used, which thwarts parallelism. Then the new solution G^{m+1}_{ij} is given by

$$G^{m+1}_{ij} = G^m_{ij} + a \cdot R_{ij}, \tag{9}$$

where a would be $\frac{1}{4}$ for no overrelaxation, but is usually between $\frac{1}{4}$ and $\frac{1}{2}$. The nonparallel part of Eq. (8) can be isolated by rewriting it

$$\begin{aligned}R_{ij} &= G^m_{i+1,j} + (G^m_{i-1,j} + a \cdot R_{i-1,j}) + (G^m_{i,j-1} + a \cdot R_{i,j-1}) \\ &\quad + G^m_{i,j+1} - 4 \cdot G^m_{ij} - F_{ij} \\ &= (G_{i+1,j} + G_{i-1,j} + G_{i,j-1} + G_{i,j+1} - 4 \cdot G_{ij} - F_{ij})^m \\ &\quad + a \cdot R_{i-1,j} + a \cdot R_{i,j-1} \\ &= r_{ij} + a \cdot R_{i-1,j} + a \cdot R_{i,j-1},\end{aligned} \quad (10)$$

where

$$r_{ij} = G_{i+1,j} + G_{i-1,j} + G_{i,j-1} + G_{i,j+1} - 4 \cdot G_{ij} - F_{ij}. \quad (11)$$

Now field r in Eq. (11) can be calculated in parallel, and only equation Eq. (10) resists vectorization. Thus the nonparallel kernal has been isolated to equation Eq. (10) with preceding and following vectorizable equations (11) and (9), respectively. On the ASC an assembler-code subroutine to implement Eq. (10) runs about 4 times faster than the FORTRAN scalar code. The entire solution when vectorized and customized for Eq. (10) runs 5–10 times faster than the comparable scalar code.

V. ESTIMATING PARALLELISM IN MODELS

The type of programming adjustments described in the previous section are found to a greater or lesser degree in all the models of the Laboratory. The role of the program-conversion effort at the time of ASC delivery was to introduce these kinds of changes to the repertoire of Laboratory programs, which had been running on a Univac 1108 or on an IBM 360/91. The main business of the laboratory is its research—not extracting the absolute maximum performance from a computer system. So the thoroughness with which such special coding is implemented varies from model to model. That notwithstanding, how much vectorization has been achieved?

Let me first define vectorization to be how much of an equivalent scalar program can be executed with parallel (or vector) instructions, achieving the maximum performance of a machine. Unfortunately, such vectorization cannot be measured directly. What can be measured directly is how much of the time a vectorized program is executing vectors on the ASC. It is also possible to constrain the compiler from creating vector instructions, so that the overall speed of a vectorized program relative to its scalar counterpart can be measured.

We have attached a high-speed pen-chart recorder to signals in the ASC that indicate when a vector instruction is is progress in a particular pipe.

This time in vector mode has been recorded for a wide spectrum of Laboratory programs. Since the recorder has only two channels, we could examine only two pipes at a time; pipe 1 and pipe 4 were selected. Each program displays a unique signature. From these recordings two broad generalizations are apparent. First, pipe 4 is just as busy as pipe 1. This means that the compiler is almost always successful in creating four concurrent vectors to keep all pipes busy. Second, the overall character of a program signature is determined more by vector length than by the degree of vectorization. This dependence stems from the vector startup time of the ASC. To the extent that this dependence is an artifice of the machine, although other machines might well display a similar effect, it is desirable to eliminate its effect when determining vectorization.

While the compiler is effective at assuring four concurrent vectors, there is usually a dependence between successive sets of four vectors. This dependence requires that all four vectors terminate before the first vector of the next set may be initiated. With a startup time of 25 cycles, the fourth pipe will not be executing its vector until 100 cycles after the first vector is initiated. Then the first pipe, will have to wait 100 cycles for the fourth pipe to finish. Thus the maximum observable fraction of time in vectors as a function of vector length is given by

$$f_{max} = L/(L + 100), \qquad (12)$$

where L is the length of each vector running in a pipe. If vectorization is achieved only in a horizontal direction, L will be on the order of 100, and f_{max} will therefore be limited to 0.5. If vectorization is achieved in a vertical slab, L is on the order or 10×100, in which case $f_{max} = 0.91$. Because the grid size and geometry vary considerably from model to model, the recorder plot for each model exhibits an envelope defined by the vector lengths characteristic of that model. The range of L is typically between 20 and 2000.

When one looks at the recording under the length-dictated envelope, the trace indicates that the ASC is running vectors about 30–50% of the time, when vector startup is excluded. Consider how to calculate the level of vectorization of a model v. Let the fraction of time running vectors be designated f. The fraction of a model that still runs in scalar mode is $1 - v$, while the vector part will run (nominally) 30 times faster than it would in scalars, or $v/30$. Thus we can write

$$f = \frac{v/30}{1 - v + v/30} = \frac{v}{30 - 29v}, \qquad (13)$$

or, rewriting to find v,

$$v = \frac{30f}{1 + 29f}. \qquad (14)$$

For f in the range 0.3–0.5, v has the range 0.93–0.97. In other words, it is estimated that only 3–7% of a geophysical model is not carried out in parallel.

Note that the overall speedup V one can realize by vectorizing a model on the ASC can be estimated by

$$\frac{1}{1 - v + v/30},$$

neglecting vector-length effects, and by

$$V = \frac{L}{(1 - v + v/30) \cdot (L + 100)} \tag{15}$$

if vector-length effects are included. For a range of programs from $v = 0.93$ and $L = 20$ to $v = 0.97$ and $L = 1000$, the expected range of V is 1.7–15. Thus a large range in speedup due to vectorization can be expected with programs, depending on the typical vector lengths and vectorization achieved with the programs.

VI. CONCLUSION

The geophysical models of the Laboratory exhibit a high innate level of parallelism. The principal problem achieving such parallelism on the ASC lies not in the physical and mathematical processes of the models so much as in the style of the FORTRAN code and its ability to exploit the computing parallelism available. With the acquisition of the ASC, we provided for a substantial program-conversion effort to carry out this restructuring of most Laboratory programs. The high level of parallelism achieved, estimated to be about 95%, testifies to the success of this strategy.

ACKNOWLEDGMENTS

I wish to thank Leith Holloway for his suggestions and review of this chapter. And I also thank Howard Frazier, my supervisor and long-time friend, and Dr. Joseph Smagorinsky, the Director of the Laboratory, for their continuing encouragement.

Experiences with a Floating Point Systems Array Processor

Kenneth G. Wilson

Newman Laboratory of Nuclear Studies
Cornell University
Ithaca, New York

I. INTRODUCTION

A. *Preamble*

This article on the Floating Point Systems Array Processors (the AP-120B, the AP-190L, and to a lesser extent the FPS-100 and FPS-164) is basically a report of the experiences at Cornell with an AP-190L attached to an IBM 370-168 host computer. This report is addressed to someone who might be considering the acquisition of an AP (shorthand for any Floating Point Systems Array Processor) for a range of scientific-type computations, which is the purpose of the AP at Cornell. There is not much emphasis in this report on the technical problem of rearranging algorithms for specific calculations to run efficiently on the AP. The reason for this is that the parallelism of the AP is a relatively unimportant consideration. If a program fits the obvious limitations of word length, memory size, and limited access to peripherals that the APs suffer from, an AP provides spectacularly cost-effective computing—almost always better than any other computer, from microcomputers to a Cray-1. This occurs *regardless of how the algorithm is structured*, or what it is doing. Unfortunately, the APs have never won acceptance for what they can do in the world of FORTRAN-based large-scale computations. They cannot do everything; it is not always easy to get them to do anything. Thus the problem with the AP is not how to make it go fast, but only how to reach the stage where one has an AP to work with and then how to get a program into it and running; the rest is gravy.

To obtain more detailed technical information than is supplied in this article, one can write to Floating Point Systems, Inc., at P.O. Box 23489, Portland, Oregon 97223, or call them at (503) 641-3151.

B. *History of the Cornell AP Project*

In 1976 I was looking for a cheap computer that would carry out floating-point computations very rapidly, namely, at roughly the speed of the IBM 370-168. This search was triggered by the previous work of Kunz (1976) of the Stanford Linear Accelerator Center (SLAC), who designed a cheap emulator for the 16-bit integer instructions of the 370-168.

Erich Knobil (at that time working for the Laboratory of Nuclear Studies at Cornell) suggested Floating Point Systems, a company I had never heard of, in Oregon. On looking through the Hardware manual, I found that the FPS AP-120B Array Processor was the computer of my dreams. It was faster than the 370-168; its raw speed is about the same as the CDC-7600. For example, a floating-point multiply required 0.5 μsec to complete, namely, 3 machine cycles. A new multiply could be started every cycle, so that machine was capable to performing 6 multiplies per microsecond. For comparison, I rarely achieved more than 1 multiply per microsecond on the 370-168. Above all, it was cheap: less than $30,000 for the arithmetic unit itself, to which one had to add the cost of memory and an interface to the outside world (namely, a host computer). A complete unit with a reasonable amount of memory cost only $50,000–150,000.

The term "Array Processor" creates an image of thousands of microprocessors working in a Rube Goldberg arrangement. Not so: the AP-120B is a *single* computational unit. The term "Array Processor" is a marketing term and does not give an accurate impression of what the AP is or what it can do. To find out, read on.

I called some of the FPS customers. They were ecstatic about the AP-120B. The AP-120B, they said, was highly reliable, and performed exactly as described in the manual.

I was looking for a fast computer because I was reaching an impasse in my own research. My research centers around the statistical behavior of *fields*: functions of space and time which fluctuate statistically. This is a problem with an infinite number of degrees of freedom: the value of the field at each space–time point is a separate degree of freedom. I had reached the limits of

Experiences with a Floating Point Systems Array Processor

the analytic methods (mostly diagrammatic expansions) available for studying these problems. I had made considerable progress in defining a general framework (the "renormalization group approach") for studying fields (see, e.g., Wilson and Kogut, 1974). Now I wanted to develop practical methods for using the framework. This meant, in practice, that I would have to treat numerically problems involving many thousands of degrees of freedom. The only numerical procedures available for this kind of problem were various forms of Monte Carlo integrations and simulations. They involve running relatively simple programs for enormous amounts of time; the time is needed to cycle through all the degrees of freedom many thousands of times in order to collect enough statistics to have reliable estimates of averages and correlations. At the time (1976) the use of Monte Carlo methods and simulations was largely restricted to a few people at the Lawrence Livermore Lab and a few other places where large amounts of computer time were accessible.

The main problem with the AP-120B was a lack of software. It was supplied with an assembler, a simulator, and an executive program to manage the movement of data and programs in and out of the AP. There was also a library of simple programs. There was no FORTRAN compiler.

For the traditional market that Floating Point Systems was serving, a FORTRAN compiler was unnecessary. The AP-120Bs were being used in data reduction applications such as image processing, speech processing, seismic data reduction, etc., where a single program would run day and night for years performing operations such as fast Fourier transforms (FFTs) very rapidly on enormous quantities of data. These programs were written in assembly language in order to achieve maximum efficiency, and there was no demand for FORTRAN. Furthermore, an AP-120B was always attached to a host minicomputer, such as a PDP-11. Only the time-critical subroutines would be executed in the AP; the rest of a data analysis program (written in FORTRAN) would execute in the host, with frequent data transfers taking place between the host and the AP.

For my own work I foresaw a different arrangement. An AP with a lot of memory (the Cornell AP has 96,000 words of memory) would be attached directly to Cornell's IBM 370-168. A FORTRAN compiler for the AP would allow me to put large programs entirely into the AP, with data transfers taking place only at the beginning and end of the computation. Assembly language routines would still be used where time was critical. It was essential to keep data transfers to the 370 to a minimum, mainly to avoid the high cost of any use of the 370. Then I could do program development on the 370, which was fast and easily accessible to me and had some good program development facilities: a primitive but reasonable time-sharing system (VM-CMS system), plus the WATFIV debugging compiler for FORTRAN.

I proposed this idea to Doug Van Houweling, the Assistant Director for User Services at Cornell, and he was enthusiastic. Geoffrey Chester, a condensed-matter physicist with a long-time interest in computing, was also eager. A consortium of five research groups was organized which agreed to pay two-thirds of the cost of an AP-190L, while one-third was paid by the Computing Center. Doug then established a FORTRAN compiler project at Cornell. Through the efforts of John Williams (of the Computer Science Department) and myself but above all Donna Bergmark (a computer science graduate student and then a staff member of the Computing Center), a FORTRAN compiler, called APTRAN was built. It began to be useful in the fall of 1978. The AP-190L itself arrived a few months earlier. By June 1979 enough people had programs working so that the AP was saturated. In the following September two heavy users (John Wilkins, another condensed-matter physicist, and myself) left on sabbaticals; since then the AP has operated at roughly 50% of its capacity.

The AP has been invaluable for Monte Carlo simulations in traditional statistical mechanical applications and quantum field theory, for band structure calculations in solids, for simulation of galaxy evolution, and a black hole calculation (see, e.g., statistical mechanics: Tobochnik and Chester, 1980; Shenker and Tobochnik, 1980; galaxy evolution: Farouki and Shapiro, 1980; black hole evolution: Shapiro and Teukolsky, 1979, 1980; quantum field theory: Wilson, 1980; band structure in solids: Wood, 1980, Richter, 1980.

II. SCIENTIFIC COMPUTING BEYOND THE CDC 7600

A. *Background*

Between about 1960 and 1972 the size and cost effectiveness of computers was increasing at a fairly rapid rate. During this period the rule was the bigger the better. Control Data Corporation brought out first the 6600 and then the 7600, with the 7600 in particular having astonishing capabilities but of course for a multimillion dollar cost. Only the big national laboratories, such as Los Alamos, Lawrence Livermore, and the Lawrence Berkeley laboratories were able to afford the cost of the 7600. Meanwhile, IBM was introducing the 360 family of software-compatible computers, and many universities, driven mostly by demands for social science and administrative computing, were buying 360s. Then the 370 series came along, upgrading the 360 series.

With the 7600 and the 370 series, the growth of capability and cost effectiveness of large mainframe computers largely ceased. For almost ten

Experiences with a Floating Point Systems Array Processor 283

years the power of the 7600 was unchallenged, except for the Cray-1 supercomputer (see Section II.B). The costs of IBM computers have been decreasing, but they showed only modest increases in capacity from the introduction of the 370-168 in the early 1970s until the just-announced 3081. Meanwhile a new generation of special purpose computers have developed which supply considerable increases in cost effectiveness or speed for specific programs. These special purpose computers include the supercomputers such as the Cray-1, array processors like the AP-120B, and for very specific applications mini- and microcomputers are often very effective, too.

In this Section I shall try to give my perspective on the various alternatives for obtaining cheap large-scale scientific computing. The reader must understand that my direct experience comes from using the AP-190L and benchmarking (in the spring of 1979) a VAX 11-780 of Digital Equipment Corporation. I have never had an opportunity to use supercomputers like the Cray-1, and my report on them is based on hearsay.

B. *Supercomputers (Cray-1, etc.)*

The mythical laurel wreath for "the world's most powerful computer" currently (end of 1980) belongs to the Cray-1. Most programs execute two to three times faster on the Cray than on the CDC 7600. In addition, the Cray-1 has a vector arithmetic processor which can add or multiply two vectors roughly ten times faster than the 7600. The Cray-1 FORTRAN compiler will recognize inner loops that can be rewritten in terms of vector instructions and use the vector unit to execute these loops. However, there are many types of programs that cannot be easily reduced to vector instructions. For example, the program that I am currently working on performs matrix multiplies by a table lookup; this is not a "vectorizable" procedure. Even when programs can be written in a form which is vectorizable, many Cray users have been unwilling or unable to spend the time necessary to put their programs in vectorizable form. Also there is usually a lot of overhead associated with any vector computation, so that the increased speed of vector computations is usually not noticeable in the overall speed of execution of an entire program unless the vector lengths are fairly large (at least 20 elements usually). There are other supercomputers besides the Cray-1, for example, the CDC 200 series.

The Cray-1 is expensive; a fully supported Cray-1 costs, very roughly, of order $10 million. Only a few national laboratories and large industrial firms have been able to afford a Cray. Most university researchers have no way to access a Cray-1 even if they wanted to, except through very expensive commercial networks.

The Crays owned by the government have been strictly limited to individual government agencies and their grantees carrying out research in specific areas: the Cray-1 at NCAR is for work in atmospheric sciences only; the machines at Livermore are accessible outside Livermore only for Department of Energy grantees working in plasma physics. The politics of large-scale computing has prevented a more rational approach of supplying low-cost computing to *all* governmental researchers instead of supplying free computing to limited areas of research.

While the Cray is the speed champion among larger computers (at the present time), it is not always the most *cost-effective* computer, due to the very high cost of a Cray installation.

C. *Mainframes*

Mainframe computers include the IBM 370 series, the CDC 7600 and Cyber 70 series, the big Univac computers, and the PDP 10 of Digital Equipment Corporation. Most universities have a mainframe; they are usually very expensive to use and often overloaded. Their speed is reasonable but not spectacular. They are usually aimed more at administrative, social science, and instructional computing than for hard-science computing. They tend now to be more appropriate for serving many users accessing a common large database than for large-scale scientific computing.

D. *Large Minicomputers*

Recently a number of large minicomputers have appeared, such as the VAX 11-780 of Digital Equipment Corporation and the IBM 4341, which have speeds from roughly 1/6 (the VAX) to 1/2 (the 4341 Group 2) of the IBM 370-168, and which offer large real memory capacity combined with virtual memory time-sharing operating systems. University science departments are acquiring these minicomputers at a fairly rapid rate; they are currently the most popular solution to hard-science computing problems. A locally owned VAX serving a small user community usually supplies more computing power *per user* than a mainframe or even a Cray-1 except for vectorizable programs. However, the raw speed of these minicomputers is still unimpressive. The benchmarks run at Cornell showed that programs written entirely in FORTRAN and compiled through APTRAN ran in the AP-190L at a speed 3 times faster than the VAX; had proper optimization been available, these programs would have been 10–20 times faster in the AP than in the VAX. This is despite the fact that an AP, by itself, is considerably cheaper than a properly supported VAX. A 4341 is faster than the VAX, but also more expensive. In summary, a large minicomputer is a reasonable first step in meeting the needs for scientific computing, but, as is discussed below, it needs to be supplemented with an Array Processor or access to a Cray for really demanding computations.

E. *Small Minicomputers or Microcomputers*

Single small minicomputers or microcomputers are much too slow and have too little memory for reasonable scientific computations such as a Monte Carlo simulation or solving a partial differential equation. It may happen that some day an array of microprocessors will be a useful computer for large calculations; such an array has been assembled by ICL Computers Ltd. in their DAP (Distributed Array Processor). A version of the DAP has been installed at Queen Mary College in London. I have no experience with it. Unfortunately, they make the mistake already made in the ILLIAC IV; namely, all the DAP microprocessors execute the *same* instruction stream in lockstep. This is likely to limit the usefulness of the DAP.

F. *Attached Processors*

The subject of this report is the use of a fast second computer, such as the AP-190L, attached to a slower or more expensive host computer, in order to carry out large-scale computations. The host could be a large minicomputer like the VAX or 4341 or it can be a mainframe like the 370-168. The benefits of this arrangement are that the AP-190L can provide the computing power of a CDC 7600 at a modest cost, and the full speed of the AP can be realized, at least in principle, for a much greater variety of problems than for a supercomputer like the Cray.

An AP is under the worst circumstances faster than either a VAX or 4341, and in the best of circumstances it should be about 20 times faster. It can in some cases be only a factor of 4 slower than a Cray-1, and when good optimizing compilers are developed it should always be within a factor of 10 of the Cray. A Cray-1 will usually have hundreds or thousands of users (to justify the expense of a Cray installation), whereas a small Array Processor is often justifiable for a single user, and a large installation (an AP-120B or AP-190L with $\sim 100{,}000$ 38-bit words of memory, or an FPS-164 with up to 1.5 *million* words of 64-bit memory) needs to serve only a small community of users (~ 5–50). The computing power *per user* delivered by an AP will normally far exceed any alternative. Furthermore, there is no compulsion to saturate an AP in order to justify its expense; it is reasonable to have a somewhat less than saturation load in order to have fast turnaround of AP jobs and convenient access for debugging runs. Since the AP can be attached to a number of minicomputers or mainframes (including the VAX, the 370 series, and the PDP 10, but not any CDC machine), an AP can often be attached to an existing computer, thus avoiding the need to buy printers, tape units, disks, etc., which often cost more than the computer itself.

The disadvantages of the AP are of three types. First, there are obvious hardware limitations. These involve word length, memory size, and a limited access to peripherals. These limitations will be reviewed in later sections. Second, the idea of using the APs for large-scale computations is rather recent and the software support for this is still somewhat primitive; the present software situation, especially regarding program development, is reviewed in Section III.E, but this will improve with time. Finally, it is not always easy to persuade users to switch to an AP even when it is very much in their interest to make the switch. The world of large-scale computing is filled with people who do not like to compute, do not want to understand the computers they are using, and are very unwilling to leave their current computing niche. People who like computing and understand computers have mostly gravitated to minicomputers and microcomputers.

There are other Array Processors on the market, for example, by Analogic Corp.; CSPI, Inc.; and Datawest. As far as I know, no other manufacturer currently supplies a FORTRAN compiler. A more expensive multiprocessing system by Denelcor, Inc., is discussed in Section V.C.

III. THE AP-190L INSTALLATION AT CORNELL

A. *Overview and Hardware Configuration*

The AP-190L at Cornell has been in operation since June 1978. It is attached (through a block multiplexor channel) to the central Cornell computer, namely, an IBM 370-168 running the VM operating system. The 370-168 is itself off-campus; it is usually accessed by low-speed on-line remote terminals, with printed output being directed to high-speed printers at selected on-campus locations. I actually work mostly from a 1200 baud home terminal (a Teleray CRT plus a 1200 baud TI 810 printer) which I use at nights and weekends when the 370 is not so crowded and the 370 rates are reduced. The AP-190L has 4096 64-bit words of Program Source memory (a separate very fast memory for programs), 96,000 38-bit words of Data memory (divided into $1\frac{1}{2}$ "pages") and 4500 38-bit words of "Table memory" (another separate memory in the AP: 2500 words of preset constants, 2000 words of reusable memory). There are no peripherals attached directly to the AP; all communication with the AP takes place through the 370.

The AP-190L is the same as the original AP-120B except that it has an interface to an IBM computer instead of a minicomputer. Both have a 38-bit floating-point word length (28-bit mantissa, 10-bit exponent). Both use 16-bit integers for standard address computations. (The forthcoming FPS-164 will have the more customary 64-bit floating-point word length, plus 32 bits for standard integers and a 24-bit address space.) Both the AP-120B and AP-190L have a maximum program source memory of 4096 words, which translates into of order 300–500 lines of FORTRAN. This was thought to be a serious limitation, but the forthcoming "segmenting loader" (see below) will solve this limitation. Both can accommodate at least 3 pages of data memory (200,000 words) and probably more; with the segmenting loader *all* this memory can be used for programs (about 3000 lines of FORTRAN per page) while only the first page is directly usable for FORTRAN data storage (using the present FORTRAN compilers). The limit on Table memory is much smaller.

B. *Purpose: Cost-Effective Computing*

The purpose of the AP is to provide very cheap large-scale computing to a small user community. The host 370 runs thousands of jobs each day for thousands of users, with typical job times ranging from less than a second to about an hour for very long jobs. The AP serves a community of about 50 users who run jobs ranging from 15 minutes to hundreds of hours. Long jobs (greater than 15 minutes) have to be run in 15-minute segments to allow all users reasonably timely access to the AP; users are instructed how to arrange for automatic submission of successive 15-minute jobs, so that a user running a many-hour job has only to check once a day or so that the job resubmission has not been stopped due to a system failure or unexpected result.

Production runs on the AP-190L typically cost $8–10 per hour to members of the consortium. This cost includes $5 per hour for use of the AP, $1.80 per hour for connect time to the 370, the rest being 370 CPU charges associated with host program execution and data transfers to and from the 370. This assumes that data transfers take place only at the beginning and end of each 15-minute AP run. The data transfers include double transfers disk → 370 → AP at the beginning of the run and AP → 370 → disk at the end of the run, as well as some output to the printer. This also assumes that there are no extensive computations in the host program. These charges are to be compared with charges of $500–1500 per hour for CPU use on the 370 itself. Casual (nonconsortium) users are charged about $40 per hour for AP use. The above costs are costs as of January 1, 1981. There is no question of the success of the AP in providing cost-effective computing; costs to consortium members have been reduced anywhere from a factor of 10 to a factor of several hundred versus using the 370, even when the capital costs of the AP are taken into account. The AP rate includes hardware maintenance and the cost of one software support person. More software support is provided out of 370 charges; see Section III.C.

It should be noted that the AP-190L is not always *faster* than the 370-168. Many users use FORTRAN programs exclusively, using the APTRAN compiler to generate AP code. Such programs run about a factor of two slower on the AP than on the 370-168, but this loss of speed is irrelevant; the consortium members are all limited by cost, not by execution time. On the other hand, users who make use of assembly language subroutines usually find the AP is about twice as fast as the 370-168, and my own program was eight times faster on the AP than on the 370. Even if a program is slower in the AP than the 370, the *turnaround* time is often faster due to heavier loading of the 370.

C. *Software and Software Support*

At the time the AP-190L was ordered there was no FORTRAN compiler available from Floating-Point Systems, and no assurance that they would build one. Hence a software support group was formed inside Cornell Computer Services, which is presently headed by Alec Grimison. The support group developed the "APTRAN" FORTRAN compiler, and is now completing the "segmenting loader" described below. The APTRAN compiler supports a subset of the FORTRAN language; no logical, complex, or double precision variables, no read or write statements, no equivalence statements, and some other restrictions. These restrictions are specific to APTRAN; the current Floating-Point Systems compiler AP-FORTRAN has fewer restrictions, and the forthcoming compiler for the FPS-164 should implement much of the new 1977 standard for FORTRAN. However, one benefit of APTRAN over AP-FORTRAN is that integers in APTRAN are stored in 38-bit floating format, allowing integers of size up to 100 million. In AP-FORTRAN integers are stored and manipulated as 16-bit integers, which leads to problems for integers greater than 30,000. As with any compiler, the APTRAN compiler has continued to exhibit bugs although most programs compile correctly and there are usually ways to reprogram when necessary to avoid compiler bugs. Fixing these bugs is one of the more painful duties of the support staff, partly because it takes a lot of effort just to separate compiler bugs from bugs in users' programs, and partly because bug fixes often generate more bugs than are fixed.

Because the AP-190L has multiple users, a scheduler was written (by Larry Chace at Cornell) to queue requests to use the AP and to assure fair sharing of AP time among the users. The APEMAN scheduler operates in two modes. It can attach the AP to a specific on-line user, who then has up to 15 minutes to use the AP as he sees fit. Alternatively, the scheduler can itself log on a given user and execute a prespecified list of commands belonging to the user. This second mode is a batch mode and is used in particular when long jobs are being executed in 15-minute segments. In this mode the last command, after completion of the 15-minute AP run, is a command to the scheduler to set up another 15-minute run (unless the job is finished). The scheduler accepts job requests with two priorities, high and low. This means that on-line users requesting AP time at high priority can jump ahead of a queue of low-priority batch jobs. Unfortunately, an on-line user who wants to make a short debugging run must still wait until the current AP job has finished executing, since the AP-190L has no provision for interrupting the current job. It is for this reason that the consortium and the Computing Center agreed on a time limit of 15 minutes per job, even though this requires extra programming in order to break up long jobs into 15-minute segments and carry out the necessary checkpointing (saving of data).

The new loader is being designed to load either APTRAN programs, assembly language programs, or FPS AP-FORTRAN programs. For programs that overflow the program source memory, the loader will segment them automatically. "Segmenting" means breaking the program up into short segments, often just the code between two consecutive branches or between a branch and the next branch target. However, assembly language and library routines will often be left unbroken. All segments will be stored in data memory in the AP. A supervisor routine will be loaded into Program source memory. The supervisor will transfer segments from data memory to program source memory as they are needed. When a given segment finishes executing and branches to another segment, the branch will be executed only if the next segment is already in memory; otherwise, the branch will transfer to the supervisor, which will load the next segment from data memory.

The segmenting loader will not be required for the FPS-164, which will automatically use data memory for both data and programs.

Aside from working on APTRAN and the loader, the AP support staff also provides general consulting on AP use and handles problems that arise with the FPS subroutine library; the less frequently used programs in this library often have bugs in them. The AP support staff presently constitutes two programmers reporting to Alec Grimison, although both have other duties besides general AP support.

D. *Experience: Hardware Reliability*

The AP-190L hardware has worked well since its arrival. The amount of down-time due to AP problems has been far less than down-time for the 370.

There were some problems with the interface to the IBM channel because we had only the second IBM interface built by FPS; the prototype interface was eventually replaced and very few problems have occurred since. The mean time to failure is now several months; the principal problems that have occurred appear to have been present all along, unnoticed until a new feature was used. The hardware rarely fails once it is working. When a hardware failure occurs, the AP is usually down for several days while arrangements are made to have the problem fixed.

E. *Experience: Program Development*

Program development for the AP usually takes considerably longer than for the 370. Debugging of AP programs is usually done using the AP itself rather than the very inefficient simulator, and the time required to find and eliminate a bug is usually longer than for a pure 370 program, plus there are more potential sources of bugs. The need to coordinate format changes from single- or double-precision IBM words to 38-bit FPS words is sometimes a problem. This is presently done by commands to the loader, specifying for each argument or common block whether floating-point data are single or double precision in the IBM. It is often important to use double precision in the IBM since single-precision 32-bit words in the IBM have less precision than the AP words. The APTRAN compiler and the FPS Math library can be sources of bugs. Hand-coded assembly language programs take a long time to debug.

In addition, a certain amount of time is usually spent arranging for the separation of programs into routines which will run in the AP and routines which stay in the host, and arranging if necessary for the breakup of program runs into 15-minute sections with checkpointing of data between sections.

Once a program is prepared and debugged, using the AP is a pleasure, especially being able to make runs of 50 hours or more without worrying about the cash outflow. In fact, one usually has to run a program for hundreds of hours before the production costs equal the debugging costs. This is especially important for Monte Carlo computations where it is important to accumulate extra statistics to be sure that the statistical errors are under control.

The problems of program development should decrease with time. The compilers and math library routines will be more heavily debugged; as compiler optimization improves there will be less reason to consider assembly language coding. It is hoped that as more use of the APs, especially the FPS-164, develops that an attempt will be made to provide FORTRAN equivalents of the library routines so that to the maximum extent possible

Experiences with a Floating Point Systems Array Processor 291

complete host + AP programs can be checked out by running them entirely in the host. I find that a lot of my problems are due to bugs that I introduce into my programs when converting them to the joint host + AP environment; I could eliminate these bugs a lot faster if I could run entire programs *after* conversion through the WATFIV debugging FORTRAN compiler.

F. *Experience: Run-Time Problems*

In production running there are two problems to be concerned about: roundoff errors and hardware errors. With the 38-bit word length of the AP-120B and AP-190L one has to be on the alert for roundoff problems, especially when making the very long runs that the AP makes possible. In my own case the roundoff errors are mostly not a problem since they usually are indistinguishable from the randomness of the Monte Carlo procedure itself. However, there was one problem that arose; I was working with unitary matrices, but the matrix multiplications introduced roundoff errors on the unitarity constraint. These roundoff errors accumulated very rapidly. The solution was simple; the matrices were checked frequently and arbitrarily adjusted to restore unitarity before the roundoff error accumulated. The errors introduced by the adjustment procedure, in the form of the matrix *after* unitarization, were absorbed in the randomness of the Monte Carlo procedure and no longer accumulated. In general, any computation which has a conservation law which is subject to roundoff error will have to be checked for roundoff problems. In addition, matrix diagonalizations are subject to severe roundoff problems; in one calculation (not my own) the accuracy of the eigenvectors after diagonalization was only about two decimal places and an iterative procedure had to be used to improve this accuracy. Roundoff problems will be much less severe in the FPS-164.

Hardware errors can be of two types: reproducible and random. Reproducible errors occur when some part of the hardware has failed; for example, a memory bit may become frozen. Such errors will usually be detected by the FPS diagnostic routines, which should be run regularly, but between diagnostic runs these errors can be encountered, although I have had few problems of this type. Random errors are nonreproducible; they can occur due to incipient hardware failures or simply random events like cosmic rays resetting a memory bit. It is wise especially on very long runs to build in as much protection as possible against errors. In my case, since unitarity had to be checked anyway, I set up an error exit anytime the violation of unitarity was larger than could be accounted for by roundoff error. In addition, at the end of each 15-minute run I checked all constants against their value at the beginning of a run. I found random errors occurring every 10 hours or so of

AP computing time; they were random in the sense that each 15-minute run that showed an error was restarted and then had no problem. I do not know whether the errors were due to hardware problems or were intrinsic to the AP. I know of no other AP user who has noticed any such problem. It is also advisable to make qualitative tests on crucial results.

Another peculiar problem that arose was that in the host IBM computer, negative zero is distinct from positive zero. Unfortunately, the AP could not accept negative zeros transferred from the 370. Normal 370 computations do not produce negative zeroes, but sometimes uninitialized variables passed to the AP created problems. The fix for this problem was to initialize every array passed to the AP completely.

G. *Experience: Sociological Problems*

We expected that a lot of users would be drawn to the AP once the FORTRAN compiler was working and people became aware of the cheap computing available on the AP. This did not happen. There have been some casual users of the AP, and one new user joined the consortium. However, most of the use comes from consortium members. The Materials Science Center has a membership in the consortium and some casual use has come through this group. The reason for this is not fully understood, but we know some contributing factors. Some potential users would have problems with the short word length; it is hoped that these users will be attracted to an FPS-164. Some potential users already have their own minicomputer systems, and minicomputer owners take an oath never to spend a cent outside their own system. Some potential users presently working on the 370 know too little about computers or computer programs to be able to manage the changes needed to use the AP. In fact, such users are extremely reluctant to make any changes whatsoever in their programs or how they are run: they mostly still use an antiquated Batch system on the 370.

IV. FPS ARRAY PROCESSORS AND PARALLEL COMPUTING

A. *Parallelism and Superfast AP Code: The Assembly Language Programmer's View*

The speed of the Floating-Point Systems Array Processors is achieved through three means. First of all, through clever design all operations including the floating-point multiply require only a few machine cycles (the multiply and a memory fetch require 3 machine cycles; all other operations require only 1 or 2 cycles). The machine cycle times is fast but not superfast: 1/6 μsec. These speeds are achieved with a minimum of use of expensive high-speed components. Second, all multicycle operations are "pipelined"; that is, a new multiply can be started every cycle, even though each multiply takes 3 cycles to complete. A new memory fetch can be started every cycle or every other cycle depending on whether "fast" or "standard" data memory is installed in the AP. However, the memory fetches must be to separate "banks" of memory; otherwise, the second fetch must wait until the first is completed. Third, the AP is divided into separate functional units: a floating-point multiply unit, a floating-point add unit, a fixed-point arithmetic unit, a branch unit, a data memory fetch and store unit, etc. A crucial feature of the AP is that instructions can be issued *simultaneously* to all the different units of the AP, and a number of data moves can also be specified. Conventional fast and expensive computers like the 370-168 or the CDC 7600 also have separate functional units, but they sit idle most of the time because instructions can be issued only to one unit at a time. When experienced programmers write assembly language programs for the AP, the inner loops which use up most of the computing time of the program will typically have about 2 instructions per instruction cycle, leading to an effective instruction rate of about 12 instructions per microsecond; for comparison the maximum instruction issue rate for the CDC 7600 is 50 instructions per microsecond, a rate which is rarely achieved in practice. Outside inner loops, where most of the code tends to be data moving rather than arithmetic, the AP code is less dense and not as fast as the 7600.

The parallelism of the AP is quite different from the vector supercomputers like the Cray-1 or the CDC 200 series. The Cray and the CDC machines also have pipelined floating-point units, but they are restricted to simple vector operations in that successive adds or multiplies can only be carried out on successive components of a vector. That is, in the Cray or CDC machine, a single vector add or multiply instruction is issued, and then successive components of the input vectors arrive in successive cycles at the floating-point unit; this process continues until all components of the vectors have been fetched. In the AP, new add and multiply *instructions* can be issued every cycle and there is no need for any connection whatsoever among the arguments used in successive multiplies and adds. This makes the AP much more flexible than the supercomputers. In the AP one can put independent computations in parallel regardless of whether they involve the same instruction sequence and simple vector memory references. As a result, good parallelism in the AP can often be achieved by rearranging the calculations of a few FORTRAN statements in an inner loop to be in parallel or putting several steps of the loop in parallel (called loop folding). These statements can include branches and unrelated memory references. The net result is that the AP can be programmed to compute at full speed on a much greater variety of problems than the vector supercomputers.

There are three other features of the AP that help in achieving high computational speeds. One feature is the large number of registers (64 floating-point registers) which can be used to minimize memory references, which are slow on the AP. The second feature is the independent random access table memory, which allows up to two memory fetches per instruction instead of one. One can move a short block of data, such as one row of a matrix, to Table memory and then access it in parallel with data from Data memory. A third feature is the existence of a number of independent data paths so that the arithmetic units can keep going at full speed. Full AP parallel speed requires six inputs and three outputs every cycle (two inputs for the floating multiplier, two for the floating adder, and two for the address computation unit). There are enough separate data paths between pairs of functional units, and between units and registers, so that the full parallel speed of the AP is often achieved in practice. These features help the AP achieve 7600 speed despite the lower instruction speed.

In the vector supercomputers there is a fair amount of startup time associated with each vector operation, so that fairly long vectors must be used to achieve very high speeds. In the case of the AP much more complicated computation can be encompassed inside an inner loop; one can easily have of order 100 arithmetic operations in a single inner loop. It is then not so necessary to have the loop repeated many times to achieve high speeds. However, it usually turns out that startup effects dominate a short loop (say, a loop executed three or four times) unless the loop is inside a set of nested loops with relatively little startup code required in front of the innermost loop. The startup time is especially devastating when the startup code for an inner loop is coded in FORTRAN and only the loop itself is coded in assembly language. With present compilers each line of FORTRAN translates to 10 or 12 AP instructions, or more, whereas an assembly language program with 100 arithmetic operations will probably require only 50–100 AP instructions.

When the AP is running at full parallel speed (an add and a multiply every cycle), it is almost a factor of 10 faster than the 370-168 and should be faster than even the newly announced 3081; it is usually faster than the CDC 7600; to do much better requires a Cray-1 or similar supercomputer, also running at full parallel speed. Many Monte Carlo simulations, for example, can be programmed to run at full speed on the AP. The AP is more cost effective for such programs by as much as a factor of 20 compared to the Cray-1 or the VAX.

B. *Parallelism in the AP: The FORTRAN Programmer's View*

With the currently existing FORTRAN compilers (APTRAN and AP-FORTRAN) the FORTRAN programmer cannot detect the parallelism

Experiences with a Floating Point Systems Array Processor

of the AP. To the FORTRAN programmer, the AP is simply another computer, but with a rather spectacular cost effectiveness for *any* kind of program. When the AP is compared with a 370 or a VAX, the same relative speeds are observed for programs which are heavy on floating-point arithmetic, or on subroutines of the APTRAN compiler itself. The APTRAN compiler subroutines are written in FORTRAN, for reasons of portability; they mostly involve moving data around and making comparisons. The reason these benchmarks are independent of the program type is simple: the AP programs that are generated by the FORTRAN compilers spend most of their time fetching data from Data Memory. As a result there is very little parallelism in these programs: an assembly language programmer would be horrified at their inefficiency. However, the output of a conventional compiler for a conventional computer like the 370-168 also mainly involves memory fetches and stores. Thus it is not surprising that benchmarks of the relative speeds of FORTRAN programs on the AP versus conventional computers turn out to be largely independent of the type of program being run.

So far there has been no serious attempt to have compilers generate optimized parallel code for the AP. The primary goal for the present compilers was that they produce usable code at the earliest possible date, with the speed of the code being of secondary importance. As a result, the present compilers (APTRAN, and AP-FORTRAN as of June 1980) do not even have sensible register allocation. As the register allocation is improved, and serious attempts to achieve parallel code are mounted, the speed of FORTRAN programs will increase, especially programs where the bulk of the time is spent in reasonably well-organized inner loops.

When full optimization is achieved, my estimate is that suitable inner loops will be speeded up by a factor of 5 or more, relative to present (as of June 1980) compilers; for startup code the speedup will probably be only a factor of 2 at most. The fastest inner loops will be those with roughly equal numbers of floating point adds and multiplies, and with less than one explicit memory reference per add + multiply pair, the other three inputs being either intermediate results of compound arithmetic statements or variables that the compiler can assign to a register for the duration of an inner loop. This means, in particular, that when full optimization is available FORTRAN programs should *not* be reduced to separate loops describing only one vector operation each; a simple loop carrying out a vector add or multiply is too memory intensive (two memory fetches and one memory store per add or multiply) to run at full speed on the AP. The inner loop should probably contain (if possible) more like 20–100 arithmetic operations, for maximum speed. However, simple loops can be brought closer to optimal speed in some cases by judicious use of Table memory, which might someday be accessible either to the compiler or directly to the FORTRAN user. Fast inner loops also must not contain recursive arithmetic (recursive arithmetic arises when each add or multiply uses the result of the immediately preceding add or multiply. It is very difficult to put recursive arithmetic in parallel.)

At present the FORTRAN programmer who is dissatisfied with the speed of FORTRAN programs on the AP has two options. The first is whenever possible to use routines from the FPS Math Library; these are hand coded and run faster than FORTRAN, but especially in the case of the simple vector operations one has to be aware of start-up costs and the fact that these routines do not make optimal use of the AP. For more complex but standard subroutines, such as an FFT (fast Fourier transform), one of course should use the Math Library routine, if available. Whenever a Math Library routine is used one should test the output frequently to be sure the routine is working and producing precisely the output you expect for the *full range of possible inputs* to the routine. (This is a precaution the wise computer user takes for *any* "canned" subroutine supplied for *any* computer!) The second option is to hand-code the time-critical inner loops, or hire/beg/browbeat someone else to do it. The unfortunate consequence of using either option is that the program cannot be fully tested using the host FORTRAN and related debugging aids, and as a result the debugging time is likely to be extensive.

C. *The AP and the World of Parallel Computing*

The AP will not seem like a particularly parallel computer when compared to the Cray or yet to appear supercomputers. Certainly, the parallel features of the AP are and will probably continue to be invisible to FORTRAN programmers. In fact, the FPS-164 in particular will probably seem a lot more similar to the CDC 7600 than to an explicitly parallel supercomputer. (Programs for the 7600 can often be speeded up by recognizing the parallelism of the 7600 architecture, but how many FORTRAN programmers know or care to know how to do this?) If users of large-scale scientific computing can be persuaded to use APs, especially in circumstances where an AP is locally accessible to the user while Crays and the like have to be used remotely, then the capabilities of the AP will define a range of problems which do not have to be run on less cost-effective supercomputers like the Cray. In particular there will be no point in restructuring a program to take advantage of the parallel architecture of a Cray-1 or a CDC 205 if the same program can in any case be run more cheaply and just as effectively on an AP.

If AP use grows, it will be possible to set up Array Processor "farms" where a reasonable size host has many APs attached to it, creating an aggregate computing power rivaling a Cray or better. Finally, and most speculatively, one would like to learn how to build a cluster of APs operating at full speed off a single common memory, so as to achieve Cray-1 speeds on problems requiring an enormous amount of memory at a reasonable cost. (The cost of the memory used in the AP is very much cheaper than the cost of memory used in the current generation of supercomputers.) The main problem in attaching several APs to a common memory is setting up the switching gear and multiple memory banks needed to allow all the APs to be accessing memory simultaneously with a minimum of delays due to multiple accesses to the same memory bank. Denelcor, Inc., has announced a multiple computer system of just this type, allowing up to 16 processors and 128 memory banks, but I have no experience with it. The memory and processing units of the Denelcor HEP are expensive for what they can do. However, the multiple processor to multiple memory switch of the HEP, *if it works to specification*, is of vital importance to the future of supercomputers. It could be used to address up to 100 million words of normal MOS memory (which is used in the AP). More-cost-effective processors similar to the FPS-164 could be designed to attach to the Denelcor switch.

V. EXAMPLES OF OPTIMAL PROGRAMMING FOR THE AP

In this Chapter I discuss a few specific problems and how they are coded to achieve maximum speed on the AP-120B or AP-190L. I shall not give actual programs since they would (at present) have to be in the assembly language of the AP and I do not want to explain details of assembly language. Timing is always given in terms of machine cycles. Each cycle is 1/6 μsec.

A. *Recursion and Ways to Avoid It: Polynomial Evaluation*

The standard procedure for efficient evaluation of a polynomial

$$f = a_0 + a_1 x + a_2 x^2 + \cdots + a_N x^N \tag{1}$$

on a computer is through a series of steps:

$$f = a_N, \tag{2}$$

$$f = a_{N-1} + xf, \tag{3}$$

$$f = a_{N-2} + xf, \tag{4}$$

$$\vdots$$

Each step involves one add and one multiply. Each step uses the outcome of the previous step; furthermore, within each step the add uses the result of the multiply. Since a multiply takes 3 cycles to complete on the AP, and an add requires 2 cycles, each step requires 5 cycles to complete. No parallelism is possible for this algorithm; the speed is therefore an add and a multiply every 5 cycles. This is five times slower than full speed.

One can use a modified algorithm which is faster: one can, for example, accumulate the even powers of x and odd powers of x separately, leading to a factor-of-2 increase in speed for a typical step, but more start-up time is required. The solution adopted in the FPS Math Library routine VPOLY is to assume that the user will want to evaluate f for several different values of x. VPOLY evaluates f for sets of five different values of x in parallel; the inner loop for this parallel evaluation proceeds at maximum AP speed.

B. *Unitary Matrix Multiply*

A subroutine that I had to construct for maximum speed involved the multiplication of two complex unitary 2×2 matrices of determinant 1 [in other words, two elements of the $SU(2)$ group]. I present this program as an example of code that runs the AP at nearly maximum speed but would run at slow scalar speed on any vector computer like the Cray. It also illustrates a feature of the AP: it *loves* arithmetic on complex numbers because this causes a high ratio of arithmetic operations to memory fetches. The matrices I was interested in have the form

$$M_1 = \begin{vmatrix} a & -b^* \\ b & a^* \end{vmatrix}, \tag{5}$$

$$M_2 = \begin{vmatrix} c & -d^* \\ d & c^* \end{vmatrix}, \tag{6}$$

$$M_3 = \begin{vmatrix} e & -f^* \\ f & e^* \end{vmatrix}, \tag{7}$$

and the program computes

$$M_3 = M_1 M_2. \tag{8}$$

The elements a, b, c, d, e, f are all complex. The program computes e and f given a, b, c, d; the formulas are

$$e = ac - b^*d, \tag{9}$$

$$f = bc + a^*d. \tag{10}$$

In terms of real numbers, there are four quantities to compute (the real and imaginary parts of e and f), and each calculation involves 4 real multiplies and 3 real adds, giving 16 multiplies and 12 adds in total. In the entire program I was building there were sequences of 3–5 matrix multiplies each;

Experiences with a Floating Point Systems Array Processor

when a matrix multiply in the middle of a sequence was performed, the only memory references required were to fetch c and d. The inputs a and b (the output of the previous multiply) were already in registers and the outputs e and f were to be left in registers. Thus the bottleneck in this subroutine is the multiplier; the routine is constructed to start a multiply on every cycle. In order to be able to start the multiplies on entering the routine, the matrix elements c and d are fetched during the previous matrix multiply; in fact, the first 3 multiplies for the current matrix multiply are also started at the end of the previous matrix multiply. The program is laid out in Table I, listing for each cycle the actions that are performed including the actions taken in the prior matrix multiply. There are also two saves to registers that take place in the succeeding matrix multiply. In Table I, the real and imaginary parts of a, b, c, etc., are denoted a_1 and a_2, b_1 and b_2, etc.

There are two sets of floating-point registers in the AP, denoted in the table as $X0$, $X1$, etc., for one set and $Y0$, $Y1$, etc., for the other set. There are a third set of registers for holding addresses and address constants. When the multiplies for the matrix multiplies begin the elements of M_1, namely, a_1, a_2, b_1, and b_2, are stored in registers $X0$, $X1$, $X2$, and $X3$; the elements of M_2 (c_1, c_2, d_1, and d_2) are stored in $Y0$, $Y1$, $Y2$, and $Y3$. The address calculations and memory fetches needed for the *next* matrix multiply take place in parallel with the multiplies and adds for the current matrix multiply; this is called "loop folding."

I shall not discuss the addressing in detail. There was a list of relative addresses stored in memory, one for each matrix M_2. The relative address had to be fetched from memory and added to a base address to give the absolute address of the first element of M_2.

The adds are displaced by 3 cycles from the multiplies, reflecting the fact that 3 cycles are required to carry out a multiply. The output of the multiplier is used in the adder as soon as the multiply is complete; this meant it was not necessary to save the multiplier output in a register. This was important because one of the restrictions of the AP is that only one X register and one Y register can be used for input to the adder or multiplier on each cycle. The multiplies all use two register inputs (always one X register and one Y register); therefore, no registers can be used as inputs to the adder. Thus the adder inputs are always either a multiplier output, an adder output, or zero (which happens to be an allowed input not requiring a register). The adds are organized as interlaced sums for e_1 and e_2 followed by interlaced sums for f_1 and f_2; with the 2-cycle add time on the AP, this arrangement allows each add, say for e_1, to use the output of the previous add for e_1.

Another problem that had to be avoided was that when the matrix elements for the next matrix multiply are saved in the registers, they erase the matrix elements needed for the current matrix multiply. Thus, for example,

TABLE I

Program for SU(2) Matrix Multiply

A. Previous Matrix Multiply (preparation for current multiply)

Cycle	Action
1	Fetch relative address of matrix M_2
4	Save relative address of M_2
5	Compute absolute address of M_2 (add base address to relative address)
6	Fetch d_2 (an element of M_2)
8	Fetch c_1
9	Save d_2 in register $Y3$
10	Fetch c_2
11	Save c_1 in register $Y0$
12	Fetch d_1
13	Save c_2 in register $Y1$
14	Multiply $a_1 c_1$ (a_1 is in register $X0$)
15	Multiply $a_1 c_2$; save d_1 in register $Y2$
16	Multiply $a_2 c_2$

B. Current Matrix Multiply

Cycle	Action
1	Multiply $a_2 c_1$; add: zero $+ a_1 c_1$
2	Multiply $b_1 d_1$; add: zero $+ a_1 c_2$
3	Multiply $b_1 d_2$; subtract: $a_1 c_1 - a_2 c_2$
4	Multiply $b_2 d_2$; add: $a_1 c_2 + a_2 c_1$
5	Multiply $b_2 d_1$; subtract: $(a_1 c_1 - a_2 c_2) - b_1 d_1$
6	Multiply $a_1 d_1$; subtract: $(a_1 c_2 + a_2 c_1) - b_1 d_2$
7	Multiply $a_1 d_2$; subtract: $(a_1 c_1 - a_2 c_2 - b_1 d_1) - b_2 d_2$ $(=e_1)$
8	Multiply $a_2 d_2$; add: $(a_1 c_2 + a_2 c_1 - b_1 d_2) + b_2 d_1$ $(=e_2)$
9	Multiply $a_2 d_1$; add: zero $+ a_1 d_1$; save e_1 in $X0$
10	Multiply $b_1 c_1$; add: zero $+ a_1 d_2$; save e_2 in $X1$
11	Multiply $b_2 c_1$; add: $a_1 d_1 + a_2 d_2$
12	Multiply $b_2 c_2$; subtract: $a_1 d_2 - a_2 d_1$
13	Multiply $b_1 c_2$; add: $(a_1 d_1 + a_2 d_2) + b_1 c_1$
14	Add: $(a_1 d_2 - a_2 d_1) + b_2 c_1$
15	Subtract: $(a_1 d_1 + a_2 d_2 + b_1 c_1) - b_2 c_2$ $(=f_1)$
16	Add: $(a_1 d_2 - a_2 d_1 + b_2 c_1) + b_1 c_2$ $(=f_2)$; RETURN

C. Next Matrix Multiply (completion of current multiply)

Cycle	Action
1	Save f_1 in $X2$
2	Save f_2 in $X3$

[a] The actions listed for the previous multiply also occur simultaneously with the current multiply except that they are preparatory for the next multiply. Likewise for the actions on the next multiply.

one sees from the table that c_1 is erased on cycle 11, and therefore all multiplies using c_1 must be started by cycle 11. The last cycle using c_1 is cycle 11; fortunately, the AP hardware is designed so that the save of the new c_1 only takes place at the end of cycle 11, whereas the transfer to the multiplier takes place at the beginning of the cycle.

One final note: there were several matrix multiply subroutines in my program; the other routines involved the multiplies $M_1 M_2^+$, $M_1^+ M_2$, and $M_1^+ M_2^+$, where M_2^+ is the Hermitian conjugate matrix to M_2. A sequence of matrix multiplies involved some start-up followed by a sequence of subroutine calls, each to one of the multiply routines. Each subroutine call required one cycle for the call itself, plus 16 cycles to execute the subroutine. During the subroutine call cycle the adder and the multiplier were left inactive, so that, for example, the subroutine call cycle did not count among the 3 cycles needed to comply a multiply.

When one first sees the instruction set for the AP, the problem of producing fast parallel code in assembly language seems overwhelming. My experience is that in practice it is not very difficult until one starts building programs of 50 or 100 cycles using lots of registers. The basic procedure for getting started on writing an assembly language program is to figure out which units are likely to be the bottleneck, and pick one of them (for example, for the program described here one could start with the floating-point multiplier). Then one simply writes down a proposed sequence of multiplies, one per cycle; if one does not have any reason for ordering the multiplies, then one orders them at random. I find it easier to start by picking the cycles on which a given set of multiplies are to be completed. One can then start writing in the instructions that use the multiplier output; one can start inserting the memory fetches so that they are available when needed, etc. Of course, at first one will very rapidly find conflicts developing, for example, needing to make references to two X registers as inputs on one cycle. But in my experience one learns fairly rapidly how to avoid these conflicts. I find very often that I can build a very tightly compressed program where on almost every cycle some disastrous conflict is barely avoided. The fact that one can do this is a tribute to the remarkably well designed instruction set of the AP. This instruction set seems rather strange, especially the rules for what *cannot* be done, until one realizes that in almost every case the rules are based on the way that the 64-bit instruction word is laid out. For example, only one X register can be referred to as input on one cycle simply because there is no room in the 64-bit instruction word to specify a second X register location.

The subroutines described here accounted for half the execution time in the program calling them. As a result the whole program was enormously faster in the AP (a factor of 8) than in the host 370-168. The entire program was run for ~ 400 hours of AP time.

C. Use of Nonarithmetic Instructions: A Random Number Generator

The AP has a good set of logical operations which act in a sensible manner on floating-point words (except for an "FEQV" instruction, whose behavior is not satisfactory due to subtle roundoff effects and should be avoided). These logical operations are useful for packing and unpacking; one especially nice application is for random number generation. The usual method for generating random number sequences is to take the nth member of the sequence, multiply it by a suitably chosen large number, and then take the low-order bits of the product modulo the word size of the computer. Unfortunately, the multiplier should be of order 48 bits in length in order to be able to generate very long random sequences by this method. (If the multiplier is n bits in length, then the "random" sequence will be periodic with period 2^n, at most; and if n is not large enough, then this periodicity can be disastrous). Because of the short word length (27-bit mantissa, plus sign and exponent) of the current APs, a 48-bit multiply has to be carried out in double precision, which takes many cycles. (I wrote a program to carry out this form of random number generation which required about 50 cycles per random number; by rewriting the program to generate many random numbers in the sequence in parallel, one could reduce the time to about 20 cycles per number.) Using logical operations one can generate random numbers at a rate of one every 4 cycles (6 cycles if standard data memory is used) and with a periodicity as large as one wants; the specific algorithm described below has a periodicity of 2^{250} numbers.

The "feedback shift register algorithm" (Lewis and Payne, 1973; Kirkpatrick and Stoll, 1981) for generating random numbers requires that a list of at least P numbers be already present in storage. Then the Nth number in the sequence (with $N > P$) is given by

$$R(N) = R(N - K) .XOR. R(N - P) \tag{11}$$

where .XOR. is the exclusive OR operation applied bit by bit to the numbers $R(N - K)$ and $R(N - P)$. Usually the XOR operation is applied to integers, in which case the exclusive OR is applied first to the 1's bit of $R(N - K)$ and $R(N - P)$, then the 2's bit of $R(N - K)$ and $R(N - P)$, etc. A specific choice for K and P is $K = 147$ and $P = 250$; other choices are possible (Zierler, 1969).

One often wants to use the random numbers in floating point format, say to generate fractions $R(N)$ lying between 0 and 1. The AP has two operations: AND and OR (actually denoted FAND, FOR), which act on two floating-point mantissas exactly as if the floating-point numbers were rewritten as binary integers and fractions and then the operation applied to every bit

Experiences with a Floating Point Systems Array Processor 303

location separately. The AND and OR (as well as XOR) all have the desirable feature that if the two input bits are 0, then so is the output bit; thus it does not matter which 27-bit window of bits is covered by the floating-point mantissas (the AP automatically aligns the mantissas with each other before carrying out the logical operation). The truth tables for AND, OR, and XOR are in Table II. To generate $R(N - K).XOR. R(N - P)$ on the AP one simply forms $R(N - K).OR. R(N - P) - R(N - K).AND. R(N - P)$. The only restriction on this operation is that the initial table of random numbers be strictly limited to positive fractions which are multiples of 2^{-27}, so that all mantissas of these numbers can be aligned with a single window covering the fraction bits 2^{-27}–2^{-1}. The inner loop is laid out in Table III. The loop is used to update an entire table of numbers, but I show only the operations needed to update one entry. As usual these operations take place over several passes through the loop. Registers $X0$ and $Y0$ are used for scratch storage. The logical operations each take 2 cycles to complete. The actions listed for pass $N - 2$ and $N - 1$ are of course repeated on cycle N but are setting up for the calculations of $R(N + 2)$ and $R(N + 1)$, respectively.

To update a full table of L random numbers ($L \geq P$) requires that the inner loop be interrupted twice. When numbers at the beginning of the table are updated, the inputs come from the end of the table, namely, one uses $R(N + L - K)$ and $R(N + L - P)$ rather than $R(N - K)$ and $R(N - P)$.

TABLE II
Result Tables for Bitwise AND, OR, and XOR

$X.AND. Y$	$Y = 0$	1
$X = 0$	0	0
1	0	1

OR	0	1
0	0	1
1	1	1

XOR	0	1
0	0	1
1	1	0

TABLE III

RANDOM NUMBER UPDATING LOOP OF RANDOM NUMBER PROGRAM[a]

Cycle	Pass $N-2$
1	Increment address of $R(N-1-P)$ to $R(N-P)$ and fetch $R(N-P)$
2	Increment address of $R(N-1-K)$ to $R(N-K)$ and fetch $R(N-K)$
3	
4	Save $R(N-P)$ in $X0$
	Pass $N-1$
1	Save $R(N-K)$ in $Y0$
2	
3	Compute $R(N-K)$ .OR. $R(N-P)$
4	Compute $R(N-K)$ .AND. $R(N-P)$
	Pass N
1	Save OR output in $X1$
2	Subtract: $(R(N-K)$ .OR. $R(N-P)) - (R(N-K)$ .AND. $R(N-P))$
3	Decrement loop counter by 1
4	Increment address of $R(N-1)$ to $R(N)$; save $R(N)$ (output of subtract) in memory; branch back to beginning of loop if loop counter > 0

[a] All steps listed occur in every pass through the loop, but only the steps needed to update a particular element $R(N)$ are shown. Timing is valid only for fast data memory.

The first K numbers are updated in this way. The next $P - K$ numbers are updated using $R(N - K)$ and $R(N + L - P)$. Then the last $L - P$ numbers are updated as originally discussed. The tests for the interrupt should of course not be placed inside the inner loop; instead an outer loop of three steps is programmed which initializes the relevant addresses and the inner loop counter for each of the three steps. The net result is a program with a total computing time of ~ 65 cycles startup time plus 4 cycles per random number. The program assumes "fast" data memory is installed on the AP.

D. *Branching*

Branching in inner loops in the AP is very simple and fast; a branch requires 1 cycle, although setting up a comparison requires 1–3 cycles prior to the branch. The comparison and the branch itself can be in parallel with other instructions. A simple list search with a branch when the desired item is found is a short loop, often only 2 or 3 cycles per element on the list. In contrast, branching in the middle of a vectorized loop on a supercomputer is awkward because no instructions are being issued while the loop is executing.

There are primitive procedures for handling certain types of branches on supercomputers, but they look not so pleasant to use; I have not actually tried them.

VI. THE TWO-MACHINE ENVIRONMENT

The traditional procedure for using the combination of a host computer and an array processor is to have a main program which runs in the host computer, and subroutines or sets of subroutines which run in the AP. Subroutines which run in the AP are called by ordinary subroutine calls. Thus if MMULT is a subroutine which is to perform a matrix multiply of two matrices A, and B, both of size $N \times N$, and C is to be the product matrix, then the host program includes the FORTRAN CALL statement

$$\text{CALL MMULT } (A, B, C, N)$$

When this CALL statement is executed, the program branches to a systems program which loads the AP with the program for matrix multiply as well as the data for the matrices A and B and the constant N. Then the system program starts the AP and waits until the AP computation is finished. When the AP computation is complete, the system program retrieves the matrix C and returns to the user's host program. If the AP computation involves subroutines called from MMULT, all necessary subroutines are loaded into the AP along with MMULT. If the AP routines include COMMON blocks or initializing DATA statements, the data for the COMMON blocks and the data statements are transferred along with the data for A and B.

It is not very difficult to write "CALL MMULT" into one's program. The difficulties come earlier. In order to establish MMULT as a subroutine running in the AP, one goes through the following steps:

1. Compile subroutine MMULT (and all other subroutines called by MMULT) using a special "cross-compiler" which produces object code for the AP. At Cornell the present cross-compiler is APTRAN.

2. "Load" the AP routines using a special load command. At Cornell, the command is APTLOAD. This command causes all the AP object modules to be linked together and stored in a data file. In addition, a host subroutine is generated with the name MMULT; this host subroutine includes the calls to all the systems programs needed to perform data transfers to and from the AP and to start the AP.

3. Load the host program routines, plus the host subroutine MMULT using the normal host loader.

The program is now ready to run. This procedure is somewhat more complicated than for conventional two-machine systems; for example, the CDC 7600 which always has a "front end" computer interfacing to terminals and printers. For the CDC 7600 an entire job is submitted to the 7600 and one is less aware of the operations at the front end. The original array processors had too little memory to hold an entire job, which is why the traditional procedure is to split a program between a host and the AP. However, a large AP-190L and especially an FPS-164 are large enough to hold entire programs; what remains to be done is to implement READ and WRITE statements in the current AP compilers along with other I/O statements. (The I/O would actually be transferred back to the host for execution, but the system would handle all the details of this.)

Some of the problems that arise in using the host–AP combination of computers have already been alluded to. The principal problems that arise due to having two computers to deal with instead of one are as follows (in no particular order):

(a) word length incompatibilities;
(b) host-to-AP data transfers: problems and costs, also the inaccessibility of host peripherals to AP programs;
(c) various error problems;
(d) scheduling a single-user AP for a multiuser host;
(e) problems of fast peripheral storage for the AP.

A. *Word Length*

The word length of the AP, 38-bit floating-point numbers, causes some minor problems because it falls between the normal word lengths for many hosts, namely, 32 bits or 64 bits. My own practice when programming the IBM 370-168 has been to use double precision (64 bits) exclusively. The roundoff errors when using 32-bit words are often intolerable, and I rarely need to use 32-bit words in order to save memory space. Unfortunately, the original loader written at Cornell assumed that by default floating-point words from the host would be single precision, and only with some restrictions could double-precision words be transferred. However, it is often important to be able to control precisely the values of AP words, in which case double-precision host words are essential to control all 38 bits of an AP word; in addition, the extra precision of the 38 bits is often important to maintain. I sometimes ran into minor problems, which always took time to fix, because I was using a mixture of single-precision words and double-precision words in the 370 in programs that were calling the AP. I should have stuck to my rule of using double-precision exclusively.

B. *Host – AP Data Transfers*

It is very easy to wipe out the benefits of the AP by having a program which spends all its time sending data back and forth to the host. The classic example of this is a host program which calls a sequence of vector routines in the AP, with data being transferred to and from the AP for each vector call. It is easy to avoid this particular trap; one has a driver program in the AP which contains all the vector calls; the host program makes one call to the driver instead of the whole sequence of vector calls. I have had no problems of this kind; my programs run uninterrupted in the AP for 15 minutes, and in this case the overhead of data transfers at the beginning of the end of the run are negligible (often less than 2 seconds of host CPU time). However, there are problems associated with host–AP transfers that are more painful to deal with. References to disk files or the printer on the host computer, in fact all FORTRAN READ and WRITE statements, must *at present* take place in the host program. Thus any program which makes frequent references to disk files, or which generate printer output frequently, must be broken into AP calls which take place only in between the READ and WRITE statements, and it may not be so easy any more to keep the data transfer times and costs low. One can collect data inside the AP to be printed later when the AP call is finished, in order to avoid frequent WRITE statements. This, however, is more inconvenient to program; one has to add code to the AP program to save the data, make sure it is transferred to the host, then put the necessary WRITE statements in the host program. This is also the way one has to deal with ERROR stops in an AP program, or with debugging output that is being generated when testing a program running in the AP. A FORTRAN compiler with I/O statements implemented would solve these programming problems, but there could still be excessive host–AP transfers.

Roughly speaking, a program that spends over a second uninterrupted in the AP should save money over running in the host; 15-minute AP runs are fantastically cheap. If running times of less than a second are contemplated between data transfers, then the data transfer costs have to be looked at carefully. The costs to be studied include the actual costs in time and dollars of host CPU and I/O operations; in addition, one has to worry about idle time in the AP while it is waiting for the transfer operations to complete. The idle time problem is more acute when the host runs under a multiuser time sharing system, so that the AP must wait its turn to have the data transfers take place.

C. Errors

One particularly annoying feature of the AP is that it is difficult to do very much when the AP goes into an infinite loop. One has to either study the program until a cause is found, or run larger and larger subsets of the program until the infinite loop appears, in order to localize the source of the loop or set breakpoints on the full program in the AP which cause only a subset to be executed, again running larger and larger subsets until the source of the loop is found. None of these options is particularly quick or easy.

The AP is designed to run a program to completion regardless of arithmetic overflows, illegal memory references, and the like. No matter how awful the results, the AP will always do *something*. If the program never reaches a normal termination as a result, then the scheduler stops the AP when a user-specified time limit is exceeded. Arithmetic overflows cause an error flag to be set, and an error message will be printed when the AP program finishes executing and returns to the host. One minor problem that arose is that whenever a 3-cycle multiply is in progress, a *new* multiply is started every cycle whether or not the programmer has requested one, and these unwanted multiplies can sometimes cause arithmetic overflows. This problem will not occur on the FPS-164. There are ways to control the problem in the AP-120B or AP-190L, at some cost in efficiency. Attempts were made to fix this problem for programs compiled by APTRAN but the fixes did not work, and no one has gotten around to fixing the fixes. Unexpected overflows can also occur at the end of a folded inner loop if some calculations are initiated (but not completed) that extend beyond the last allowed loop index.

D. *Scheduling the AP*

There is inevitably a conflict between users who want to use the AP for long production runs and users who want frequent short shots at the AP for debugging purposes. The solution to this problem for standard computers is to implement a time-sharing operating system which allows both production runs and debugging runs to execute simultaneously, or, more precisely, to alternate at millisecond intervals. The AP-120B and AP-190L are not set up for time-sharing; the FPS-164 has some hardware features to support time-sharing, but building a time-sharing operating system is an expensive and time-consuming proposition. At Cornell a compromise was adopted that seems to work reasonably well: no single AP job is allowed to run more than 15 minutes. This is implemented by a Gentleman's Agreement, rather than in the scheduler. (The "APEMAN" scheduler is described in Section III.C.) The 15-minute rule means that users may have a wait of order 5–10

Experiences with a Floating Point Systems Array Processor

minutes (on the average) to complete a debugging run even if the run itself consumes only a few seconds of AP time. On the other hand, users making production runs have to arrange their programs so that they stop at 15-minute intervals, return to the host, and then the host program saves all necessary data on disk to be carried over to the next run. In summary, program development is made more difficult both by the need to build in the data checkpointing and by the slow response time for debugging runs.

E. *Problem of Fast Peripheral Storage*

The customary source of extra scratch memory on a computer is a disk. In the case of the AP, one can use the disks on the host, but this is slow, awkward, and expensive if frequent disk references are required. Floating Point Systems supplies a disk which attaches directly to the AP, but I have no experience with it. Questions one must ask when considering this disk include

1. What is the error rate for transfers to and from disk?
2. What is the software available for accessing the disk?

If a really large peripheral memory is required, a disk is the only solution. However, disks are an awkward solution to the scratch storage problem for any fast computer. It takes milliseconds to locate a file on disk, and the data transfer rates are slow. Semiconductor memory is now quite cheap, and much more suitable as scratch storage. I think one could make a disk substitute out of semiconductor memory that would have 5–10 times the memory actually on the AP, for a reasonable cost. To date no such product is available.

VII. PRACTICAL PROBLEMS OF AP OWNERSHIP

There are some practical problems that arise when considering whether to obtain an AP, which I will discuss briefly. First of all, an AP requires a host computer, and one must decide which (if any) of the existing computers to use as a host, or else obtain one. Customarily the AP-120B has been attached to small minicomputers (PDP 11s and the like); this is a cheap and appropriate arrangement for traditional Array Processor applications. Some simple problems (especially in molecular dynamics) can use this arrangement; the FPS-100 can be an especially cheap Array Processor for this case. This is not the right solution for general scientific computing, such as for the AP use at Cornell. The host computer must be powerful enough to make possible both debugging runs and compilations of AP FORTRAN programs, and the small minicomputers have too little memory and are too slow either for debugging or for running the AP FORTRAN compilers, present and proposed. Instead, one should consider only large minicomputers like the VAX or 4341 with virtual memory operating systems or else mainframe systems to be the host. The second problem is to determine who will be using the AP if one is obtained. It is perfectly possible to go out and buy an AP and then discover that no one will use it. At the time an AP is purchased, one must have an identifiable user or set of users whose problems and programs fit the constraints of the AP being obtained (the constraints of memory size, word length, limited access to peripherals, and limited and not always debugged software), and who will take the trouble to use the AP when installed. At Cornell the formation of a consortium in advance who helped to pay for the AP and therefore had a serious commitment to using it was a crucial factor in the success of the AP project. There have been new users who have in some cases benefited considerably from having the AP available, but they provide a rather small part of the total computing load on the AP. To be explicit about what I am saying, one does *not* follow the traditional avenue of "evaluating the AP." Instead, one evaluates the potential *users* of the AP. Is the AP suitable for each user? Given the current state of software available for program development and AP operation, will the user actually get his programs converted to the AP and run them on the AP? The third problem one faces with an AP project is to determine how the project will be kept going once started. For example, who will arrange for maintenance of the hardware and software when problems arise? Somebody has to know enough about the AP hardware and software to pinpoint problems as they arise and then must badger FPS personnel to fix the problem. Who will provide consulting to users who have become stuck because of some problem and do not know how to proceed? Someone must know enough about the AP to be able to recommend procedures for locating problems, and preferably must know enough about assembly language programming for the AP to be able to fix trivial problems in Math Library routines. At Cornell, most of these functions are handled by the AP support group in the Cornell Computing Center. In many installations there is a "chief user" who handles these tasks. Usually the user community is small and a single chief user can handle these functions without a lot of lost time. The only exception is at the time of installation of the AP, when a fair amount of effort may be required to make sure that hardware problems are fixed and especially to bring the software into full operation despite the idiosyncrasies of the host operating system.

One problem that, in my experience, does *not* occur with an AP is instant saturation. The added difficulties of program development for the AP combined with its awesome capacity and limited user base means that the

AP is unlikely to become overloaded over long periods of time, with the resulting frustration of slow turnaround. This is of course a blessing. As long as the AP is handling a reasonable load, using, say, 25-50% of its capacity, it is providing more cost-effective computing than any alternative, and no one is likely to complain that it is underutilized. Keeping the load on the AP below saturation also means it is much more accessible when needed for program debugging.

One logical way to set up an AP is to follow Cornell's example and attach it to a mainframe computer in a computing center, often an IBM 370 series machine. There are some problems to be considered in this case. First, there is no interface to CDC machines (6600, 7600, or Cyber 70 series). Second, the approach to a computing center director with the proposal to buy an AP has to be made cautiously. The director typically

1. has never heard of the AP or Floating Point Systems, Inc.;
2. is heavily dependent on service personnel from IBM (or whomever) to keep his computing operations afloat and does not want to have to deal with unknown hardware and software from an unknown company;
3. has a staff which is overworked, not very responsive to users, often have unrewarding jobs (if you do not believe this, try replacing one for a day) and would resent any increase in their workload the AP might bring;
4. has a motto of "equally good (or bad) service for everyone except Accounting" and will not see the merits of supplying enormous computing power to a few users of his center.

There are exceptions to this somewhat grim picture; the Cornell Computing Center staff have been very helpful in their support of the AP project, and all are agreed that it has been a success. Clearly, a little research is in order into the character of a computing center and its staff before suggesting an AP to them.

The second problem with a mainframe is that one has to make sure it has an operating system that one can live with for program development, and make sure it is not so saturated that one gets no turnaround on test jobs. The 370 at Cornell has a time-sharing operating system (VM/CMS) that is primitive but workable; it is often saturated during the day, but I work at night when computing time is cheaper and more available.

Even if one adopts the politically more feasible approach of attaching an AP to an existing departmental large minicomputer, one must be sensitive to the problems of maintenance and consulting that the computer support staff will face if an Array Processor is brought in; often some kind of division of responsibility will have to be arranged between the support staff and the chief user; if there are enough users of the AP, then it may be necessary to obtain additional staff to handle the consulting and maintenance problems. At Cornell there are roughly 40 users, and even after discounting special software development efforts it takes roughly $1\frac{1}{2}$ support staff to keep everyone going.

One final note: if one uses a multiuser time-sharing host for an AP, make sure that the time-sharing system is a "paging" system and not a "swapping" system. In a swapping system a host program can be locked into physical memory during the whole time that an AP subroutine is executing, even though the host program is only waiting for the AP to finish.

VIII. CONCLUSIONS

The AP-190L installation at Cornell has provided a number of physical scientists with a very sizable increase in computing power over what would have been available to them elsewhere. This has made it easier for some to continue ongoing projects; others have carried out projects (usually simulations) which otherwise would not have been possible. This does not mean that the AP has unique capabilities that other computers do not have; it is just that computing time is very much cheaper on the AP than on any alternative computer and the Cornell users of the AP all have limited computing budgets.

Program development has been a serious problem. To improve this problem, the main need is to be able to debug AP programs, in their *final form*, in the host computer using the host debugging aids. This means avoiding assembly language routines or Math Library routines in the AP, or if they are necessary, having completely functionally equivalent FORTRAN routines for testing in the host. It is especially important that it be easy immediately to rerun *entirely* in the host any program that runs into trouble in the AP. This will be much easier to arrange when the FORTRAN compilers for the AP have good code optimizers. In addition, either a larger user base or a concerted testing effort is required to bring the compilers, loaders, and library software for the AP into a mature, debugged state.

Now that Floating Point Systems has a strong compiler effort, the Cornell support group will try to concentrate on bringing in an FPS compiler and making it usable at Cornell, rather than continuing to develop APTRAN. It is hoped that the FPS-164 compiler will serve this purpose. The Cornell group will make every effort to work with FPS to maximize the extent to which ultimately the standard FPS compilers can be used without modification for both the 164 and the 190L. In addition, the Cornell group will be exploring methods to achieve fully optimized code from the compilers, including novel approaches such as Monte Carlo optimization methods.

One effect of my own programming efforts for the AP has been to become much more aware of the limitations of the FORTRAN language itself. The facilities for packing and unpacking bits are abysmal, the enticements to write unreadable code are overwhelming, and all this becomes painfully apparent whenever one tries to make wholesale changes in code in order to achieve more efficient running times on the AP. Alternatives to FORTRAN, especially when they come with a good program development system, need to be explored. The obvious candidate at the moment is the UNIX (trademark of Bell Telephone Laboratories, Inc.) operating system and the C language; for the future the Defense Department's ADA language should be watched.

The FPS-164 should overcome many of the hardware limitations of the AP-120B and AP-190L, and with its present maximum memory size of 1.5 million words it should provide relief even for many users of overloaded Cray-1s. As larger memory chips become available the maximum memory size of the FPS-164 will also increase. The biggest problem, however, is to win acceptance for the AP for what it can do in the excessively conservative world of large-scale computing, where people seem more interested in either raw speed or software compatibility with present computers than in cost-effectiveness, if they are interested in anything at all except to resist change.

ACKNOWLEDGMENTS

I would like to thank the many people at Cornell and at Floating Point Systems who have helped make the AP project at Cornell a success. I am especially grateful to Doug Van Houweling for organizing the AP project in the Computing Center and Geoffrey Chester for organizing the consortium of users; to Alec Grimison for his able leadership of the AP project; and to the members of the project for their continued efforts in maintenance, software development, and consulting help. Above all, I am grateful to Donna Bergmark, who almost single-handedly, often under trying circumstances, developed APTRAN into a usable production compiler. This was the first working compiler for the AP, and virtually all projects carried out on the AP at Cornell have relied upon it.

REFERENCES

Farouki, R., and Shapiro, S. L. (1980). Rep. No. 740 and 743, Center for Radiophysics and Space Research, Cornell University, Ithaca, New York.
Kirkpatrick, S., and Stoll, E. (1981). *J. Comput. Phys.* **40**, 517.
Kunz, P. (1976). *Nuclear Instrum. Methods* **135**, 435–440.
Lewis, T. G., and Payne, W. H. (1973). *J. Assoc. Comput. Mach.* **20**, 465–468.
Richter, R. (1980). Ph.D. Thesis, Physics Dept., Cornell University, Ithaca, New York.
Shapiro, S. L., and Teukolsky, S. A. (1979). *Astrophys. J.* **234**, L177–L181.
Shapiro, S. L., and Teukolsky, S. A. (1980). *Astrophys. J.* **235**, 199–215.

Shenker, S., and Tobochnik, J. (1980). *Phys. Rev. B: Condens. Matter* **22**, 4462.
Tobochnik, J., and Chester, G. V. (1980). Rep. No. 4226, Materials Science Center, Cornell University, Ithaca, New York.
Wilson, K. (1980). *In* "Recent Developments in Lattice Gauge Theories" (G. 't Hooft *et al.*, eds.). Plenum, New York.
Wilson, K., and Kogut, J. (1974). *Phys. Rep.* **12C**, 75–199.
Wood, D. M. (1980). Ph.D. Thesis, Physics Dept., Cornell University, Ithaca, New York.
Zierler, N. (1969). *Inform. and Control* **15**, 67–69.

A Case Study in the Application of a Tightly Coupled Multiprocessor to Scientific Computations

Neil S. Ostlund
Peter G. Hibbard
Robert A. Whiteside

Department of Computer Science
Carnegie-Mellon University
Pittsburgh, Pennsylvania

I. INTRODUCTION

Computational physicists, chemists, biologists, etc., need hardware and software facilities that are capable of solving very large numerical problems. Processors which are capable of executing several operations simultaneously are frequently a more cost-effective way of supplying these needs than are serial computers. For example, the pipelined vector processor Cray-1 (Russel, 1978), which employs a relatively limited degree of parallelism, has proved to be a practical and successful architecture. There are many different feasible multiprocessor designs, however, and the design space has been only partially explored. It is possible that better candidates for very high performance scientific computation in the next decade or so will be multiprocessors which execute many instructions (perhaps thousands) simultaneously.

In this paper we report on a collection of experiments designed to understand the potential of a particular general-purpose tightly coupled multiprocessor: Cm* (Fuller et al., 1978). Since the hardware is experimental and not highly optimized, our experiments have more of the nature of software design explorations and feasibility studies than performance measurements. For this reason, we spend some time explaining what are the basic issues which need to be addressed when designing programs for this particular class of machines. We present a collection of "standard" decompositions of programs, and we describe several problems that have been examined on Cm*. We indicate why some of the architectural characteristics of Cm* may become increasingly important in future generations of parallel computers.

In Section II, we describe the hardware and software characteristics of multiprocessors. In Section III, we describe experiments in numerical computation that are being carried out on Cm*. Finally, in Section IV, we describe our initial conclusions of this continuing work. The appendix contains details of the parallel programming of three specific numerical applications that are considered here.

II. TIGHTLY COUPLED MULTIPROCESSORS

In this section we describe the hardware organization of tightly coupled multiprocessors, and the design of a particular example—Cm*. We then describe some of the software issues raised by this class of computer.

A. *Hardware Organization*

Flynn (1972) has presented a taxonomy of parallel computers which is particularly useful for describing the features of multiprocessors. Parallel machines in this taxonomy are characterized by the number of independent paths to memory along which data can move in parallel, and by the number of independent instructions which can be executed in parallel. The conventional serial computer has a single *instruction stream* and a single *data stream*. It is characterized as a SISD (single instruction stream, single data stream) computer.

A computer which executes a single instruction stream, but performs operations in parallel on several data items to produce several results, is characterized as SIMD (single instruction stream, multiple data stream). A block diagram for such a machine is shown in Fig. 1. The processing component of a SIMD machine consists of a single control unit (CU) and many processing elements (PEs). The CU fetches and decodes the stream of instructions from a program. An instruction is executed either by the CU or is sent simultaneously to all the PEs to be executed by each in parallel. PEs

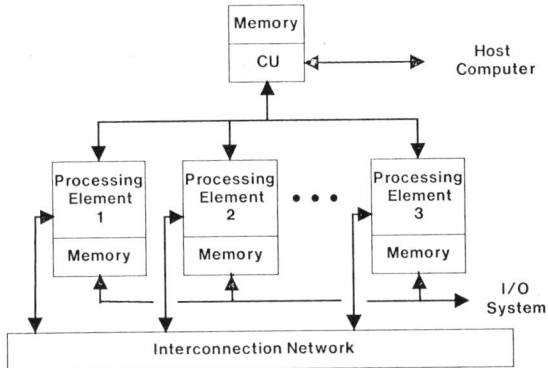

Fig. 1. Block diagram of a SIMD architecture.

may be active or inactive during an instruction execution as a result of control signals from the CU or as a consequence of data-dependent conditions which arise during the execution of instructions. The memory component of an SIMD computer may be a single memory accessible to the CU and all the PEs, but more generally it is partitioned as shown, in order to allow simultaneous access by each PE. The switching structure of an SIMD computer permits data to be broadcast from the CU to all the PEs. Data may also be shipped from the PEs to the CU. In order to allow the direct exchange of data between PEs, an interconnection network is sometimes provided. Full PE interconnection is unlikely because of cost. A typical interconnection scheme is four connections per PE, one to each neighbor in a two-dimensional array. Connections to an I/O system and a host computer are normally provided also. The archetypical example of an SIMD machine is ILLIAC IV (Barnes *et al.*, 1968).

A computer which has several instruction streams that operate simultaneously upon independent data streams is characterized as MIMD (multiple instruction stream, multiple data stream). A block diagram for such a machine is shown in Fig. 2. There are n processing components in such a computer, each of which has an instruction decoder and an arithmetic and logical unit (ALU) entirely independent of the other processors, so that the processors can decode and execute instructions simultaneously. The memory component of a MIMD computer consists of m memory modules, each of which may be accessed independently. The crosspoint switch allows any of the n processors to access any of the m memories and thus any of the $n \times m$ crosspoints may be a potential connection. In order to allow parallel access, the switch must allow n of the crosspoints to be in use simultaneously. In addition to the structure presented here, there will normally be a simple

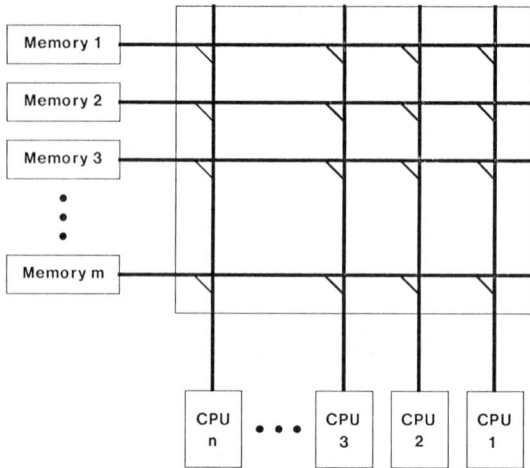

FIG. 2. Block diagram of a MIMD architecture.

interconnection network to allow control signals to pass between the processors and an I/O system. The archetypical example of an MIMD machine is C.mmp (Wulf and Bell, 1973).

A comparison of the different design criteria which have led to SIMD and MIMD machines is instructive. The processors of an MIMD machine must each contain an instruction decode unit, an ALU, and memory addressing hardware; potentially they are more complex than the PEs of an SIMD machine, which use the single centralized instruction decode unit. However, the MIMD processors can be (more or less) off-the-shelf general-purpose processors, whereas the PEs must be specially designed. Current trends in hardware technology are widening the gap between highly integrated off-the-shelf computers and specialized one-of-a-kind processors, and are therefore causing an MIMD machine to be more cost-effective than a more highly engineered SIMD machine. There are other reasons than pure cost-effectiveness for preferring a MIMD architecture. After dispatching an instruction, the CU of the SIMD machine might be required to wait for control signals from each PE indicating the completion of the instruction before it can dispatch the next instruction. There is a large amount of distributed control exerted at every instruction cycle of the machine. An MIMD machine need not exert distributed control at every cycle since each processor operates independently. In fact, it might exert control only very infrequently. Distributed control is expensive, both in terms of logic and in terms of the time taken to generate, receive, and act upon the control signals. It imposes a limit to the number of PEs that it is feasible to incorporate in a

Application of a Tightly Coupled Multiprocessor

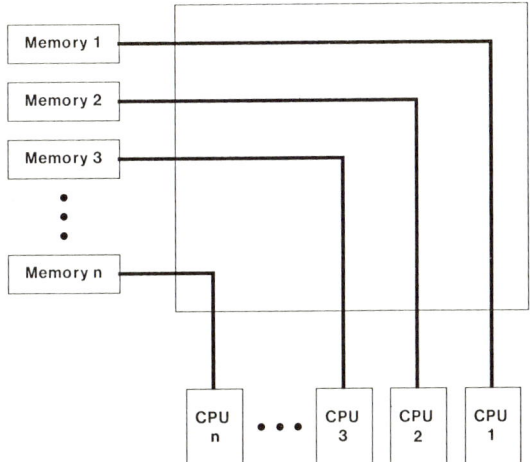

FIG. 3. Block diagram of n independent computers.

single machine structure, and it is unlikely that SIMD machines much larger than ILLIAC IV will be developed in the future.

The MIMD architecture which has been presented does, however, suffer from a potentially severe limitation. In order to obtain n-fold parallelism, there must be at least n memory units, and the switch must contain $O(n^2)$ crosspoints. Thus the cost of the switch will increase as n^2, and for suitably large n it will dominate the cost of the computer. Furthermore, the switch poses a reliability problem since its failure renders the whole machine inoperable. Refinements of the basic MIMD architecture have generally addressed themselves to ways of avoiding an n^2 switching structure.† We take here an argument which leads to the switching structure of Cm*; other switching structures are certainly possible (Wu and Feng, 1979).

A collection of n computers, each composed of a processor and a memory unit, can be represented as in Fig. 3. In this view it is clear that the crosspoint switch is a device which converts a collection of computers with low access costs to local memory and infinite access costs to nonlocal memory into a collection of computers with the same access cost to any memory. However, experiments with programs running on MIMD machines have shown that the memory references generated by an individual processor are localized. That is, they generally fall into a single memory unit, with only occasional

---

† For modest numbers of high-performance processors, switch costs form only a small fraction of the total machine cost, and architectures using a full crosspoint are feasible. Such an example is the S-1 computer being developed at Lawrence Livermore (McWilliams et al., 1977), which will have a C.mmp structure with 16 processors each of Cray-1 power.

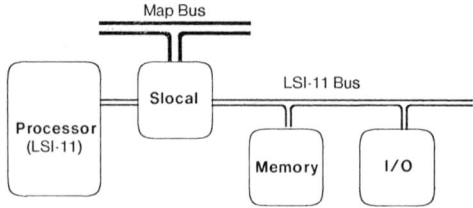

Fig. 4. A computer module of Cm*.

references being made to other memory units (Raskin, 1978). The crosspoint switch, therefore, gives more flexibility to their addressing structures than most programs require; consequently, it is worthwhile exploring structures that trade off the flexibility of the switch against its cost. A switch that takes Fig. 3 as a starting approximation to the memory access paths may be better starting point than a switch which starts from Fig. 2.

The Cm* switching structure illustrates this. The individual components are processor-memory pairs connected together with a high bandwidth communication channel. Off-the-shelf LSI-11s are used for Cm*, with the LSI-11 bus providing the communication channel. In order to connect together the component processors, a custom-designed switch, the Slocal, is attached to the LSI-11 bus; the processor–switch–memory combination is termed a computer module (Cm) and is shown in Fig. 4. A collection of Cms is connected together, through their Slocals, by a map bus, which is presided over by a special-purpose microcoded mapping processor, the Kmap, as shown in Fig. 5. This collection is known as a cluster. Finally, several clusters may be connected together by connecting their Kmaps by intercluster buses as shown in Fig. 6.

Each word of memory has a unique physical address, as given by the cluster number, Cm number within the cluster, and word within the local memory. Each processor is able to address any word in memory by exercising the

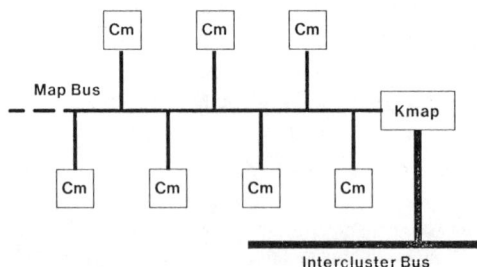

Fig. 5. A cluster of Cms.

Application of a Tightly Coupled Multiprocessor

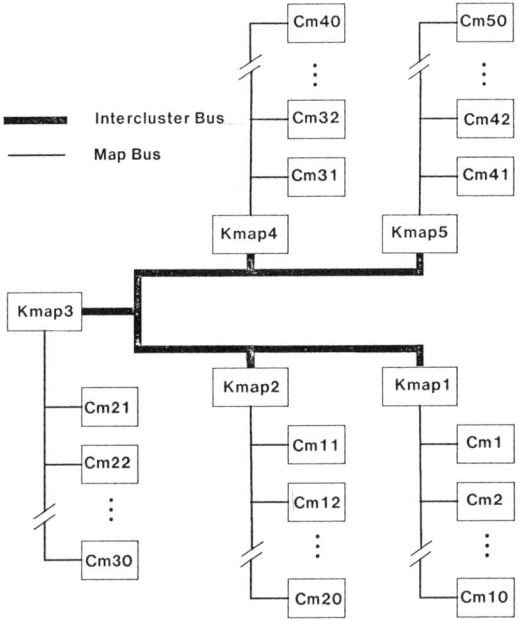

FIG. 6. A multicluster configuration of Cm*.

appropriate access paths through the local LSI-11 bus, map bus, and intercluster bus. In order to avoid the need for processors to use system-wide addresses, mapping is performed by the Slocals and the Kmap, thus allowing processors to use virtual addresses.

To access a word of memory, the LSI-11 processor places a virtual address on the LSI-11 bus. Depending on the physical location of the word that corresponds to that virtual address, one of several actions can take place, as follows:

(a) The Slocal inspects the virtual address to determine if it corresponds to a physical address within the local memory. If it does, the corresponding location within the local memory is accessed. There is sufficient latitude in the LSI-11 addressing mechanism for the necessary lookup and address translation to be done without slowing down the access. Thus, if memory accesses are completely local, the LSI-11 operates at full speed. Since each Slocal is operating independently, n accesses can proceed simultaneously.

(b) If the Slocal finds that the virtual address generated by the LSI-11 processor is not in local memory, it generates an information packet containing the addressing information, which is sent to the Kmap; the LSI-11 processor is then held up. The Kmap inspects the virtual address in the packet which has been sent, and if it corresponds to a physical address within the cluster, an information packet and request to access memory is sent to the appropriate Slocal. The destination Slocal accesses memory invisibly to the processor in the Cm by cycle stealing. A return package is generated, routed through the Kmap back to the source Slocal, and the LSI-11 processor is permitted to continue. Such a nonlocal access takes three times as long as a local access.

(c) If the Kmap finds that the virtual address sent to it corresponds to a physical address in some other cluster, the packet is redirected towards that cluster, where it is serviced as for an intracluster access. This imposes an additional threefold delay on access.

There are several interesting features of such an addressing structure. First, the hierarchical organization takes advantage of locality of reference, and introduces only as much switching and address mapping actions as are necessary to translate the address and access the memory. This is in contrast with the canonical organization (Fig. 2), which imposes the same switching and mapping overhead irrespective of the locality of reference. Second, the switching hardware costs grow much less than n^2: each Cm requires a Slocal which adds a cost linear in the number of Cms, and each cluster requires a Kmap, adding a cost linear in the number of clusters. Third, the amount of memory in each Cm, the number of Cms in each cluster, and the number of clusters in the machine are not fixed by the design. Suitable values for these can be chosen after observing the access and performance characteristics of programs, whereas for the canonical MIMD machines it is fixed.† Finally, the switch is distributed, so it is no longer a central point for failure. It is possible to partition the machine into separate parts, each with the functional characteristics of the whole, by closing down some of the intercluster links.

It is necessary to trace the history and motivation for the design of Cm* to be able to understand the implementation, and to interpret the results of our experiments in the appropriate light. The design of Cm* started in response to lessons learned in the development of C.mmp (Wulf and Bell, 1973), and in order to take advantage of trends in integrated manufacturing style. Rather than design processor–switch–memory modules specifically tailored to the machine, and thereby lengthen the design period of the project by about 2 years, it was decided to use currently available components. In choosing LSI-11s several restrictions were accepted:

(a) The LSI-11 is slow. Average memory access time is about 3 μsec and typical instructions take about 6 μsec. Floating point operation times are very slow—a 32-bit floating-point add requires more than 42 μsec.

† Thus, potentially at least, one can have as much switching structure as one needs or can afford.

(b) The address space of the LSI-11 is only 64K bytes, which complicates the addressing of large data structures.

(c) The physical memory of LSI-11s is limited to 256K bytes. If the programs residing in one Cm need to access more physical memory than this, they are forced to access it nonlocally.

In contrast to the Cms, the Kmaps were custom built and tailored to their application—packet switching and address translation. Each Kmap is a fast 150-nsec horizontally microcoded special-purpose machine. Eight hardware microprocess contexts and several queues are maintained in hardware to allow a Kmap to handle up to eight access requests simultaneously. Since there was considerable research interest in determining the appropriate functional characteristics of a Kmap, they were given disproportionately large computing power. Currently, this is being used to experiment with different addressing structures (such as simple address mapping and capability-based mapping), and for determining which functions (such as message passing) it is appropriate to centralize into the Kmaps. Each Kmap possesses 8K words of microstore, but the simplest address mapping microcode requires only 250 instructions.

B. *Software Organization*

The factors which lead to the hardware organization of MIMD machines in general, and Cm* in particular, are fairly clear: the availability of cheap LSI and VLSI components and the need to impose a communication hierarchy on the machine. Whether the resulting architecture is of value for practical problems is less clear, for several reasons. First, on an MIMD (and a pure SIMD) machine one must ensure that the whole program fits the architecture well, rather than a small part of it. Second, because the program structure for SIMD and MIMD machines is radically different from that for pipeline and vector machines, few software tools are available. As it happens, a large amount of culture has grown up around ILLIAC IV, but there is little experience with MIMD machines. There are two reasons to be encouraged in our attempt to develop practical programs for MIMD machines, however. First, a great deal of theoretical work has analyzed classes of problems for MIMD machines and has shown encouraging results (Kung, 1976, 1980) for the speedups possible with simple computer models. Second, the basic structure of an MIMD machine is similar to that presented by a multiprogramming operating system, which maps processes consecutively onto a single physical machine, rather than simultaneously onto several machines. Thus many of the basic structures needed to build software for

MIMD machines have been exposed and understood in different contexts, and the issues which are raised concern more the matching of program characteristics to the machine performance than basic issues of programming paradigms.

Our analysis in this section examines the basic principles of software organization for MIMD machines. We concentrate on two factors: what are the architectural features which affect most strongly the performance of programs, and what organization should be imposed on a problem to decompose it for a multiprocessor?

1. *Program Structure*

MIMD programs, which need to control and coordinate several simultaneous instruction streams, are inherently more complex than those for serial machines and for other classes of parallel machines. In order to manage this complexity, it is necessary to impose a well-defined organization onto programs. The appropriate organizational unit is the process, corresponding in part to a program on a serial machine, through which runs a single thread of control. A program for an MIMD machine then comprises a collection of processes which (potentially) execute simultaneously.

It is not necessary for the number of processes to be equal to the number of processors, since it is possible to map processes onto processors either at load time or at run time using standard scheduling techniques. It is, in fact, an advantage for a program not to require some fixed number of processors in order to execute because the actual number of processors available may not be known until run time. In addition, some algorithms can use this flexible run-time mapping to reduce the total execution time by dynamically allocating processing resources to the critical path of the computation.

Communication between processes is done either by means of *messages*, which are sent between processes only when the sender is ready to send and the receiver is ready to receive, or by means of *shared variables*, which may be read from and written to by several processes. Shared variables map directly into accesses to the shared memory, while message-passing requires implementation in terms of shared variables. It might seem, therefore, that shared variables are a more natural way of expressing communication in MIMD machines, and this is indeed the case for some algorithms. There are, however, other algorithms for which message passing is more natural and imposes a stronger discipline on the program. When communication is via shared variables, the ability to enforce *mutual exclusion* is required; a process must be able to acquire exclusive access to a data structure in shared memory. For a number of different reasons, a process may want to lock out another process while it manipulates a particular data structure.

Application of a Tightly Coupled Multiprocessor

In a multiprocess program, it is usually necessary for a process to be able to synchronize with one or more other processes. That is, before any process involved in the synchronization action can proceed past the synchronization point, all of the processes must be at their corresponding synchronization point. This generally involves some processes waiting for others to reach the synchronization point. Synchronization is required to coordinate the activities of several processes, and to ensure the orderly transfer of information between them. A message causes synchronization as well as the transfer of information.

2. Measures of Performance

We measure performance either by the *speedup* ratio t_1/t_n, where t_1 is the time for the program to execute using 1 processor, and t_n the time to execute using n processors, or by the processor *utilization* $t_1/(n \times t_n)$, which measures the fraction of time the processors spend performing useful work. Running a parallel program on a single processor will usually involve some overhead which is not present in the corresponding serial algorithm. Thus, it is probably better to use the execution time of a *serial* program for t_1 rather than the execution time of a parallel program running on only one processor. The algorithms for these two choices of t_1 may well be different, but the measure of how much better a multiprocessor performs is more concrete when comparison is made with the serial algorithm.

Measuring performance by the speedup and utilization reveals three interesting aspects:

(a) For most problems and decompositions, the speedup ratio rises (generally less than linearly) to a maximum as the number of processors increases, and then starts to fall, so that adding more processors slows down the program. Some programs continue to speed up all the way out to 50 processors.

(b) The average processor utilization, as measured against a uniprocessor executing a serial algorithm, is sometimes low. This indicates that the multiprocessor hardware and software organization is imposing high overheads, and this is frequently quoted as illustrating the essential inefficiency of multiprocessors. However, we need to be aware of how these overheads arise: they arise because an MIMD machine has little hardware to assist in the distributed control of computations. Control must be accomplished by software, and therefore is prominent in our performance figures. More hardware is devoted to distributed control in SIMD machines and (generally) in highly optimized serial machines than in present multiprocessors. This hardware is not generally visible to the programmer. In order to obtain an improvement in performance over a strictly serial machine it will be necessary to devote machine hardware resources to coordinating and synchronizing the activity of concurrent operations. Of course, if one has a multiprocessor of strictly limited size (because of architectural properties) such as ILLIAC IV (limited to 64 PEs) or S-1 (limited to 16 processors), then one needs to chose algorithms which have a reasonably high processor utilization. For Cm*, which may be made indefinitely large for nearly linear cost, one should be more concerned with chosing decompositions which continue to speed up with very large numbers of processors, rather than with decompositions which have a high processor utilization.

(c) Very surprisingly, there are algorithms which yield more than linear speedup on a multiprocessor, even compared to the best decompositions on uniprocessors. At first sight this appears paradoxical, since one expects overheads to occur in the multiprocessor case which do not exist in the uniprocessor case. Closer examination shows, however, that this need not be so. An example is given below.

3. *Basic Algorithmic Decompositions*

There are several factors which affect how a program is best decomposed into processes. A part of the research reported here has been to investigate the influence of these factors. The factors are

(a) The communication hierarchy imposes different accessing cost to memory. It is, therefore, worthwhile to localize the words accessed by a process to the Cm executing the process; this is not possible if the information is scattered or if there is too much to fit into the memory of a single Cm. In this case it is better to maintain local copies of the information accessed most frequently, provided the program does not require the sharing of this information. As it happens, the memory accessing pattern of the LSI-11 architecture imposes some very clear requirements: about 80% of all accesses are to code, 10% are to stack data, and 10% to stored data, measured over a wide variety of programs. Thus it is clearly advantageous to have the code and stack local to the Cm executing the process, and in general this can always be trivially arranged. †

(b) Memory contention and Kmap contention. The Cm/Kmap ratio has been chosen so as to allow the Kmaps to perform mapping functions for inter-Cm references without saturating, if the Cms are generating a reason-

---

† The code for very large processes may not fit into the memory of a single Cm. If this is the case, it will be necessary either to copy code into a Cm prior to executing it or to move the locus of control in a process from one processor to another. None of the problems we have studied so far on Cm* has required this additional complication.

Application of a Tightly Coupled Multiprocessor

able mix of local and global references. As it happens, the Kmap is sufficiently fast that only pathological programs cause Kmap contention, and no special measures need be taken to avoid it. Contention for a single memory module is a problem, however. If a number of Cms repeatedly access the same memory, as might happen if the memory contains a variable shared between several processes, severe degradation results. Thus it is necessary to avoid decompositions which require a single shared variable to be accessed repeatedly, and if a decomposition requires several shared variables, these should be spread out over several Cms.

(c) Scheduling costs. If there are more processes than processors in a program decomposition or if synchronization occurs frequently, it is necessary to alter, during run time, which process is being executed by which processor. This involves computational overhead, because the internal state (registers, stack, etc.) of one process needs to be saved, another process then selected for execution, and the internal state of that process loaded into the processor. To a first approximation, the fewer such context swaps which need to occur, the lower the overhead and the faster the computation. One effect of adding more processors is to reduce the number of necessary context swaps—this is one reason why more than linear speedup can be observed for some programs. It is also possible to reduce the number of context swaps by adjusting the amount of work for each process; since one's intuitions as to the execution time of processes are frequently wrong, fine tuning a program can sometimes dramatically improve its performance. However, rescheduling is not always to be avoided; in some cases it is worth expending moderate effort in deciding how to allocate processors to processes. These cases are generally those in which it is not possible to determine, before the program is run, where processing resources should be allocated. By defering scheduling decisions until run time, better estimates of the processing needs can be obtained, and a better allocation performed. Such rescheduling strategies are therefore adaptive to the instantaneous needs of the decomposition.

(d) Synchronization costs. Synchronization involves one process (and one processor) waiting for another, and thus wasting time and slowing down the computation. One way of avoiding an idle processor is to schedule it onto another process. However this will only be useful if the expected wait time is longer than the context swap time, and if there are other useful tasks for the processor to perform. In general, a strong driving factor in the choice of decompositions is to reduce synchronization costs since synchronization costs are usually the dominating factor in limiting the performance of a multiprocessor program (Oleinick and Fuller, 1978).

We can look for decompositions suitable for MIMD machines in several areas. Since we can program an MIMD machine so that each process executes the same task and synchronizes after each operation, an MIMD machine can execute any SIMD algorithm. However, because software synchronization on an MIMD machine costs far more than hardware synchronization on an SIMD machine, we would not expect the performance to be good. Alternatively, we can program an MIMD machine to act like a pipeline machine, by arranging a sequence of processes to perform successive parts of a computation, with partial results being passed down the sequence of processes using messages. Again, the high overheads of message passing (combined with the implicit synchronization involved) lead to poor performance. SIMD-like and pipeline computations become suitable for MIMD machines only if the amount of computation performed by the processes is large compared to the cost of interaction. More generally, we would expect an MIMD computation to involve a collection of processes more diverse than the identical processes that execute on an SIMD machine, synchronizing much less frequently than at the operation level, and with a less regular communication pattern than is exhibited by a pipeline computation.

Our aim then is to increase the amount of computation performed by each process relative to the number of synchronization actions: in the limit we wish to approach a decomposition in which no synchronization is performed, and the algorithm is "asynchronous." Unfortunately, as we reduce the amount of synchronization, our algorithms change from being "deterministic" to being "nondeterministic." By "deterministic" here we mean both "predictable" (we may predict the state of the machine after each operation) and "repeatable" (executing the program will always produce exactly the same sequence of internal states). Thus an asynchronous algorithm is "nondeterministic" because slight variations in the speeds of the individual processors cause unpredictable interleavings of operations, and therefore unpredictable sequences of internal states. MIMD algorithms must have an appropriate termination state in spite of the possibility that they reach that state from one of several different paths. A suitable algorithm is one which permits a wide variety of paths to the final state (since this implies that there need be little synchronization between the processes). It would appear that devising such algorithms and demonstrating their correctness under any execution sequence would be extremely hard; fortunately, this is not always the case.

C. Cm* Software Environment

Cm*, being composed of LSI-11 processors, is potentially capable of executing most PDP-11 software. However, the operating system support for

Cm* differs radically from that for uniprocessor PDP-11 configurations, and much of the software has to be specially written for the machine.

The software support which is currently available includes basic software to load and to debug the Kmap microcode, and to load and initiate programs on the computer modules. Two experimental operating systems provide a higher level of support—StarOS (Jones *et al.*, 1977, 1979) is a capability-based operating system and Medusa (Ousterhout *et al.*, 1980) is a message-based operating system. The experiments described here are run using either the basic software or the Medusa operating system. Programs are generally written in Bliss-11 (Wulf *et al.*, 1971), which generates highly optimized code.

III. CASE STUDIES

The unique aspect of Cm* is that it provides a working, functioning machine that can be used to test out and experiment with new parallel algorithms. Previous work on parallel algorithms for MIMD machines consists almost entirely of theoretical work. With any scientific discipline there can be a major difference between the results predicted by theory and those actually obtained in the laboratory. Cm* has 45–50 functioning processors, and the experimental laboratory that it represents provides one of the first opportunities for scientists, whether computer scientists or physical scientists, to test the effectiveness of new parallel algorithms for the solution of important computational problems—problems that have to date only employed serial algorithms for their solution. The case studies given here thus represent, in many ways, the very first attempts to use a multiprocessor for solving some of the numerical applications of interest to many members of the community of physical and biological scientists.

While Cm* provides an excellent vehicle for exploring communication and synchronization problems associated with parallel algorithms and MIMD machines, it is important to keep in mind that Cm*, itself, does not constitute a practical computing system for solving numerical problems. Even 50 LSI-11s executing a parallel numerical algorithm with perfect efficiency (a speedup of 50 over a single processor) cannot be considered a high-performance scientific computer. This is not to say, however, that the knowledge gained by implementing and experimenting with numerical algorithms on Cm* is not of interest for high-performance scientific computation. On the contrary, experiments on Cm* can lead the way in indicating how hundreds of high-performance single processors, operating in parallel, could be efficiently applied to the solution of an important scientific problem. For example, knowledge such as that gained from Cm* will be necessary in the future to make full use of MIMD supercomputers such as the 512 processor system proposed by Burroughs (1979) for the NASA aerodynamic simulation facility or the 16 processor S-1 system (McWilliams *et al.*, 1977).

The results presented here represent two types of experiments. The first type of experiment involves attempts to devise and test parallel algorithms for a few simple, model problems of universal importance. The second type of experiment involves attempts to investigate MIMD algorithms and architectures as a practical solution to a specific, very large computational problem. The prime example of the first type of experiment is ordinary matrix multiplication. At first glance this might seem a trivial problem hardly worth including in a publication proposing to represent the state of the art in computational physics. MIMD machines hold great promise for vastly extending the scale of the computations that physicists and other scientists can perform; however, to make effective use of a MIMD architecture may require a complete reformulation of the original problem. One cannot assume that present serial formulations of problems carry over to an asynchronous parallel environment. If, as may be the case, MIMD machines represent the wave of the future, it will be necessary to rethink the methods, algorithms, procedures, etc., of current computational physics, in light of the architecture of a specific MIMD machine. The results given here for a few standard benchmark problems represent a step in that direction.

The second set of results presented here represent an attempt to experiment with the applicability of a MIMD architecture, like that of Cm*, to a particular application area. We have chosen to investigate as an application the statistical mechanical calculation of the macroscopic properties of a liquid from the microscopic interactions of its constituent particles. Such computations consume a nonnegligible fraction of the computational facilities available to the physics community. If MIMD machines are to have a substantial impact on computational physics, chemistry, or biology, then it must be shown, here and subsequently by others, that MIMD architectures can provide more cost-effective solutions to such important problems than can existing architectures. We could, equivalently, have chosen to study any one of a number of large computational problems (for example, electronic or nuclear structure calculations, trajectory calculations of chemical kinetics, quantum scattering calculations, or calculations of fluid flow from the Navier–Stokes equation). Eventually, it will be necessary to show how these and the other resource-consuming calculations of computational physics map onto MIMD architectures. The calculations of the structure of liquids that are described here were chosen because the Monte Carlo and molecular dynamics methods that they use are relevant to many other areas. For example, molecular dynamics calculations directly relate to studies of the internal motion of large biological molecules such as enzymes or DNA.

A. Standard Benchmarks

1. Matrix Multiplication

Although several theoretical analyses of matrix operations have been presented previously, few have considered the simplest case of matrix multiplication. Though other matrix operations may be more important in scientific applications (inversion and diagonalization, for example), the design of efficient algorithms on MIMD machines is sufficiently different from that of serial machines that the careful reexamination of simple problems is useful. The models of parallel computation used in previous analyses (Sameh, 1977) have typically focused on arithmetic operations, neglecting communication and synchronization costs. In particular, it has usually been assumed that (i) an unlimited number of processors is available and (ii) no memory access penalties are incurred. The first assumption is generally not a problem, since it is often straightforward to employ the same or similar parallel algorithms when only a limited number of processors is available. The second assumption is more dubious, however, and in particular it *cannot* be corrected by simply adding some constant memory access time to that necessary to perform an arithmetic operation. An analysis of the matrix multiplication problem will illustrate this.

We wish to compute the matrix product $C = A \times B$, where A, B, and C are $N \times N$ matrices. The elements of C are given by

$$C_{ij} = \sum_{k=1} A_{ik} \times B_{kj}. \tag{1}$$

The straightforward serial algorithm for evaluating the matrix product in Eq. (1) involves N^3 multiplications and N^3 additions. However, it is easy to see that all of the N^3 multiplications are independent and could be performed in parallel, suggesting the following steps in an algorithm employing N^3 processors:

1. Each of the N^3 processors retrieves from memory an element of A and an element of B to be multiplied.
2. Each processor multiplies its elements.
3. The summation of the products is performed.
4. The resulting final values of C are stored into memory.

In the absence of memory costs, steps 1 and 4 are free. Since all of the multiplications are done in parallel, the time necessary for step 2 is $O(1)$. In the absence of synchronization costs, the summation of the products in step 3 can be performed in at best $O(\log_2 N)$ time. Thus, neglecting memory and synchronization costs, the summation step 3 is the bottleneck, and matrix multiplication can be performed in $O(\log_2 N)$ time using N^3 processors.

The importance of including communication costs in the analysis of the algorithm are apparent when memory contention is included in the model. Note that in step 1, each element of the arrays A and B must be read by N different processors. It is true of most real machines that each memory location can be accessed by at most one processor at a time. Thus, the time necessary to perform the "fetch" operation in step 1 is $O(N)$, and this dominates the overall calculation. Indeed, the assumption here that each memory location is simultaneously addressable is rather optimistic: for a single-port memory, the time necessary for step 1 is $O(N^3)$! It is apparent, therefore, that memory access patterns must be included in a useful model of parallel computation.

The best decomposition of matrix multiplication on Cm* will not be the one described above. It will be influenced by several factors. First, there will not be an arbitrary number of processors available. For matrices of reasonable size there will not be N^3 or even N^2 of these available. Second, accesses to nonlocal shared memory are expensive, even when there is no contention. The problem gets much worse, of course, if many processors make frequent accesses to shared memory. Thus, an efficient algorithm on Cm* will have a high locality of memory reference: the amount of work that can be done locally for each access to nonlocal memory should be maximized.

A simple decomposition scheme proposed by Stone (1975) involves assigning the computation of each column of the output matrix to a different processor. Thus, on Cm*, each processor would copy its assigned column of B to local memory, then retrieve each row of A, and compute its assigned elements of C. Finally, the computed vector of C must be stored into shared memory. Note, however, that for each element of the array A that is accessed, only two arithmetic operations are performed (one multiplication and one addition).

An alternative way to decompose the problem is to partition the output matrix C into square submatrices and assign to each processor the task of computing one of these blocks. Thus, if the number of processors is K^2, then the size of each submatrix is $(N/K) \times (N/K)$. To compute its submatrix contribution to the product matrix C, each processor must retrieve a block of size $(N/K) \times (N/K)$ of A and a corresponding block of B for a total of $2(N/K)^2$ nonlocal memory references. However, once these are retrieved, the processor can perform $(N/K)^3$ multiplications and $(N/K)^3$ additions locally. Thus each processor performs (N/K) operations for each nonlocal memory reference using this scheme, as opposed to only two operations for the previous one.

These two methods for dividing the work among available processors differ only in the way that the work is partitioned. If memory reference costs are neglected in the model, the two schemes are equally good: using K processors, the model predicts a speedup of K for each. However, since locality of memory reference can decrease both access time and contention, the submatrix partitioning scheme is the superior one on a real MIMD machine like Cm*.

A pseudocode listing of the matrix multiplication algorithm is shown in the appendix. If we assume $N \times N$ matrices and K^2 processors (where N is divisible by K), then we divide the $N \times N$ matrix up into K^2 blocks each of size $(N/K) \times (N/K)$ and have a process (processor) responsible for generating its block of A and B and for calculating its block of C by accessing blocks from other processes. A process must make its blocks of A and B available to other processes, but we assume here that a particular block of C will only be used by the process which calculates it. Thus the calculation performed by a process can be represented by

$$(C_{RS})_{\mu\nu} = \sum_{T=0}^{K} \sum_{\lambda=0}^{N/K} (A_{RT})_{\mu\lambda}(B_{TS})_{\lambda\nu}, \qquad (2)$$

where R, S, and T are indices labeling blocks of the full matrix.

We have recently implemented this matrix multiplication algorithm on Cm*, and the results of preliminary benchmark tests are summarized in Table I. This table presents both the speedup t_1/t_n, and the processor utilization $t_1/(n \times t_n)$, where n is the number of processors used and t_n is the time necessary to perform the multiplication using n processors. We have used

TABLE I

SPEEDUP AND PROCESSOR UTILIZATION [IN BRACKETS] FOR MATRIX MULTIPLICATION BENCHMARKS

Matrix size	Number of processors				
	1[a]	1	4	9	16
24	1.0	0.93 [0.93]	2.78 [0.69]	4.58 [0.51]	5.12 [0.32]
30	1.0	0.95 [0.95]	3.03 [0.76]	5.55 [0.62]	
36	1.0	0.96 [0.96]	3.24 [0.81]	6.24 [0.69]	8.93 [0.56]
42	1.0	0.95 [0.95]	3.32 [0.83]	6.68 [0.74]	
48	1.0	0.95[b] [0.95][b]	3.46 [0.87]	7.06 [0.78]	10.98 [0.69]

[a] This is a *serial* algorithm on one processor and is used as the reference time t_1 for computing the processor utilization.

[b] Due to technical difficulties this value was not measured, but rather is estimated from results for smaller matrices.

for the reference time t_1, the time necessary to perform the multiplication using the normal serial algorithm on one processor. Thus the third column of the table compares this time to that necessary for the *parallel* algorithm on one processor. Apparently, the extra overhead in the parallel algorithm on one processor as compared to the serial algorithm on one processor is about 5%. As expected, the parallel algorithm is best for larger matrices, since each processor can then perform more local operations for each nonlocal memory reference.

2. Partial Differential Equations

Parallel algorithms for a model problem related to the practical solution of large sets of partial differential equations (PDEs) have previously received fairly extensive study (Baudet, 1978; Raskin, 1978; Jones and Gehringer, 1980). The problem that has been investigated is Laplace's equation in a plane,

$$\partial^2 U/\partial x^2 + \partial^2 U/\partial y^2 = 0, \tag{3}$$

subject to fixed values on the boundary (Dirichlet's problem). Taking finite difference approximations to the above derivatives directly leads to a set of equations for U on a two-dimensional grid. That is,

$$U_{ij} = \tfrac{1}{4}(U_{i-1,j} + U_{i+1,j} + U_{i,j-1} + U_{i,j+1}), \quad 1 \le i \le n, \quad 1 \le j \le m, \tag{4}$$

where the values of U on the boundary, that is, U_{1j}, U_{i1}, U_{nj}, and U_{im}, are fixed.

Iterations of the above equation can be performed in numerous ways. It is useful to distinguish a completely synchronous solution from a completely asynchronous solution. If we assign a subset of the points $\{U_{ij}\}$ to a process, then the completely synchronous solution has a process calculate, locally, values of $U_{ij}^{(r)}$ (for the rth iterate) from values $U_{ij}^{(r-1)}$ [for the $(r-1)$th iterate] that are stored in shared memory. After all processes have computed their new values (locally), then the shared variables are updated. Only when all values for the rth iteration have been updated does each process goes on to the $(r+1)$th iteration.

Completely asynchronous iterations, on the other hand, proceed with any process updating immediately any shared variable U_{ij} with the value it has just computed. Similarly, any process accesses directly the latest and "best" value of the shared U_{ij} variables on the right-hand side of Eq. (4), without regard to the iteration number of a particular point. Experimental results have shown (Baudet, 1978) that asynchronous iterations perform far better than synchronous iterations. The cost of synchronizing parallel processes is perhaps the major detriment to linear speedup. If available or known, asynchronous algorithms are to be preferred, provided they converge† to the correct result.

B. Simulations of Molecular Motion

The second type of case study that we report here involves an exploration of possible parallelism in simulations of the structure of liquids. This particular problem was chosen as representative of general problems involved in the theoretical study of molecular motion. The basic problem can be stated as follows: given the microscopic interactions between particles (the potential function), predict the static and/or the dynamic properties of a collection of such particles. The collection may include a macroscopic number ($\sim 10^{23}$) of particles, in which case some form of periodic boundary condition (Metropolis et al., 1953) must be employed, or a relatively small finite number of particles representative of a large biological molecule, a microscopic "drop" of a liquid, a hydration cluster, etc. Generally, the number of discrete particles that can be treated is less than 1024. For example, in current simulations of liquid water (Pangali et al., 1980; Mezei et al., 1977), it is common to have 216 H_2O molecules in the central bounding box. One of the hopes of parallel computation that is considered here involves increasing by orders of magnitude the number of individual particles that can be dealt with in a particular simulation.

One of the intriguing aspects of computing molecular motion on a multiprocessor is the possibility of performing some sort of mapping of the three-dimensional structure of the physical system to be simulated onto the three-dimensional structure of the computation machine performing the simulation. The importance, for efficient multiprocessor performance, of maximizing local computation and minimizing global communication has an analog in the rapid decay with distance of intermolecular potentials. Molecules "communicate" mostly with their nearest neighbors. In the same way, the aperiodic nature of most physicochemical systems that one wishes to simulate should map much more effectively onto an asynchronous MIMD multiprocessor than onto a synchronous ordered SIMD array of processors. The initial experiments in the simulation of molecular motions that are described below constitute relatively mundane translations of the usual serial algorithms and take no cognizance of such possible physical mappings. The "best" parallel algorithms of the future will not likely be quite so constrained.

---

† Proof of convergence for asynchronous iterations is more difficult than for synchronous iterations.

To relate the properties of a collection of particles to the individual interactions between particles, two types of averaging are common—either ensemble averaging or time averaging. The experiments described below correspond to these two procedures. The first of these, the ensemble average of classical statistical mechanics, is represented in our calculations by the Monte Carlo (Metropolis) algorithm (Metropolis et al., 1953). The second of these, the time average—related to the ensemble average by the ergodic hypothesis—is represented in our investigations by the molecular dynamics (Verlet) algorithm (Verlet, 1967). These are well-known serial algorithms. Our initial investigations constitute relatively straightforward translation of these serial algorithms to the parallel environment provided by Cm*. A much better procedure would be to invent new and unique algorithms that are optimal for the specific physics (ensemble average or time average) being simulated and programmed. In particular, our present parallel algorithms for these two problems are *synchronized* algorithms. As the discussion involving the solution of partial differential equations indicated, it is generally preferred on multiprocessors to have asynchronous rather than synchronous algorithms. Unfortunately, in the present situation, it appears that the generation of asynchronous algorithms requires a rather substantial revision of the way the physics of the problem is approached. For example, an asynchronous molecular dynamics algorithm requires averaging the interactions between two particles occupying different positions in space *and time*! This anomaly indicates how constrained our modes of thought have been by the sequential model of computation. We have not yet made a serious attempt to find asynchronous algorithms for the ensemble average or time average problems.

Prior to describing the particular Monte Carlo and molecular dynamics calculations, we first need to describe the physical system being simulated.

1. *Structure of Liquids*

The equilibrium macroscopic properties of a liquid can be obtained from a knowledge of the interactions between the N constituent particles of the liquid, by performing averages of certain microscopic quantities $X(\mathbf{r}_1, \mathbf{r}_2, \ldots, \mathbf{r}_N)$. Particles here are considered to be spherical with center-of-mass coordinate vectors $\mathbf{r}_1, \mathbf{r}_2, \ldots, \mathbf{r}_N$. The quantity X might be the total potential energy (leading to the thermal energy), the standard deviation in the potential energy (leading to the heat capacity), the total dipole moment (leading to the dielectric constant), etc. In the canonical ensemble, the quantity $\langle X \rangle$ is obtained by averaging over all configurations of the N particles,

$$\langle X \rangle = Z^{-1} \int X(\mathbf{r}_1, \mathbf{r}_2, \ldots, \mathbf{r}_N) \exp \frac{-E(\mathbf{r}_1, \mathbf{r}_2, \ldots, \mathbf{r}_N)}{kT} d\mathbf{r}_1 \, d\mathbf{r}_2 \cdots d\mathbf{r}_N, \quad (5)$$

where the partition function is

$$Z = \int \exp \frac{-E(\mathbf{r}_1, \mathbf{r}_2, \ldots, \mathbf{r}_N)}{kT} d\mathbf{r}_1 \, d\mathbf{r}_2 \cdots d\mathbf{r}_N \qquad (6)$$

and E is the potential energy of the configuration $\{\mathbf{r}_1, \mathbf{r}_2, \ldots, \mathbf{r}_N\}$.

According to the ergodic hypothesis (Farquhar, 1964), an ensemble average is equivalent to a long-term time average,

$$\langle X \rangle = \frac{1}{\tau} \int_0^\tau X(\mathbf{r}_1, \mathbf{r}_2, \ldots, \mathbf{r}_N; t) \, dt, \qquad (7)$$

where the particles move according to the classical Hamiltonian equations of motion. That is, the motion of the N-particle system is obtained by integrating Newton's equations. The computational identity of the ensemble average and the time average, as well as the relative computational efficiency of the two procedures is of considerable interest (Rao et al., 1979).

In the usual procedure, the potential is approximated to be pairwise additive,

$$E(\mathbf{r}_1, \mathbf{r}_2, \ldots, \mathbf{r}_N) = \sum_{i>j} E_{ij}(\mathbf{r}_i, \mathbf{r}_j) = \sum_{i>j} E_{ij}(\mathbf{r}_i - \mathbf{r}_j), \qquad (8)$$

the integration is replaced by a sum over a finite (but large) number (M) of configurations,

$$\int \Rightarrow \sum_{\text{configuration}}^{M}, \qquad (9)$$

and surface effects are minimized by surrounding a central "box" of N particles by mirror image boxes such that when a particle exits from the box a second particle enters from the opposite edge to replace the first particle. In the computations considered here, we have not used these periodic boundary conditions and instead have considered a relatively simple system consisting of a "drop" of a finite number of positive and negative ions. These ions interact according to the potential

$$E_{ij}(r) = B_1(r^*/r)^9 + z_i z_j B_2/r, \qquad (10)$$

where $z_i = \pm 1$ is the charge on ion i and the parameters B_1, r^*, and B_2 were chosen (Ramanathan and Friedman, 1970) to approximate Cs^+ and Cl^- in the gas phase. This model potential has been used previously to investigate a new serial Monte Carlo algorithm (Rossky et al., 1978).

2. The Monte Carlo Metropolis Algorithm

The Metropolis (Metropolis et al., 1953) algorithm is well known and will only be sketched here. We assume for simplicity that we are averaging the total configuration energy ($X \equiv E$) to obtain the thermal energy $\langle E \rangle$. We also assume that some initial configuration and initial value of the configuration energy are available. Subsequently, we randomly move particles to generate a Markov chain of configurations. The basic bottleneck in these calculations or probably any other such particle simulation method involves evaluating the $O(N^2)$ interaction energies E_{ij} (or, equivalently, their derivatives with distance, the interparticle forces F_{ij}). In the Metropolis scheme, particles are moved one at a time, the new configuration being either accepted or rejected with a probability that is based on the Boltzmann factor for the move. If the binding energy of one of the particles is defined as

$$B_i = \sum_{j \neq i} E_{ij}, \tag{11}$$

then the energy of the system after a move is given in terms of the energy prior to the move by

$$E^{\text{new}} = E^{\text{old}} + B^{\text{new}}_{\text{moved particle}} - B^{\text{old}}_{\text{moved particle}}. \tag{12}$$

The bottleneck of the computation for each move (we assume insufficient memory to store all N^2 interactions) involves the computations required to compute the binding energy of one of the particles, following and preceding its actual move. This computation of the binding energy involves evaluating the interaction of the moved particle with each of the other particles and thus involves $O(N)$ computations. The basic idea of the present parallel algorithm is to use K processors to reduce the complexity of this step (the calculation of the binding energy of one of the particles) to $O(N/K)$. The interactions can be evaluated in parallel with no communication between processors. However, the contributions calculated by individual processors must be added together according to Eq. (11). In addition, the computation is synchronized at each move since it is necessary to know the total energy of the new configuration before it is decided whether or not to accept the new configuration and go on to the next move.

The program that has been implemented (for details, see the appendix) consists of a master process which generates and evaluates the moves, and an arbitrary number of slave processes that are responsible for evaluating the interactions E_{ij} and summing the interactions to form the binding energy $B_{\text{moved particle}}$. The master process first of all reads appropriate parameters describing the computation and then dynamically creates as many slave processes as there are processors remaining. The slave processes initially just spin, waiting for a signal from the master process to start their continual computation of interactions.

The procedure in the master process responsible for obtaining the binding energy of a configuration first initializes the required binding energy (a variable in shared memory) to zero and then releases all the slaves to begin their computation of individual contributions to the binding energy. When all the contributions $E_{\text{moved particle}, j}$ ($j \neq$ moved particle) have been computed and summed to give $B_{\text{moved particle}}$, then control returns to the master process. The slave processors compute interaction energies E_{ij} between $i \equiv$ moved particle and $j \equiv$ my particle. The values of j can be determined dynamically by inspecting a shared variable describing the number of interactions yet to be evaluated or values of j can be preassigned to each slave process by the master process. Slave processes add the values of the interaction they have calculated to the running sum (a shared variable) representing the binding energy. A protected shared variable can be used to keep track of the number of interactions that have been evaluated (it is updated when a slave adds a new interaction to the binding energy), and the slave process evaluating the last interaction signals the master process that the binding energy of the moved molecule is available. If there are no more interactions to be evaluated, a slave will wait for the master process to signal the next round of interaction computations. A qualitative view of the parallel computation of binding energies is shown in Fig. 7.

Experimental results for the speedup of the above algorithm on Cm* are shown in Figs. 8 and 9. These results represent only first attempts at programming the problem, and one might expect considerably better results with a modified and "fine-tuned" algorithm.

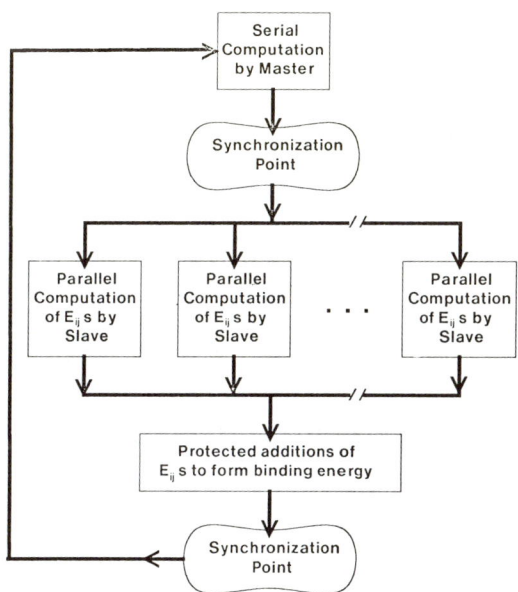

FIG. 7. Parallel computation of binding energies.

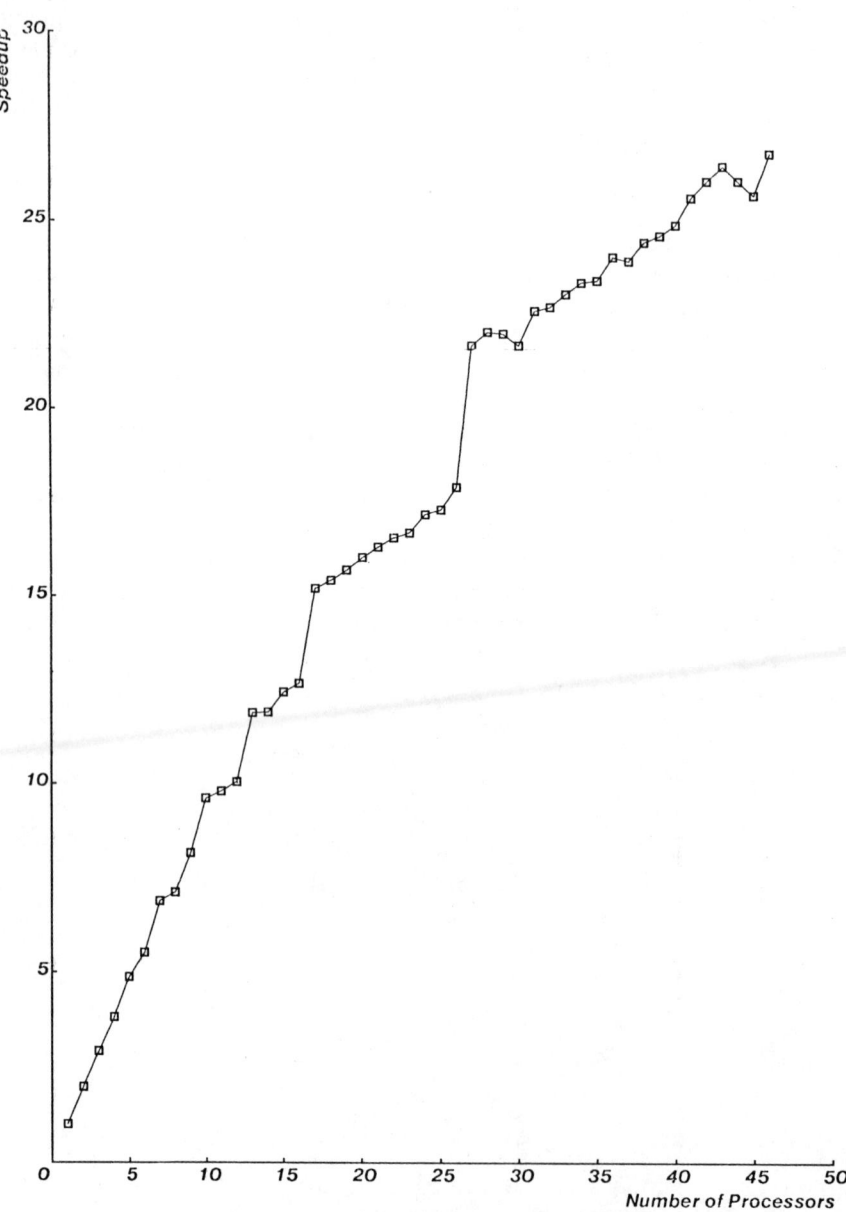

FIG. 8. Speedup of the Monte Carlo algorithm.

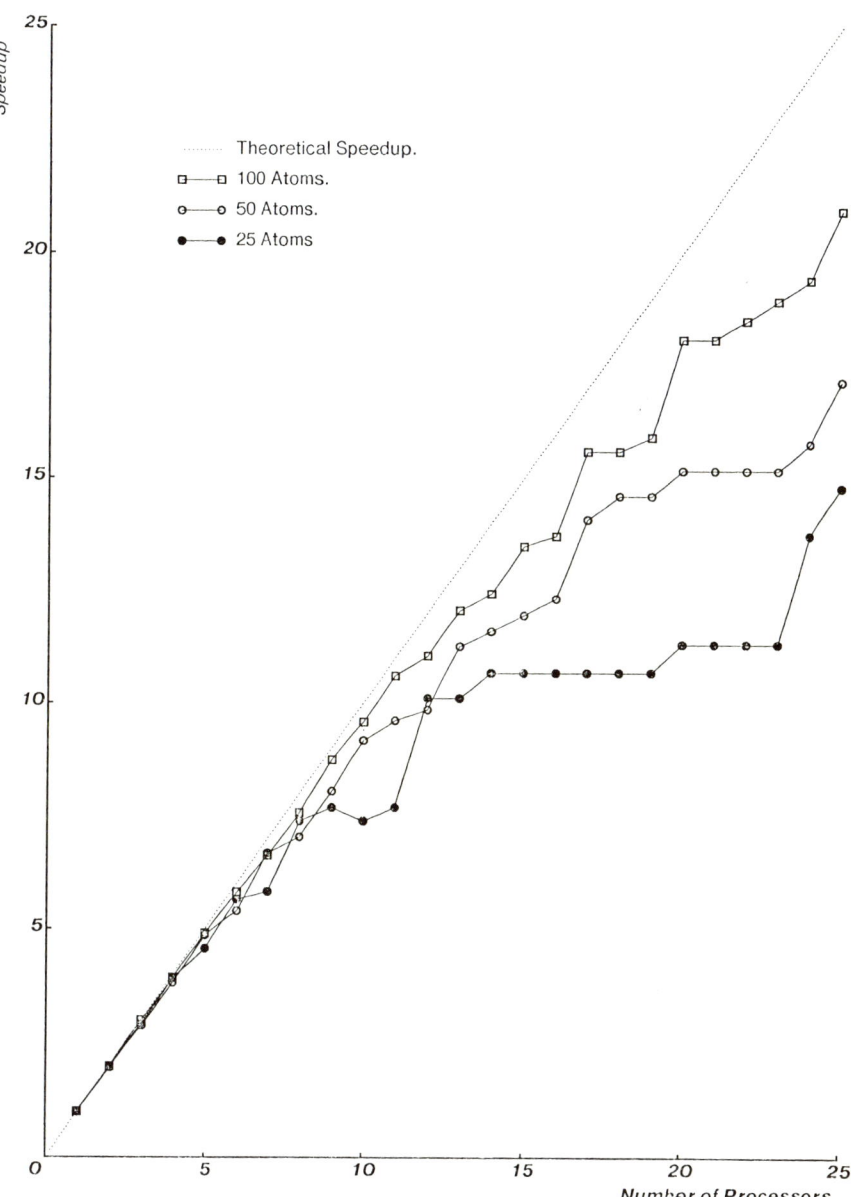

FIG. 9. Speedups for Monte Carlo with various numbers of atoms.

The results in Fig. 8 are for a system of 50 particles. The result for 50 processors, therefore, represents a "worst-case" locality where each processor calculates on the average only a single interaction prior to adding its contribution to the shared variable representing the binding energy. While the speedup in Fig. 8 is adequate, depending on one's point of view, there do exist a number of problems with the algorithm. First, for every interaction that a slave process calculates, it must enter a *critical section*† to obtain the index of the particle for which it is calculating the interaction with the moved particle and also when updating the binding energy and the variable describing the number of contributions to the binding energy that have been calculated. If the number of particles is the same as the number of processors, all this overhead of global communication is necessary. However, a more likely real-world case would have much fewer processors than the number of particles. In this case, sums of interactions can be computed locally by a processor for the subset of the total number of particles for which it is responsible, prior to any global communication, thus reducing the amount of global communication. That this is the case is illustrated in Fig. 9. These results were obtained under a slightly different operating environment which allowed only up to 25 processors to be used by applications programs. The several curves, representing speedups for the algorithm with varying numbers of particles, illustrate the improved performance as the number of particles increases.

For the present algorithm, the computation of the sum of the interactions is also a bottleneck. The sum involves $O(N)$ computations per "move" rather than the $O(N/K)$ computations being strived for. The algorithm can be rearranged to effect an $O(\log_2 N)$ summation, but in practice it was not found that this improved the speedup much. The major problem with the algorithm is that at the lowest level (assuming that the number of particles and processors is identical) a processor computes only a single interaction before global communication is required. The Metropolis algorithm is likely to be effective for large numbers of processors only if the number of particles is much larger yet ($N \gg K$), and a processor computes approximately N/K interactions locally before any global communication is necessary. With this slight revision, the present algorithm should be effective in giving linear speedup for large numbers of processors provided that contention for read-only access to the shared-variable coordinates is not a problem. The limitations to obtaining linear speedup that have been observed, to date, involve synchronization and global communication, not contention in reading shared variables.

---

† A critical section is a piece of code which can only be executed by one process at a time. The appendix shows how it is implemented on a multiprocessor.

3. The Molecular Dynamics Verlet Algorithm

The Verlet algorithm (Verlet, 1967) is basically the finite difference form of Newton's equations. One generates an initial set of velocities for the particles based on a Maxwell distribution for temperature T. Then, given an initial set of coordinates at time $t - \Delta t$, the velocities can be used to predict a set of coordinates at the later time t. With coordinates for these two initial times ($t - \Delta t$ and t), a simple algorithm that will predict the coordinates for all future times is then

$$\mathbf{r}_i(t + \Delta t) = -\mathbf{r}_i(t - \Delta t) + 2\mathbf{r}_i(t) + \mathbf{F}_i(t)/m_i(\Delta t)^2, \qquad (13)$$

where $\mathbf{F}_i$ is the total force on a particle of mass m_i.

Consider the problem of finding the average energy. Each of the time steps generates a new configuration, the energy of which is

$$E = \sum_{i \in \text{particles}} B_i/2. \qquad (14)$$

If B_i^A is the binding energy of particle i in the configuration A, then

$$\langle E \rangle = M^{-1} \sum_{A \in \text{configurations}}^{M} \sum_{i \in \text{particles}}^{N} B_i^A/2. \qquad (15)$$

If the computation is performed as shown in this last equation and processes are assigned to subsets of the total number of particles, then it is necessary to perform a systemwide sum of binding energies (involving all the attendant communication problems) at each new configuration. However, a simple reordering of the summation,

$$\langle E \rangle = M^{-1} \sum_{i \in \text{particles}}^{N} \sum_{A \in \text{configurations}}^{M} B_i^A/2, \qquad (16)$$

allows a process to sum its subset of binding energies *locally* and the shared-variable sum is required only once at the end of the computation. A more detailed description of the molecular dynamics algorithm is given in the appendix.

Figure 10 summarizes the results of our molecular dynamics benchmarks. The most prominent feature of the graph is the "stepped" nature of the curves. This is caused by the unequal distribution of work among the processors in the flat regions of the curve. Processors with less work to do finish sooner, but must wait idly for the remaining processors to complete their tasks. The peaks on the curves approaching linear speedup represent points at which the work to be done can be divided most evenly among the available processors.

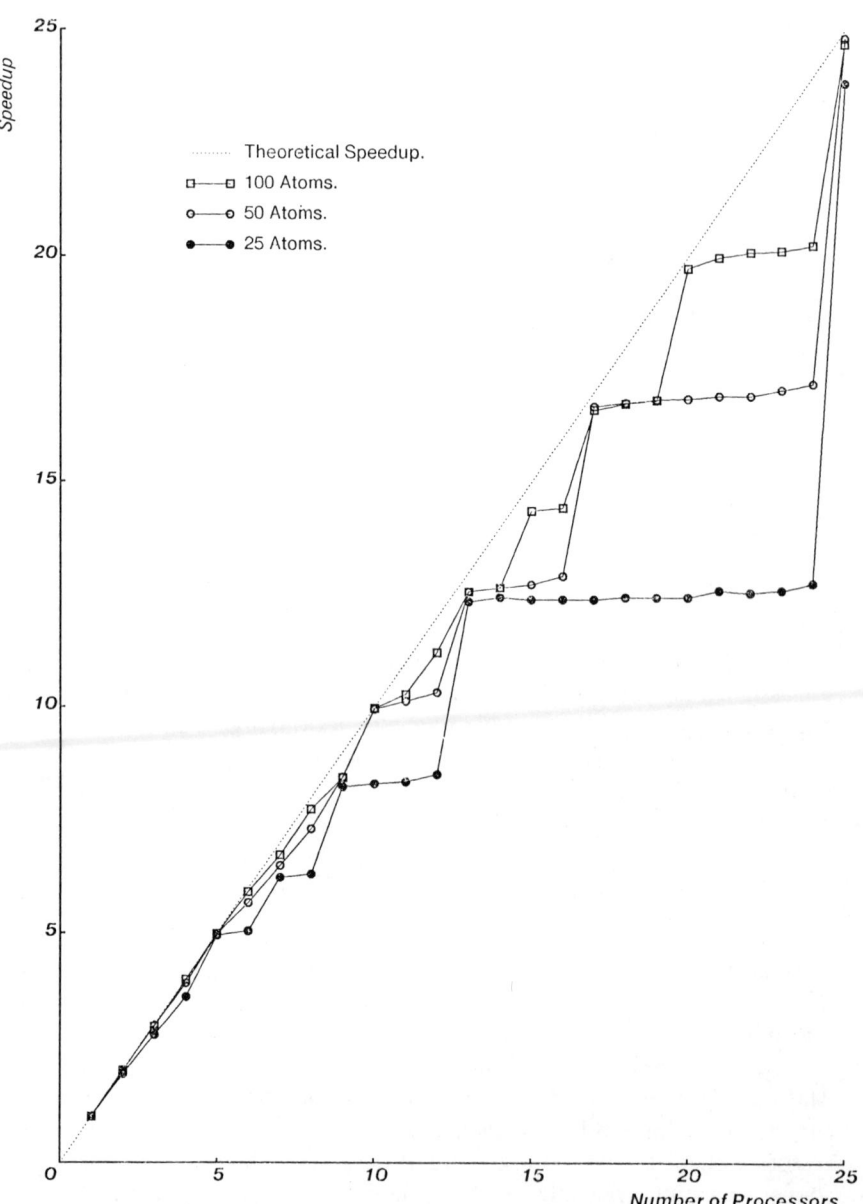

FIG. 10. Speedups for molecular dynamics algorithm.

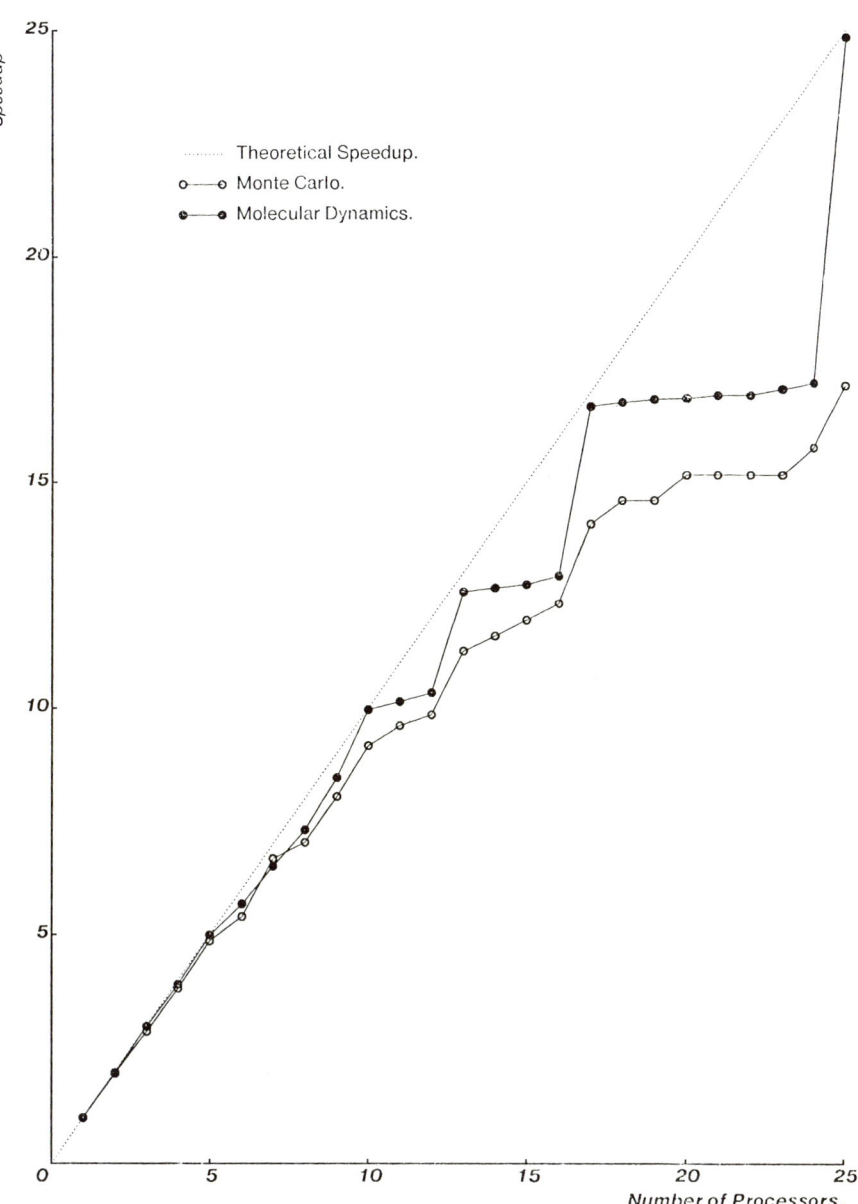

FIG. 11. Comparison of Monte Carlo and molecular dynamics for 50 atom calculations.

4. Comparison of Monte Carlo and Molecular Dynamics

The molecular dynamics algorithm is considerably superior to the Monte Carlo algorithm (Fig. 11), at least for calculations closely associated with the structure of liquids calculation described here. The ideal of parallel computation involves frequent, cheap, local computation and infrequent, expensive, global communication. The molecular dynamics algorithm totally avoids any problem of performing sums of shared variables. In addition, since particles move simultaneously rather than one at a time, the $O(N^2)$ serial computation of the energy of a configuration is converted to an $O(N^2/K)$ parallel computation, by employing K processors. In the Monte Carlo algorithm, on the other hand, the $O(N)$ serial computation of a binding energy is converted to an $O(N/K)$ parallel computation. Thus a process computes $O(N^2/K)$ interactions prior to communication, in the molecular dynamics example, as opposed to only $O(N/K)$ interactions prior to communication, as in the Monte Carlo example. For the 50-particle/50-processor example, a process in the Monte Carlo case calculates only a single interaction before communicating its contribution to the binding energy of the moved particle. In the molecular dynamics case, a process calculates all N-1 (49) interactions of a given particle, prior to the synchronization necessary to go on to a further time step.

The molecular dynamics algorithm appears to be more applicable to parallel MIMD decomposition than the Monte Carlo algorithm. Both cases, however, are synchronous algorithms. A better solution would be to find original asynchronous algorithms for the ensemble average and time average that are the basic problems of fundamental interest.

IV. CONCLUSIONS

From the hardware engineering point of view, multiprocessors may offer more attractive solutions to the problem of increasing the performance of computers than other parallel machines, since they are able to take advantage of simple regular designs which employ replicated standard components. When suitable communication structures are used they also allow extensibility and reliability. To a certain extent, though, the regularity of the design is achieved at the expense of removing specialized and centralized control, and one of the characteristics of software which executes on MIMD machines is the need to provide control using a variety of programming techniques. Inevitably this involves software overheads, both in terms of design complexity and execution costs. Whether the hardware savings compensate for the added complexity of the software and the reduced execution speed of programs will vary, of course, from machine to machine and from program to program. Our experiments show, however, that the task of designing programs for MIMD machines, while more complex than that for other machines, is quite feasible, and that reasonable decompositions exist for a variety of problems of practical scientific interest.

We have only investigated a few of the factors which influence the choice of decomposition and the performance of the resulting programs: the memory locality and accessing costs, synchronization costs, and communication costs. There are certainly other aspects which remain to be explored. In particular, we want to explore how to exploit the inherent complexity of control in multiprocessor programs to help choose performance parameters at run-time. We would hope to subsequently study a number of adaptive algorithms to see how they perform on multiprocessors under a variety of flexible scheduling policies.†

It is clear to us that our explorations have been hindered, and partly molded, by the lack of adequate programming tools. While we now have basic operating system tools to help us load programs and initiate them, the lack of good program monitoring tools, and the lack of languages which allow us to express our programs at the appropriate level of abstraction, cause severe problems. A second factor which has hindered us has been the inadequate computational power and the small address space of the LSI-11. This has made it difficult to present totally convincing proofs of the ability of MIMD multiprocessors to perform serious scientific computations.

We are convinced of the long-range viability and usefulness of MIMD machines, and of the importance of an understanding of MIMD problem decompositions. In the near future one can expect there to be implementations of MIMD machines for general-purpose scientific computations (McWilliams et al., 1977; Burroughs, 1979). Multiprocessors currently are used in special situations were reliability is a principle concern (Heart et al., 1972; Barlett, 1978; Bedard et al., 1977). Variants of MIMD machines, incorporating functionally specialized processors, are being developed, as are network machines. Similar programming considerations apply to these. MIMD software organization is likely to continue to be a fruitful area for scientific investigation.

The present investigations have hardly begun to scratch the surface of numerical scientific applications on MIMD multiprocessors. There remains a great deal of effort required to explore and develop parallel algorithms for a coming generation of multiprocessors. It is likely that many of our existing modes of thought regarding serial computation will have to be modified or abandoned to successfully use an asynchronous multiprocessor. It is also

---

† The Medusa operating system does not currently allow rescheduling of processes.

unfortunately likely that good parallel algorithms for one parallel architecture will differ rather drastically from good parallel algorithms for another parallel architecture. In order to take full advantage of the technological developments in microelectronics, computational physicists will need to come to grips with these new parallel architectures and algorithms.

APPENDIX. IMPLEMENTING PARALLEL ALGORITHMS

As indicated in the text, implementation of the parallel algorithms being discussed is accomplished by programming in the Bliss language (Wulf et al., 1971), either on the bare Cm* hardware or with the assistance of the experimental Medusa (Ousterhout et al., 1980) operating system. Medusa and its associated linker certainly allow a user to implement a parallel algorithm in a much easier fashion than by programming in Bliss the primitive machine operations that implement the communication and synchronization of processes. However, the Medusa system does still not lead to a clear way of expressing the basic algorithms that have been used. A high-level parallel language suitable for directly coding the present applications has not yet been implemented on Cm*. This leads to some difficulty in expressing in unambiguous terms the parallel algorithms being discussed. To overcome this difficulty and, in addition, to allow the reader who is unfamiliar with parallel computation to obtain some feeling for what is involved in programming a MIMD machine, we give the syntax and semantics here of a parallel language for Cm* and express our three algorithms (matrix multiplication, Monte Carlo calculations, and molecular dynamics calculations) in this language. The primitives in this language are at a lower level of abstraction than is normally convenient for expressing parallel algorithms; however, since we want to expose important implementation details, we have chosen a language fairly close to the hardware.

A. *A Parallel Language*

We describe here a slightly extended version of the C (Kernighan and Ritchie, 1978) programming language.† This language is extensively used in the UNIX (Ritchie and Thompson, 1974) programming environment, and

---

† For those not familiar with C, the major syntactic differences from FORTRAN or ALGOL that we use here involve curly braces for BEGIN and END pairs; the use of "For ($i = 0$; $i < N$; $i + +$) Statement;" to represent the execution of "Statement" for the values of $i = 0, 1, \ldots, N - 1$; and the use of !, !=, and = = for "not," "not equal," and "equal," respectively.

Application of a Tightly Coupled Multiprocessor

a straightforward superset of the language provides a convenient tool for describing our parallel algorithms.

We suppose that a task consists of a set of parallel processes. In the programs described here each process executes one particular process only; thus no dynamic rescheduling of processes onto processors is involved. A parallel language for Cm* then requires the following facilities:

(a) A way of representing a process.
(b) A way of indicating the allocation of shared memory.
(c) A way of enforcing mutual exclusion on shared memory.
(d) A way of synchronizing processes.

We discuss each of these in turn.

1. *Processes*

The decompositions described here establish a particular relation between the processes of a task; i.e. we suppose that there can be *one and only one* Master process, but several Slave processes.

```
Process Master (N_Processors)
Int N_Processors;
{
«declaration of shared variables owned by the master»

Main()
        {
        «the program text to be executed by the master»
        }
}

Process Slave_A(«parameters»)
{
«declarations of shared variables owned by the slave»

Main()
        {
        «the program text to be executed by the slave»
        }
}
Process Slave_B(«parameters»)
{
«declarations of shared variables owned by the slave»
```

Continued

```
Main()
         {
         «the program text to be executed by the slave»
         }
}
/* etc. */
```

The master and slave program texts each comprise a main program and procedures. When the task starts, a single processor starts to execute the master program text: this is the master process. It is passed, at run time, the total number of physical processors (**N_Processors**) allocated to this task by the operating system. The master process can create dynamically any number of slave processes. The tasks described here are such that the total number of processes does not exceed the total number of physical processors which have been allocated to the task.

For example, a master process might invoke a number of slave processes via the call

```
For (Instance=0; Instance<4; Instance++)
Create Slave: Instance( «parameter list» );
```

Four slave processes would be created; each would be passed an appropriate set of parameters and each would execute the same program text. The slave processes start to execute at the time of their creation. A process is referred to by its name and, except for the master process, its instance number.

2. *Memory Allocation*

In a multiprocessor like Cm*, or any other multiprocessor with large numbers of processors, locality of code and data is of principle importance to efficiency. A multiprocessor language needs to allow a programmer to specify this locality. On Cm*, at least, code is best placed local to the processor executing it, and this is easily arranged by having copies of the slave code resident in the local memory of each processor. To describe the allocation of memory for data, we assume four basic types of variables: *Universal variables* are accessible to any process, and are allocated to memory anywhere in the system; *shared variables* are accessible to any process, but are allocated to a memory module local to that process which declares the variables; *global variables* are accessible to any routine within a particular process; and *local variables* are accessible only to the routine that declares them. Only universal and shared variables are unique to the multiprocessor environment; Global and local variables are present in ordinary C, although they are not explicitly declared with these names. Global here means "global-within-the-process," not "global-within-the-task."

Shared variables may be declared at the start of any process. For example,

```
Shared Float Cartesian_Coords [3,100];
```

This declaration results in the shared variable being allocated storage space in the local memory of the processor executing the process in which this declaration occurs; the process is said to "own" the shared variable. Shared variables can be accessed inside the same process that declared them by simply referring to them. For example,

```
Cartesian_Coords [i, j] = 0.0;
```

In a different process from the one which declared them, shared variables need to be declared as external,

```
Extern Shared Cartesian_Coords [3,100];
```

and accessed using the variable name qualified by the owner as, for example, in

```
Cartesian_Coords <Master> [i, j] = 0.0;
```

When there is more than one instance of a process executing the same program text, then the variable name qualifier includes the instance number. For example,

```
Local_Coords <Slave: Which_One> [i, j] = 0.0;
```

3. *Mutual Exclusion*

Shared or universal variables can be accessed simultaneously by more than one process. Since it might be necessary for a process to have exclusive access, for a period of time, to some particular shared variable, there must be a way of ensuring that processes are mutually excluded from simultaneous access. Two operations are required: **Lock** and **Unlock**, used in the following way to surround a critical section.

```
Lock(Gate)
«the critical section: freely access shared or universal data»
«no danger of interference from another processes, provided»
«that they also surround their accesses to the shared or»
«universal data by similar Lock and Unlock operations»
Unlock(Gate);
```

The variable **Gate**, which is also shared or universal, has values of **Locked** (0) or **Unlocked** (1). The operation performed by executing **Lock(Gate)** involves waiting while the variable **Gate** has a value of **Locked**. If the variable **Gate** ever becomes **Unlocked**, then one (an arbitrary one) of the processes waiting on the lock sets **Gate** to **Locked**† and then enters the critical section. Upon exiting the critical section, **Unlock(Gate)** is executed. The **Unlock** procedure simply sets **Gate** to **Unlocked** and returns. In this way, one and only one process gains access to the data structures protected by **Gate**.

4. *Synchronization*

Finally, there must be a way of synchronizing processes. That is, it must be possible for a process to wait for a signal to be generated by another process. A simple mechanism for this involves the **Wait** primitive, which causes a process to wait until a signal arrives from another (any) process and a **Signal⟨Process:Instance⟩** primitive which causes a process to send a signal to another (particular) process. **Signal** and **Wait** may be easily implemented using the same machine operations as **Lock** and **Unlock**. For a more complete treatment of process synchronization, see (Habermann (1976).

B. *A Matrix Multiplication Calculation*

In the program below, the master process passes to each of a set of slave processes the indices of a block of the matrix C, A, and B which that process is to calculate. The slave processes first generate locally their blocks of A and B and then inform the master process of their readiness to proceed with the matrix multiplication. The matrix multiplication starts when all slave processes are ready. Each loops over the appropriate blocks, first caching them, and then multiplying them locally.

```
Process Master(N_Processors)
Int N_Processors;
{

Main( )
{
Int Completed;
Int N, K, Size, Instance;
Int Block_I, Block_J;
Float K_Squared;
```

---

† This whole operation must be an "atomic" action and the hardware must in some way allow for inspecting for the value **Unlocked** and setting to **Locked** in one action.

```
/* Compute slave initialization parameters */
Input(N);
K_Squared = Square_Less_Than(N_Processors);
K = Sqrt(K_Squared);
Size = N/K;

/* Start up the slaves */
For (Instance = 0; Instance <K_Squared; Instance++)
        {
        Your_Block_I = (Instance)/K;
        Your_Block_J = Instance-(Your_Block_I)*K
        Create Multipliers: Instance (Your_Block_I,
                                      Your_Block_J,
                                      Size, K);
        }
/* Now wait for all the slaves to create A and B */
Completed = 0;
While (Completed <N_Processors - 1)
        {
        Wait;
        Completed = Completed + 1;
        }
/* Now signal the slaves to start the multiplication */
For (Instance = 0; Instance <K_Squared; Instance++)
        Signal <Multipliers: Instance >;
}
}


Process Multipliers(Block_R, Block_S, Size, K)
Int Block_R, Block_S, Size, K;
{
Shared Float A [Size, Size ], B [Size, Size ];

Main( )
{
Extern Shared Float A [Size, Size ], B [Size, Size ];
Int Mu, Nu, Lambda;
Int Block_T, Instance_RT, Instance_TS;
Float C [Size, Size ];
Float Local_A [Size, Size ], Local_B [Size, Size ];
```

Continued

```
/* The shared values of A and B are created somehow */
Generate_Values_Of(A);
Generate_Values_Of(B);
/* Now wait until all the slaves have created their A and B */
Signal<Master>;
Wait;

/* The multiplication */
For (Mu=0; Mu<Size; Mu++)
For (Nu=0; Nu<Size; Nu++)
C [Mu, Nu]=0.0;

For (Block_T=0; Block_T<K; Block_T++)
{

/* Cache local copies */
For (Mu=0; Mu<Size; Mu++)
For (Nu=0; Nu<Size; Nu++)
        {
        Instance_RT = Block_T + Block_R*K;
        Instance_TS = Block_S + Block_T*K;
        Local_A [Mu, Nu] = A<Multipliers:Instance_RT> [Mu, Nu];
        Local_B [Mu, Nu] = B<Multipliers:Instance_TS> [Mu, Nu];
        }

/* Compute C */
For (Mu=0; Mu<Size; Mu++)
For (Nu=0; Nu<Size; Nu++)
For (Lambda=0; Lambda<Size; Lambda++)
C [Mu, Nu] = C [Mu, Nu] + Local_A [Mu, Lambda]*Local_B [Lambda, Nu];

}
}
}
```

C. A Monte Carlo Calculation

The master process performs a straightforward Monte Carlo calculation except that it uses slave processes to assist in the calculation of the binding energy of the **Moved_Molecule** (a shared variable) before and after it is actually moved. The slaves operate with two variables that are in shared memory—**Calc_To_Do**, which describes the number of jobs yet to be picked up by the slaves, and **Calc_Done**, which describes the number of jobs which

Application of a Tightly Coupled Multiprocessor 355

have been successfully completed by slaves. The master process initializes these two variables, starts all the slaves, and waits for a signal from one of the slaves indicating the completion of the last job.

A slave uses **Calc_To_Do** as an index describing **My_Molecule**; each job comprises a calculation of the interaction between **Moved_Molecule** and **My_Molecule**. When a slave accesses **Calc_To_Do** in a critical section, it decrements it for the next slave, prior to exiting the critical section. Thus, **Calc_To_Do** falls to a negative value when all jobs are picked up. Any free slave recognizes this negative value and waits till a signal is received from the Master indicating that a new set of jobs is queued. When a slave completes a job it enters a critical section to add its contribution to the binding energy and increment **Calc_Done**. The slave process which completes the last job informs the master that the binding energy is available.

```
Process Master(N_Processors)
Int N_Processors;
{
#Define Molecule_Limit 1024
Shared Int Moved_Molecule;
Shared Float Sum;
Shared Float Coordinates[3, Molecule_Limit];

Main( )
{
#Define K 1.38066E-23
#Define Unlocked 1
Universal Int Calc_To_Do;
Universal Int Calc_To_Do_Gate = Unlocked;
Universal Int Calc_Done;
Universal Int Calc_Done_Gate = Unlocked;
Int Instance;
Int Molecule, Good_Move;
Int Pass, N_Passes, N_Molecules;
Float Temperature, Boltzmann_Factor;
Float E_Initial, E_Configuration, E_Sum, E_Average;
Float BE_Old, BE_New;

Input(N_Molecules);
Input(Temperature);
Input(N_Passes);         /* a pass moves each of N
                            molecules once */
/* Start up the slaves */
```

Continued

```
For (Instance=0; Instance<N_Processors-1; Instance++)
Create Slave: Instance(N_Molecules);

Generate_Initial_Configuration(N_Molecules, E_Initial);
E_Sum=0;
E_Configuration=E_Initial;

For (Pass=0; Pass<N_Passes; Pass++)
For (Moved_Molecule=0; Moved_Molecule<N_Molecules;
                                            Moved_Molecule++)
        {
        /* Compute binding energy in parallel */
        BE_Old=Binding_Energy(Moved_Molecule);

        Displace(Moved_Molecule);

        /* Compute binding energy in parallel */
        BE_New=Binding_Energy(Moved_Molecule);

        Boltzmann_Factor=Exp((BE_New-BE_Old)/(K*Temperature));
        Good_Move=Probability_Test(Boltzmann_Factor);
        If (!Good_Move)
                {
                Undisplace(Moved_Molecule);
                BE_New=BE_Old;
                }
        E_Configuration=E_Configuration+BE_New-BE_Old;
        E_Sum=E_Sum+E_Configuration;
        BE_Old=BE_New;
        }

E_Average=E_Sum/(N_Passes*N_Molecules);
}
}


Float Binding_Energy(Molecule)
Int Molecule;
{
Extern Float Sum;
Extern Int N_Processors;
Universal Int Calc_To_Do;
```

```
Universal Int Calc_Done;
Int Instance;

Calc_To_Do = N_Molecules - 1;
Calc_Done = 0;
Sum = 0;

/* Start up all the slaves */
For (Instance = 0; Instance < N_Processors - 1; Instance++)
Signal <Slave: Instance>;
/* Wait for the last one to complete */
Wait;
Binding_Energy = Sum;
}


Process Slave(N_Molecules)
Int N_Molecules;
{

Main( )
{
#Define False 0
#Define True 1
#Define Molecule_Limit 1024
Extern Shared Int Moved_Molecule;
Extern Shared Float Sum;
Extern Shared Float Coordinates [3, Molecule_Limit];
Universal Int Calc_To_Do;
Universal Int Calc_To_Do_Gate;
Universal Int Calc_Done;
Universal Int Calc_Done_Gate;
Int More_Interactions;
Int N_Completed;
Int My_Molecule, Other_Molecule, Direction;
Float R, R_Squared, Delta, Interaction;
Float X [3];

While (True)
{
```

Continued

```
Wait;      /* Wait for a new moved molecule */
Other_Molecule = Moved_Molecule<Master>;   /* cache these */
For (Direction = 0; Direction<3; Direction++)
X [Direction ] = Coordinates <Master> [Direction, Other_Molecule ];

            More_Interactions = True;
            While (More_Interactions)
             {

            Lock(Calc_To_Do_Gate);
               My_Molecule = Calc_To_Do;
               If (My_Molecule == Other_Molecule)      /* no self-
                                                          interactions */
               My_Molecule = My_Molecule - 1;
               Calc_To_Do = My_Molecule - 1;
            Unlock(Calc_To_Do_Gate);

            If (My_Molecule >= 0)
                   {
                   R_Squared = 0;
                   For (Direction = 0; Direction<3; Direction++)
                         {
                         Delta = X [Direction ]
                            -Coordinates <Master> [Direction,
                                                    My_Molecule ];
                         R_Squared = R_Squared + Delta*Delta;
                         }
                   R = Sqrt(R_Squared);
                   Interaction = Eij(R);

                   Lock(Calc_Done_Gate);
                      Sum<Master> = Sum<Master> + Interaction;
                      N_Completed = Calc_Done;
                      N_Completed = N_Completed + 1;
                      Calc_Done = N_Completed;
                   Unlock(Calc_Done_Gate);

                   /* Last one to complete signals master */
                   If (N_Completed == N_Molecules - 1)
                         }                         Signal <Master>;

                   Else
                   More_Interactions = False;
             }
}
}
}
```

D. A Molecular Dynamics Calculation

This computation operates in a similar fashion to the previous Monte Carlo calculation except that the basic slave computation has a coarser grain. A job here (also described by the job number, **Calc_To_Do**) consists of the calculation of *all* interactions with a given molecule, not just a single interaction as in the Monte Carlo case. In addition, a job requires computation of all the forces on a given molecule and the generation of new coordinates for that molecule according to, for example, Verlet's algorithm.

The master process initializes **Calc_To_Do** and **Calc_Done**, activates all slaves, and then waits till all jobs are completed and the slaves have computed new coordinates (stored locally) for the particles. The slave completing the last job signals the Master and then the Master reactivates all slaves to update the coordinates in shared memory with the new values just computed and stored locally by the slaves. When each slave completes the updating of coordinates in shared memory, it signals the Master. The Master keeps count of the slaves signaling the completion of the updating of their subset of the coordinates. The Master then continues on to a new time step when all coordinates have been updated.

```
Process Master(N_Processors)
Int N_Processors;
{
#Define Molecule_Limit 1024
Shared Float Coordinates[3,Molecule_Limit];

Main()
{
#Define Unlocked 1
Extern Shared Float E_Local_Sum;
Universal Int Calc_To_Do;
Universal Int Calc_To_Do_Gate = Unlocked;
Universal Int Calc_Done;
Universal Int Calc_Done_Gate = Unlocked;
Int Pass,N_Passes;
Int N_Molecules;
Int Instance, Completed;
Float Temperature;
Float E_Average, E_Sum;

Input(N_Molecules);
Input(Temperature);
Input(N_Passes);    /* a pass moves each of N molecules once */
```

Continued

```
/* Start up the slaves */
For (Instance = 0; Instance <N_Processors - 1; Instance++)
Create Slave: Instance(N_Molecules);


Generate_Initial_Configuration(N_Molecules);
Generate_Initial_Velocities(N_Molecules, Temperature);
E_Sum = 0;

For (Pass = 0; Pass <N_Passes; Pass++)
        {

        /* Start the slaves calculating the interactions */
        Calc_To_Do = N_Molecules - 1;
        For (Instance = 0; Instance <N_Processors - 1; Instance++)
        Signal <Slave: Instance >;

        /* Wait for the last to complete */
        Wait;

        /* Start the slaves updating the coordinates */
        Calc_Done = 0;
        For (Instance = 0; Instance <N_Processors - 1; Instance++)
        Signal <Slave: Instance >;

        /* Now wait for all the slaves to complete */
        Completed = 0;
        While (Completed <N_Processors - 1)
                {
                Wait;
                Completed = Completed + 1;
                }
        }

For (Instance = 0; Instance <N_Processors - 1; Instance++)
        E_Sum = E_Sum + E_Local_Sum <Slave: Instance >;
E_Average = E_Sum/N_Passes;
}
}
```

```
Process Slave(N_Molecules)
Int N_Molecules;
{
Shared Float E_Local_Sum;

Main( )
{
#Define False 0
#Define True 1
#Define Molecule_Limit 1024
#Define Local_Limit 256
Extern Shared Float Coordinates [3, Molecule_Limit ];
Universal Int Calc_To_Do;
Universal Int Calc_To_Do_Gate;
Universal Int Calc_Done;
Universal Int Calc_Done_Gate;
Int More_Molecules;
Int My_Molecule, N_Completed;
Float Local_Coords [3, Local_Limit ];

While (True)
{
Wait;     /* Wait for another time step */

        More_Molecules = True;
        While (More_Molecules)
         {

        Lock( Calc_To_Do_Gate );
          My_Molecule = Calc_To_Do;
          Calc_To_Do = My_Molecule - 1;
        Unlock( Calc_To_Do_Gate );

        If (My_Molecule >= 0) Then
                {
                E_Local_Sum = Binding_Energy(My_Molecule);
                Calculate_Force(My_Molecule);
                Advance(My_Molecule);
```

Continued

```
        Lock(Calc_Done_Gate);
           N_Completed = Calc_Done;
           N_Completed = N_Completed + 1;
           Calc_Done = N_Completed;
        Unlock(Calc_Done_Gate);


        If (N_Completed == N_Molecules) Signal <Master>;
        }

        Else
        More_Molecules = False;
        }
Wait;   /* Wait until all slaves have generated new
                                                      coordinates */


/* Update new coordinates */
For (Direction = 0; Direction <3; Direction++)
For (My_Molecule «contained in subset moved this time step»)
Coordinates <Master> [Direction, My_Molecule ] =
   Local_Coords [Direction, My_Molecule ];
Signal <Master>;
}
}
}



#Define Molecule_Limit 1024
Float Binding_Energy(My_Molecule);
Int My_Molecule;
{
Extern Shared Float Coordinates [3, Molecule_Limit ];
Extern N_Molecules;
Int Direction, Molecule;
Float R_Squared, Delta, R;
Float X [3];


/* cache coordinates of My_Molecule */
For (Direction = 0; Direction <3; Direction++)
X [Direction ] = Coordinates <Master> [Direction, My_Molecule ];
```

```
For (Molecule = 0; Molecule <N_Molecules; Molecule++)
     If (Molecule! = My_Molecule)
               {
               R_Squared = 0;
               For (Direction = 0; Direction <3; Direction++)
                          {
                          Delta = Coordinates <Master>
                                            [Direction, Molecule ]
                                     - X [Direction ];
                          R_Squared = R_Squared + Delta*Delta;
                          }
               R = Sqrt(R_Squared);
               Binding_Energy = Binding_Energy + Eij(R);
               }
}
```

Acknowledgments

The authors gratefully acknowledge financial support for this research from the National Science Foundation under Grant MCS79-20698. This research was also sponsored in part by Control Data Corporation, Grant 79C13, and by the Defense Advanced Research Projects Agency (DARPA), ARPA Order No. 3597, monitored by the Air Force Avionics Laboratory under Contract F33615-78-C-1551.

References

Barlett, J. F. (1978). *Proc. Hawaii Conf. System Sci. 11th*, pp. 103–117.
Barnes, G. H., Brown, R. M., Kato, M., Kuck, D. J., Slotnik, D. L., and Stokes, R. A. (1968). *IEEE Trans. Comput.* **C-17**, 746–757.
Baudet, G. M. (1978). Ph.D. Thesis, Carnegie-Mellon University, Pittsburgh, Pennsylvania.
Bedard, C. J., Mellor, F., and Older, W. J. (1977). *Proc. Internat. Software Applications Conf. (COMPSAC)*, *1st* pp. 772–777.
Burroughs Corporation (1979). Final report, Numerical Aerodynamic Simulation Facility, feasability study. Prepared for NASA under Contract No. NAS2-9897, Tech. Rep., Burroughs Corporation, Paoli, Pennsylvania.
Farquhar, I. E. (1964). "Ergodic Theory in Statistical Mechanics" Wiley, New York.
Flynn, M. (1972). *IEEE Trans. Comput.* **C-21**, 948–960.
Fuller, S. H., Ousterhout, J. K., Raskin, L., Rubinfeld, P. I., Sindhu, P. J., and Swan, R. J. (1978). *Proc. IEEE* **66**, 216–228.
Habermann, A. N. (1976). "Introduction to Operating System Design." Science Research Associates, Chicago, Illinois.
Heart, F. E., Ornstein, S. M., Crowther, W. P., and Barker, W. B. (1972). *Proc. Amer. Federation Inform. Process. Soc.: Nat. Comput. Conf.* **42**, 529–537.
Jones, A. K., and Gehringer, E. F., eds. (1980). The Cm* Multiprocessor Project: A research review. Tech. Rep. CMU-CS-80-131, Carnegie-Mellon University Computer Science Dept., Pittsburgh, Pennsylvania.

Jones, A. K., Chansler, R. J., Jr., Durham, I., Feiler, P., and Schwans, K. (1977). *Proc. Amer. Federation Inform. Process. Soc.: Nat. Comput. Conf.* **46**, 657–663.
Jones, A. K., Chansler, R. J., Durham, I., Schwans, K., and Vegdahl, S. R. (1979). *Proc. Symp. Operating System Principles, 7th*, pp. 117–127.
Kernighan, B. W., and Ritchie, D. M. (1978). "The C Programming Language." Prentice-Hall, Englewood Cliffs, New Jersey.
Kung, H. T. (1976). *In* "Algorithms and Complexity" (J. F. Traub, ed.), pp. 153–200. Academic Press, New York.
Kung, H. T. (1980). *In* "Advances in Computers" (M. C. Yovits, ed.), Vol. 19, pp. 65–112. Academic Press, New York.
McWilliams, T. M., Widdoes, L. C., Jr., and Wood, L. L. (1977). Advanced digital processor technology base development for Navy applications: The S-1 Project. Tech. Rep. N00014-77-0023, Office of Naval Research, Washington, D.C.
Metropolis, N., Rosenbluth, A. W., Rosenbluth, M. N., Teller, A. H., and Teller, E. (1953). *J. Chem. Phys.* **21**, 1087–1092.
Mezei, M., Swaminathan, S., and Beveridge, D. L. (1977). *J. Chem. Phys.* **71**, 3366–3373.
Oleinick, P. N., and Fuller, S. H. (1978). The implementation and evaluation of a parallel algorithm on C.mmp. Tech. Rep. CMU-CS-78-125, Carnegie-Mellon University Computer Science Department, Pittsburgh, Pennsylvania.
Ousterhout, J. K., Scelza, D. A., and Sindhu, P. S. (1980). *Comm. ACM* **23**, 92–104.
Pangali, C., Rao, M., and Berne, B. J. (1980). *Molecular Phys.* **40**, 661–680.
Ramanathan, P. S., and Friedman, H. L. (1970). *J. Chem. Phys.* **54**, 1086–1099.
Rao, M., Pangali, C., and Berne, B. J. (1979). *Molecular Phys.* **37**, 1773–1798.
Raskin, L. (1978). Ph.D. Thesis, Carnegie-Mellon University, Pittsburgh, Pennsylvania.
Ritchie, D. M., and Thompson, K. L. (1974). *Comm. ACM* **17**, 365–375.
Rossky, P. J., Doll, F. D., and Friedman, H. L. (1978). *J. Chem. Phys.* **69**, 4628–4633.
Russel, R. M. (1978). *Comm. ACM* **21**, 63–72.
Sameh, A. H. (1977). *In* "High Speed Computer and Algorithm Organization" (D. J. Kuck, D. H. Lawrie, and A. H. Sameh, eds.), pp. 207–228. Academic Press, New York.
Stone, H. S. (1975). *In* "Introduction to Computer Architecture" (H. S. Stone, ed.), pp. 318–374. Science Research Associates, Chicago, Illinois.
Verlet, L. (1967). *Phys. Rev.* **159**, 98–103.
Wu, C., and Feng, T. (1979). *Proc. Internat. Conf. Parallel Process.* pp. 160–174.
Wulf, W. A., and Bell, G. C. (1973). *Proc. Amer. Federation Inform. Process. Soc.: Fall Joint Comput. Conf.* **41**, 765–777.
Wulf, W. A., Russel, D. B., and Habermann, A. N. (1971). *Comm. ACM* **12**, 780–790.

Computer Modeling in Plasma Physics on the Parallel-Architecture CHI Computer

Robert W. Huff
John M. Dawson

Department of Physics
University of California
Los Angeles, California

G. J. Culler

CHI Systems
Goleta, California

I. INTRODUCTION

Traditionally, there have been two lines of attack for understanding the behavior of physical systems. The first is the experimental approach, in which one asks questions of nature through controlled experiments on the system of interest in the laboratory; one collects data on the response. For large-scale physical phenomena one often substitutes observations for experiments. The second approach is that of determining the predicted consequences of well-established physical laws through the application of mathematical analysis. The joint application of these two types of investigation has led to the remarkable development experienced by the physical sciences over the past 400 years.

Experiments have provided data that establish the laws, while analysis provides predictions whose verification strengthens our confidence in the laws; in the case of lack of verification they point to the need for revised physical theories.

Despite the success of these methods, they have their limitations, particularly when applied to complex systems with many interacting degrees of freedom. For such systems, analysis is incapable of handling the many interactions that take place and simplifying assumptions and approximations must be made. These may or may not be justified, but in any case, it may be difficult or impossible to determine their validity. On the other hand, experiments are also often difficult or impossible (as, for example, determining the influence of mountains on weather, the evolution of stars, the evolution of galaxies); in any case, these provide limited information as to what is going on. However, we often believe that we understand the fundamental physical laws and that we could understand and predict the behavior of these complex systems if we could just carry through the required calculations.

With the advent of modern high-speed computers, a third technique for investigating the behavior of physical systems has become possible: the construction of numerical models of the physical system and the performance of numerical experiments on these models. This approach is an extension of the analytic approach in that it employs idealized models. However, the models can be much more complex than can be handled analytically. (Millions of interacting degrees of freedom can be handled, and the computer follows nonlinear motion just as easily as it does linear motion.) The results of such computations can provide us with as much detail on the time development as we desire; it can provide for more detail than is available from laboratory experiments. The correctness of assumption and approximation relevant to simplified models can be checked. The science (or perhaps art) of numerical modeling based upon our knowledge of physical laws is generally quite different from the approach applied in analytic theory. It requires its own methods and techniques, which often bear a close relation to techniques used in laboratory experiments.

Often with the aid of the results of such simulations, we can determine which effects are important and which can be neglected. Armed with this knowledge it is often possible to construct a theory to explain the system behavior. It is, in fact, when used in this way that the method is most powerful; piles of computer output are generally not of much use, but if they can be condensed into a simple theory which contains the important phenomenon, then we have learned something.

Hot plasmas are an example of physical systems which are difficult to perform experiments and observations on. They also exhibit much complex behavior, and support a multitude of waves and instabilities which interact with each other and with the particles that make up the plasma. The behavior is often turbulent and nonlinear, and the analysis is often difficult and obscure.

Computer models have proved a powerful means of investigating plasma behavior. Although a great deal has been achieved with existing computers, to tackle many problems of interest requires computers of even greater power. Parallel computation offers one means of obtaining this added power. This chapter describes our exploration of the capabilities of the parallelism of the CHI system as applied to plasma simulation.

The set of problems defining the UCLA plasma-physics computing environment is described in Section II. Although small-scale special-case problems are included, emphasis is placed upon the formulation of the large-scale simulations, especially the electrostatic particle model. Section III summarizes the design of the CHI computer system, including the scheduling of the parallel array processes by the operating system. Section IV is a detailed discussion of the programming methods used to attain maximum execution efficiency on the large-scale simulations. Finally, Section V presents some general observations regarding the performance achieved with the computer system, the programming strategies that have proven most useful, and the behavior of the various users of the system. Ongoing elaboration of the large-scale simulations is described, and related to possible expansion of the computer system on two different levels.

II. FORMULATION OF THE SIMULATION PROBLEMS

A. *Electrostatic Particle Model*

The calculations performed on the CHI computer at UCLA fall into two groups: large-scale simulations and small-scale special-case problems. The large-scale simulations typically involve 10^5–10^6 particles or gridpoints, and 10^3–10^4 timesteps. Runs are repeated many times with different input parameters, for a broad range of applications. The required code modifications are relatively minor, so that a developed code may have a life of many years. Therefore considerable programming effort is justified in order to attain high execution efficiency for these simulations.

The most extensively developed class of simulation codes on the CHI computer is the electrostatic particle model, primarily the $2\frac{1}{2}$-dimensional version with uniform three-dimensional magnetic field, although a one-dimensional version without magnetic field also exists. The term "$2\frac{1}{2}$-dimensional" refers to allowing three-dimensional motion of the particles, but restricting the electrostatic forces between them to be two-dimensional. Physically, this is equivalent to a full three-dimensional electrostatic model in which the "particles" are infinite, uniformly charged rods aligned parallel to the z axis, as illustrated in Fig. 1. The restriction to a two-dimensional

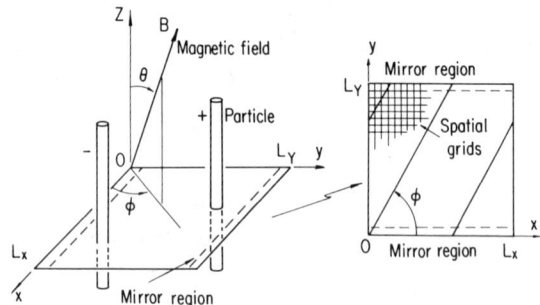

FIG. 1. Equivalent three-dimensional configuration for the $2\frac{1}{2}$-dimensional electrostatic particle simulation with uniform and constant magnetic field **B** and a mirror region.

force reduces the simulation to a tractable scale while retaining a sufficient richness of system phenomena. Because this model possesses most of the significant features found in large-scale simulations, it will be used as the principal vehicle for exhibiting these features.

The model consists of finite-size particles, moving in a uniform and constant magnetic field **B** and interacting via a self-consistent electrostatic field **E**, with the equations of motion (Kamimura *et al.*, 1975):

$$d\mathbf{v}_i/dt = (q_i/m_i)[\tilde{S} * \mathbf{E} + \mathbf{v}_i \times \mathbf{B}], \tag{1}$$

$$d\mathbf{r}_i/dt = \mathbf{v}_i, \tag{2}$$

$$\nabla \cdot \mathbf{E} = S * \rho, \qquad \nabla \times \mathbf{E} = 0, \tag{3}$$

$$\rho(\mathbf{r}) = \sum_i q_i \delta(\mathbf{r} - \mathbf{r}_i). \tag{4}$$

Here $\mathbf{v}_i$, m_i, and q_i are the velocity, mass, and charge of the *i*th particle, whose center position is denoted by $\mathbf{r}_i$. The particles have a common shape factor S, with the normalization $\int S \, d\mathbf{r} = 1$. The charge-center density is $\rho(\mathbf{r})$, while the true charge density is given by convolution with S:

$$[S * \rho](\mathbf{r}) = \int d\mathbf{r}' S(\mathbf{r} - \mathbf{r}') \rho(\mathbf{r}') = \sum_i q_i S(\mathbf{r} - \mathbf{r}_i). \tag{5}$$

The effective electric field seen by the *i*th particle is the weighted value

$$\mathbf{F}(\mathbf{r}) \equiv [\tilde{S} * \mathbf{E}](\mathbf{r}) = \int d\mathbf{r}' \, S(\mathbf{r}' - \mathbf{r}) \mathbf{E}(\mathbf{r}'), \tag{6}$$

which is the convolution of $\tilde{S}(\mathbf{r}) \equiv S(-\mathbf{r})$ with $\mathbf{E}(\mathbf{r})$. Finite-size particles are used, rather than point particles, in order to suppress some undesirable numerical and statistical effects.

The above physical equations must be converted to a form suitable for computation. Equations (1) and (2) are approximated by the time-centered difference equations

$$\mathbf{v}_i^{n+1/2} - \mathbf{v}_i^{n-1/2} = (q_i\, dt/m_i)[\mathbf{F}^n + \mathbf{v}_i^n \times \mathbf{B}], \tag{7}$$

$$\mathbf{r}_i^{n+1} - \mathbf{r}_i^n = \mathbf{v}_i^{n+1/2}\, dt. \tag{8}$$

Here dt is the timestep, and the time arguments $t^n = n\, dt$, etc., are denoted explicitly by superscripts. Although $\mathbf{v}_i^n$ appears in Eq. (7), it is not obtained in the course of iterating these equations. A usable scheme is obtained by replacing $\mathbf{v}_i^n \to \tfrac{1}{2}(\mathbf{v}_i^{n+1/2} + \mathbf{v}_i^{n-1/2})$ in Eq. (7), and solving for $\mathbf{v}_i^{n+1/2}$,

$$\mathbf{v}_i^{n+1/2} = \mathbf{R}_i \cdot \mathbf{v}_i^{n-1/2} + \tfrac{1}{2}(q_i\, dt/m_i)[\mathbf{1} + \mathbf{R}_i] \cdot \mathbf{F}^n, \tag{9}$$

where the dyadic

$$\mathbf{R}_i \equiv [1 + \Omega_i^2]^{-1}[1 - \Omega_i^2 - 2\Omega_i \times + 2\Omega_i\Omega_i] \tag{10}$$

represents a rotation about the magnetic field, with

$$\Omega_i \equiv \tfrac{1}{2}(q_i\, dt/m_i)\mathbf{B}. \tag{11}$$

In order to complete this iteration scheme, an algorithm must be derived for obtaining $\mathbf{F}^n$ from the set $\{\mathbf{r}_i^n\}$. The time superscripts will be suppressed for convenience. The algorithm is based on a solution of Eq. (3) via Fourier transforms. The derivation is simplest for an infinite continuous system, for which Eq. (3) transform to

$$i\mathbf{k} \cdot \mathbf{E}(\mathbf{k}) = S(\mathbf{k})\rho(\mathbf{k}), \qquad i\mathbf{k} \times \mathbf{E}(\mathbf{k}) = 0. \tag{12}$$

The solution for dimension $v \leq 3$ is

$$\mathbf{E}(\mathbf{k}) = -i\mathbf{k}S(\mathbf{k})\rho(\mathbf{k})/\mathbf{k}^2 \tag{13}$$

and

$$\mathbf{F}(\mathbf{r}) = \int_\infty d\mathbf{k}\, \frac{-i\mathbf{k}|S(\mathbf{k})|^2}{(2\pi)^v \mathbf{k}^2}\, \rho(\mathbf{k})e^{i\mathbf{k}\cdot\mathbf{r}}. \tag{14}$$

The infinite continuous transform is defined here as

$$S(\mathbf{k}) \equiv \int_\infty d\mathbf{r}\, S(\mathbf{r})e^{-i\mathbf{k}\cdot\mathbf{r}}, \tag{15}$$

and we have used the fact that convolutions transform into products and that $S(-\mathbf{r})$ has the transform $\tilde{S}(\mathbf{k}) = S(-\mathbf{k}) = S^*(\mathbf{k})$ for real $S(\mathbf{r})$, where * denotes the complex conjugate.

The integral in Eq. (14) is unsuitable for computation, and must be replaced by a sum over $\mathbf{k}$ values. This suggests that the simulation model be made periodic in each of the v nontrivial coordinate directions. The only change from the previous results is the replacement of Eq. (14) by

$$\mathbf{F}(\mathbf{r}) = \sum_{\mathbf{k}} \frac{-i\mathbf{k}|S(\mathbf{k})|^2}{Vk^2} \rho(\mathbf{k}) e^{i\mathbf{k}\cdot\mathbf{r}} \qquad (16)$$

and the reduction to finite continuous transforms

$$\rho(\mathbf{k}) \equiv \int_V d\mathbf{r}\, \rho(\mathbf{r}) e^{-i\mathbf{k}\cdot\mathbf{r}} = \sum_i q_i e^{-i\mathbf{k}\cdot\mathbf{r}_i}. \qquad (17)$$

The i summation extends only over the particles within the fundamental v-dimensional coordinate volume V and does not include their periodic images. The sole effect of the latter is the retention of the infinite transform in Eq. (15) for $S(\mathbf{k})$. The $\mathbf{k}$ summation extends over the v-dimensional reciprocal lattice appropriate to the periodicity in coordinate space. Although this summation is infinite, $S(\mathbf{k})$ is always chosen to vanish sufficiently rapidly for increasing k^2 that only a finite number of terms need be included. The usual form chosen is $S^2(\mathbf{k}) = \exp(-a^2 k^2)$, corresponding to $S(\mathbf{r}) = \exp(-r^2/2a^2)$, where a is the rms radius of the particles.

Equations (16) and (17) have the appearance of finite discrete transforms, which invites the use of an efficient fast Fourier transform (FFT) algorithm. However, Eq. (16) is to be evaluated at the particle positions for use in Eq. (9), and Eq. (17) involves the same positions, which do not coincide with the regularly spaced lattice or grid positions required by FFT algorithms. For FFT use, let a regular v-dimensional grid, consistent with the periodicity conditions, be established in coordinate space. As a first approximation, each particle position may be replaced by the nearest grid position in Eqs. (16) and (17). Since this approximation is known to give an unacceptable amount of numerical noise, the next approximation must be obtained. Let the ith particle be associated with the nearest grid point $\mathbf{r}_g$, with displacement $d\mathbf{r}_i \equiv \mathbf{r}_i - \mathbf{r}_g$. Then each term in Eq. (17) may be Taylor expanded about $d\mathbf{r}_i = 0$ to give

$$\rho(\mathbf{k}) = C(\mathbf{k}) - i\mathbf{k}\cdot\mathbf{D}(\mathbf{k}) + \cdots, \qquad (18)$$

where the charge and vector–dipole arrays are

$$C(\mathbf{k}) \equiv \sum_g C(g) e^{-i\mathbf{k}\cdot\mathbf{r}_g}, \qquad \mathbf{D}(\mathbf{k}) \equiv \sum_g \mathbf{D}(g) e^{-i\mathbf{k}\cdot\mathbf{r}_g} \qquad (19)$$

and

$$C(g) \equiv \sum_{i(g)} q_i, \qquad \mathbf{D}(g) \equiv \sum_{i(g)} q_i\, d\mathbf{r}_i. \qquad (20)$$

The notation $\sum_{i(g)}$ denotes a summation including only those particles having $\mathbf{r}_g$ as their nearest gridpoint. $\mathbf{F}(\mathbf{r}_i)$ may be evaluated by interpolating between values at adjacent gridpoints or by using a similar Taylor expansion in Eq. (16) to obtain

$$\mathbf{F}(\mathbf{r}_i) = \mathbf{F}(\mathbf{r}_g) + d\mathbf{r}_i \cdot \mathbf{G}(\mathbf{r}_g) + \cdots, \quad (21)$$

where

$$\mathbf{G}(\mathbf{r}_g) \equiv \sum_{\mathbf{k}} [\mathbf{kk} |S(\mathbf{k})|^2 / V k^2] \rho(\mathbf{k}) e^{i\mathbf{k} \cdot \mathbf{r}_g}. \quad (22)$$

$S(\mathbf{k})$ will always be chosen to vanish sufficient rapidly for increasing k^2 that the $\mathbf{k}$ summations in Eqs. (16) and (22) can be restricted to the first Brillouin zone, centered about $\mathbf{k} = 0$. Thus Eqs. (16), (19), and (22) are now in the form required for employment of FFT algorithms.

The number of FFT operations can be reduced if the factor $i\mathbf{k}$ in Eq. (18) is approximated by a form having a simple coordinate-space transform. One such approximation replaces Eqs. (18) and (19) by

$$\rho(\mathbf{k}) = \sum_{g} [C(g) - \mathbf{\Delta} \cdot \mathbf{D}(g) + \cdots] e^{-i\mathbf{k} \cdot \mathbf{r}_g}. \quad (23)$$

Here the vector central-difference operator $\mathbf{\Delta}$ has components defined by

$$\Delta_x f(x_g) \equiv [f(x_g + dx) - f(x_g - dx)]/(2dx), \quad (24)$$

with the transform

$$\Delta_x(k) = i(\sin kdx)/dx \approx ik, \quad (25)$$

where dx is the grid spacing. Similarly, the evaluation of $\mathbf{F}(\mathbf{r}_i)$ by interpolation is equivalent to approximating the tensor $\mathbf{G}(\mathbf{r}_g)$ by a form such as $\mathbf{\Delta F}(\mathbf{r}_g)$, and also reduces the required number of FFT operations. Higher accuracy can be obtained by including more terms in the expansion of $\rho(\mathbf{k})$ and $\mathbf{F}(\mathbf{r}_i)$ than the two terms explicitly retained above.

The simulation begins by computing the charge and dipole arrays, $C(g)$ and $\mathbf{D}(g)$, for a suitably constructed initial distribution of particle positions and velocities. It proceeds through FFT calculation of $\rho(\mathbf{k})$, to the inverse FFT calculation of $\mathbf{F}(\mathbf{r}_g)$ and possibly $\mathbf{G}(\mathbf{r}_g)$, and then advances all particle positions and velocities by one timestep via Eqs. (8) and (9). This cycle is then repeated for the desired number of timesteps. Accurate integration of the particle trajectories generally has little value per se, however, and is seldom the direct goal of simulation. The shear volume of such data for large-scale simulations exceeds both the storage capacity of presently envisioned computer systems and the data-digestion capability of the human mind. (Complete three-dimensional trajectory data for a quite reasonable million-particle 10,000-timestep simulation with 4-byte precision would occupy 240 gigabytes, or four times the capacity of the mass store being installed at the National Magnetic Fusion Energy Computer Center.) Of more direct interest and value are macroscopic quantities, such as temperature and mean velocity for subsets of particles, diffusion rate for subsets of particles, amplitude and energy content of selected Fourier modes of the electrostatic field, and spatial distribution of temperature and particle density. These quantities are essentially weighted averages of functions of particle positions and velocities, and thus consist of far fewer data elements than the particle data from which they derive. Furthermore, they typically vary more slowly in time than the individual particle data, and thus need be calculated only at selected timesteps during the simulation.

Two such macroscopic quantities, or diagnostics, are the electrostatic field energy

$$\varepsilon_F = \frac{1}{2} \int_V d\mathbf{r}\, \mathbf{E}^2(\mathbf{r}) = \sum_{\mathbf{k}} \frac{|\rho(\mathbf{k})|^2 |S(\mathbf{k})|^2}{2V k^2} \qquad (26)$$

and the particle kinetic energy

$$\varepsilon_K = \sum_i \frac{1}{2} m_i v_i^2. \qquad (27)$$

Since the total energy, $\varepsilon_K + \varepsilon_F$, is conserved for the physical system being simulated, it serves a dual purpose as a diagnostic: first, as a measure of the macroscopic accuracy of the various approximations made in deriving the simulation equations and implementing them on a given computer; and second, as a detector of computer error during a simulation run, to permit execution of user-defined recovery procedures. For these reasons, ε_K and ε_F are usually computed and stored at every timestep.

Four additional diagnostics of great flexibility are the velocity tensor and vector

$$\mathbf{T} = \langle \mathbf{v}_i \mathbf{v}_i \rangle, \qquad \mathbf{U} = \langle \mathbf{v}_i \rangle \qquad (28)$$

and the displacement tensor and vector

$$\mathbf{P} = \langle (\mathbf{r}_i - \mathbf{r}_i^0)(\mathbf{r}_i - \mathbf{r}_i^0) \rangle, \qquad \mathbf{Q} = \langle (\mathbf{r}_i - \mathbf{r}_i^0) \rangle, \qquad (29)$$

where $\mathbf{r}_i^0$ is the initial position of the ith particle and the angular brackets denote an average over some subset of particles. $\mathbf{U}$ and $\mathbf{Q}$ are just the mean velocity and displacement. The temperature in a given direction $\hat{s}$ (e.g., parallel or perpendicular to the magnetic field) for a subset of particles of the same mass m is

$$T = m\hat{s} \cdot (\mathbf{T} - \mathbf{U}\mathbf{U}) \cdot \hat{s}. \qquad (30)$$

Similarly, diffusion is obtained from the second-moment tensor $\mathbf{P} - \mathbf{Q}\mathbf{Q}$. Either of these second-moment tensors can be diagonalized, and the corresponding principal axes for temperature or diffusion can be determined.

A portion of the array $\rho(\mathbf{k})$ centered about $\mathbf{k} = 0$ is usually saved as a diagnostic, since Eqs. (13) and (26) show that a single array element is sufficient to obtain the electrostatic field amplitude $\mathbf{E}(\mathbf{k})$ and energy content for that Fourier mode. This diagnostic can also yield a good reconstruction of the spatial field $\mathbf{E}(\mathbf{r})$ because the factor $S(\mathbf{k})$ greatly suppresses the contributions from the larger-k^2 modes. Hence the latter would need to be saved only if the fine-grained short-wavelength features of $\mathbf{E}(\mathbf{r})$ were desired. This diagnostic has proved valuable for frequency analysis via an FFT in time, and is therefore usually saved every few timesteps.

A final aspect to be considered for this simulation is the modification needed for differing applications. The addition of new diagnostics is relatively simple because they execute independently of the particle-advance and force-array calculations. One modification of the particle-advance algorithm is the addition of a mirroring algorithm to restrict particle motion along the magnetic field. By this algorithm one or all of the velocity components are reversed for any particle which approaches a mirror [the (x, z) plane and its periodic images; cf. Fig. 1] with a ratio of parallel to perpendicular velocity components (usually relative to the magnetic field) less than some limit. The simplest modification of the force calculation is the addition of an external electric field for the purpose of pumping energy into the plasma. This pump field is usually of a single frequency, with a simple spatial dependence. The change from a fully periodic system to a system bounded in one direction requires a more complicated modification. The analysis is based on the scalar potential $\phi(\mathbf{r})$ satisfying $\nabla^2 \phi = -S * \rho$, and uses a solution in the form, $\phi = \phi_1 + \phi_0$, where ϕ_1 is the FFT-based solution with $\phi_1(\mathbf{k}) = S(\mathbf{k})\rho(\mathbf{k})/k^2$ and ϕ_0 is a general solution of the homogeneous Laplace equation $\nabla^2 \phi_0 = 0$. The constants in ϕ_0 are chosen such that $\phi_1 + \phi_0$ satisfies the desired boundary conditions. The determination of $\mathbf{F}(\mathbf{r}_g)$ from $C(g)$ and $D(g)$ is thus a more elaborate computation than in the case of a fully periodic system.

B. *Other Large-Scale Simulations*

Other large-scale simulations show significant features beyond those found in the electrostatic particle model. One class of codes was developed for the magnetohydrodynamic (MHD) fluid model, with equations of motion (Dawson *et al.*, 1978; Roberts and Potter, 1970)

$$\partial \rho / \partial t = -\nabla \cdot (\rho \mathbf{v}) + \varepsilon \nabla^2 \rho, \tag{31}$$

$$\partial(\rho \mathbf{v})/\partial t = \nabla \cdot (\mathbf{BB} - \rho \mathbf{vv}) - \nabla(\tfrac{1}{2}\mathbf{B}^2 + \rho T) + \nu \nabla^2(\rho \mathbf{v}) - \lambda \rho \mathbf{v} + \rho \mathbf{g}, \tag{32}$$

$$\partial \mathbf{B}/\partial t = \nabla \times (\mathbf{v} \times \mathbf{B}) + \eta \nabla^2 \mathbf{B}, \tag{33}$$

$$\partial T/\partial t = -\mathbf{v} \cdot \nabla T - \mu T \nabla \cdot \mathbf{v} + \eta \mu \rho^{-1}(\nabla \times \mathbf{B})^2 + \kappa \mu \nabla^2 T$$

$$+ \nu \mu \sum_{ij} \frac{\partial v_i}{\partial r_j}\left(\frac{\partial v_i}{\partial r_j} + \frac{\partial v_j}{\partial r_i} - \delta_{ij} \nabla \cdot \mathbf{v}\right). \tag{34}$$

The fluid mass density and velocity, ρ and $\mathbf{v}$, the magnetic field $\mathbf{B}$, and the temperature T are all spatially dependent quantities. The acceleration $\mathbf{g}$ and the parameters ε, λ, μ, ν, η, and κ are all constants, some of which are introduced to ensure numerical stability in the simulation.

As before, these equations are approximated by time-centered difference equations. A uniform coordinate grid is introduced, and the spatially dependent quantities are defined only over this grid. The spatial derivatives in Eqs. (31)–(34) are approximated by difference operators on the grid. This directly couples the grid quantities at neighboring gridpoints at each timestep of the simulation. In contrast, the particles of the electrostatic model are coupled only indirectly, via the electrostatic field, and are thus advanced independently of one another at each timestep. The topological considerations necessitated by this direct coupling are an important feature of fluid simulations. The topological complexity is greater for the full three-dimensional model than for the $2\frac{1}{2}$-dimensional model with its two-dimensional grid. Moreover, the difference equations for the fluid simulation are far more elaborate than their particle counterparts, Eqs. (8) and (9).

Another large-scale simulation was developed by a guest from the field of chemistry. This molecular-dynamics model consists of particles interacting via the two-body truncated Lennard-Jones potential, with equations of motion

$$m_i \frac{d\mathbf{v}_i}{dt} = \sum_{j \neq i} \mathbf{F}(\mathbf{r}_i - \mathbf{r}_j), \tag{35}$$

$$d\mathbf{r}_i/dt = \mathbf{v}_i, \tag{36}$$

where

$$\mathbf{F}(\mathbf{r}) \equiv \begin{cases} (Ar^{-14} - Br^{-8})\mathbf{r} & \text{for } r < R \\ 0 & \text{for } r \geq R \end{cases} \tag{37}$$

and the parameters A, B, and R are constants. Because of the two-body form of interaction, the topological complexity is much greater than in the case of fluid or electrostatic-particle simulations. The update of a particle velocity involves a sum of contributions from a significant fraction of the complete set of particles, rather than from only the nearest neighbors or a single electrostatic field. The required computation per particle is thus much greater than for the plasma simulations, even though the two-body force itself has a very simple form. Fortunately, several hundred to a thousand molecules are sufficient for a number of applications.

C. Small-Scale Special-Case Problems

This group includes many problems, but all of small scale, such that the group uses only a small fraction, probably 10–20%, as much computer time as the group of simulations. The problems range in scale from desk-calculator type, through simple processing and display of minor amounts of experimental data, to the solution of coupled differential or integral equations or the integration of some thousand particle orbits in a given spatially dependent electric or magnetic field. If programs are repeated for more than a few runs, they usually undergo frequent and extensive code modification. Thus considerable programming effort is expended on this group of problems, but there is neither motivation nor strong reason to increase this effort in order to maximize execution efficiency.

If one of these programs proves sufficiently valuable that its use is prolonged, perhaps with widening application, or if its scale is sufficiently enlarged, then effort to increase its execution efficiency would be justified. So far, although some programs have had greater use than initially expected, none has expanded sufficiently to impel such effort.

The users in both problem groups are not professional programmers, but are research scientists who do their own programming. Those in the large-scale group might be termed simulationists, as they have specialized in the use of simulations as a principal tool in developing their intuition and understanding of plasma theory. In contrast, users in the small-scale group tend to be theorists or experimentalists for whom computation is only one of many tools of analysis. Therefore, the small-scale users generally cannot be expected to have the same level of programming experience, sophistication, or addiction commonly demanded of simulationists. Because the small-scale users are several times as numerous, the importance of ensuring them convenient computer access is far greater than might be suggested by the small fraction of computer time they use.

III. DESIGN OF THE COMPUTER SYSTEM

The CHI computer system at UCLA is based upon an earlier system designed by Culler-Harrison, Inc. (now CHI Systems) and used by them in their research in signal-processing applications. An electrostatic particle simulation serving as a pilot study was performed jointly by CHI and the UCLA plasma simulation group (Kamimura *et al.*, 1975) using this earlier computer. The pilot study indicated that a significant increase in speed could be obtained if more disk drives, with a higher transfer rate, were used for intermediate storage of particle data, and if the transfer of particle data between disk and memory could be overlapped with the computations in the two floating-point and fixed-point arithmetic processors.

The structure of the improved system installed at UCLA in 1978 is shown in Figs. 2 and 3. The fixed-point Macro Processor (MP-32A) was retained as a host or control processor, with some improvements to handle an expanded multiuser environment. The essential characteristics of the MP-32A are summarized in Table I. A floating-point Array Processor (the AP-120B, manufactured by Floating Point Systems, Inc.) replaced its design predecessor (the CHI-designed AP-90). Four Calcomp Trident T80 disk drives were used to attain the desired transfer rate. In order to achieve complete overlapping among the disk transfers, floating computations, and fixed-point control operations, an independent microprogrammable Input/Output Processor (IOP) was designed to handle high-speed data transfers among the MP and AP memories and the disk drives and to remove this function from the MP. Four of these IOPs were included in the system, each originally controlling one disk drive, but each capable of controlling up to four T80 or T300 disk drives in any combination desired. The essential characteristics of the AP-120B, IOPs, and disk drives are summarized in Table II.

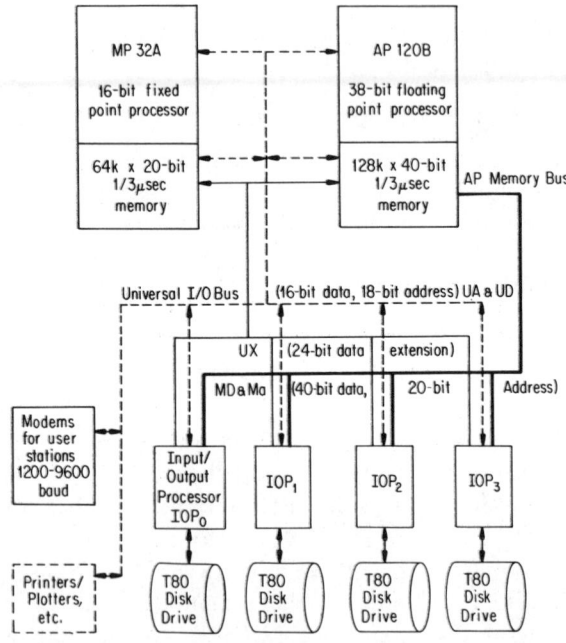

FIG. 2. Block design of the CHI computer system at UCLA.

FIG. 3. Physical arrangement of the CHI computer system at UCLA. The left rack contains the MP, the central rack contains the AP and four IOPs, and the right rack contains the four disk drives. One of the CRT terminals with attached supplementary keyboards appears at the left.

The standard data memory of the AP-120B has not been included in the system. Instead, a data memory of equivalent speed has been used that allows shared access on a DMA basis by all six microprocessors of the system: the MP, AP, and four IOPs. For disk transfers each IOP uses one memory access during each 4-μsec interval. The shared-memory controller is designed to ensure that the AP arithmetic unit has access on both sides of any IOP access. To organize the computation, the MP host interprets the application program into a sequence of array processes (of which a block transfer is a simple example) and then schedules either the AP arithmetic unit or one of the IOPs for each process of the sequence. In many cases the arrays to be used are outputs of earlier processes, so that each array referenced by a scheduled process must be locked against subsequent processing until any outstanding process upon it is complete.

The MP host maintains a scheduling queue in its own memory for each IOP. These queues are available, via a DMA facility, to the IOPs, and each

TABLE I

ESSENTIAL CHARACTERISTICS OF THE MP-32A

16-bit fixed-point processor   167-nsec cycle
    64 words scratch pad data memory
    512 words fixed instruction memory
    64 words writable instruction memory
    64 kilowords of $\frac{1}{3}$-μsec instruction and data memory (CD)
    16 × 16 multiply in 333 nsec
Schedules IOPs and Array Processor
Controls data transfer to modems (1200–9600 baud) and
    other PDP-11 compatible I/O devices over Universal I/O bus

---

queue element is a process control block that specifies the process, input and output arrays and their lengths, and a status word for process validation. When an IOP finishes a process, an end-of-process interrupt is sent to the MP, and the next process on the queue is initiated. The MP responds to this interrupt by checking the process validity, unlocking the arrays referenced by the completed process, and scheduling another process. Detected errors for a process are treated by repeating those processes that are safely repeatable (i.e., block transfers), and by passing control to a user-defined recovery

TABLE II

ESSENTIAL CHARACTERISTICS OF THE AP/IOP/DISK GROUP

Array Processor (AP-120B)

Very high speed 38-bit floating-point arithmetic unit
    Two 32-word scratch memories (Data Pad)
    2560 words 1/3-μsec fixed table memory
    128 kilowords 1/3-μsec data memory (MD)
    2048 words instruction memory

All memories can be referenced in two 167-nsec machine cycles

Typical operation times:
Vector add, multiply, subtract	1 μsec/point
Vector SQRT, divide	1.8 μsec/point
Vector EXP, LOG, SIN	5–6 μsec/point

IOPs and Disks

4 CHI I/O Processors (IOPs) and Trident T80 drives—over 16 × 10⁶ 38-bit words each
    All four can be simultaneously transferring data to/from AP–MD or MP–CD, each at 250,000 words/sec
    Any IOP can be used for memory transfer between AP–MD and MP–CD at 1 × 10⁶ words/sec
    Expandable to four T80 or T300 drives per IOP

program in other cases. This provides a high error tolerance at the system level. The management of the AP is similar, except that the scheduling queue for the AP resides in the shared memory, to which the AP arithmetic unit has greater access.

The results presented in this paper were obtained with the original 64-kiloword shared AP memory. This is the minimum size needed for a large-scale simulation, including 4 kilowords for system purposes. The original operating system was therefore designed to handle multiple users on a time-shared basis, with the complete AP memory transferred to and from disk between successive users. With 1-sec time slots, fully one third of the system time was spent in swapping between users when two or more large-scale simulations were sharing the system, even though overlapped data transfer to a pair of disks was employed. Shorter time slots would have given lower execution efficiency, while longer time slots would have given unacceptably long response time for multiple users, especially the many small-scale interactive users.

The shared AP memory was recently expanded to 128 kilowords to permit more elaborate simulations. This was accompanied by a rewriting of the operating system to provide both faster response time and more efficient time sharing. For this purpose, the shared memory is partitioned when the system is initialized, with the number and size of the partitions selected by the system manager. The small-scale interactive users, which may include simulationists who are processing diagnostic output from completed simulations, will typically need 20k–30k partitions. Thus a large-scale simulation in a 64k partition can be time sharing with two or three such users without the need for swapping between shared AP memory and disk. Alternatively, a single 124k partition can be provided for more elaborate simulations.

Four programming languages are available on the system: microlevel assembly languages on the MP host and the AP, a macrolevel assembly language used for most system programming, and the high-level Math System Language (MSL) used for 99% of all applications programming. The mathematical operations available in MSL include the elementary functions of classical analysis, both real and complex. The operands of MSL are integers, and floating-point scalars, vectors, multidimensional arrays and subarrays. The result is a natural and concise language which avoids most of the loop and subscript bookkeeping of lower-level scalar languages such as FORTRAN. It is used for on-line interactive computing of the desk-calculator type, as well as for simulation programming. MSL is interpreted at execution time, and controls the scheduling of array processors as described above. (The term "array processor" derives from the use of the AP and its predecessor in such a software environment. It is inaccurate to use it purely as a hardware descriptor for scalar devices such as the AP.)

Any special functions or AP operations not provided as part of the system can be written in AP assembly language. This is also done for critical pieces of an MSL program where maximum execution efficiency is required. The resulting AP program is then invoked from an MSL program in the same way as the existing mathematic operations.

Access to the system is provided by Tektronix 4013 terminals, each augmented by a supplementary 64-function keyboard to support the MSL operators and the library and edit utilities. The storage CRT of these terminals provides a flexible and highly interactive graphic-display capability for the real and complex operands of MSL.

IV. PROGRAMMING FOR EFFICIENCY

One characteristic of large-scale simulations is the great effort usually exerted to approach maximum execution efficiency. This section describes techniques applied toward this goal on the CHI computer at UCLA. Considerable details are given, not because of their importance per se, but rather to allow users who are more familiar with large-memory general-purpose computers to see the different sort of analysis necessary in the present context.

From the standpoint of the simulationist, the formulation of the particle model may itself be viewed as a significant gain in execution efficiency via a reduction in number of calculations, namely, by aggregating the 10^{19}–10^{24} particles of a physical plasma into 10^5–10^6 simulation particles in such a way as to leave various macroscopic quantities unchanged. The use of finite-size particles, rather than point particles, counteracts some of the adverse effects of this aggregation, such as increased collision rates. In addition, introducing an electric field to replace the N^2 two-particle force contributions in an N-particle plasma further reduces the required calculations from a number proportional to N^2 to one proportional to N only, assuming the particle advance dominates the computation. This latter technique may not be feasible for more complicated two-particle forces, such as in simulations of chemical molecular dynamics. But even in such a case, the medium-to-long-range interactions might be represented by a field, so that only the short-range forces from a particle's few neighbors need be computed individually.

There are several techniques which need to be mentioned only briefly, since they are generally worthwhile regardless of which particular computer is used to implement the simulation. Since $\rho(\mathbf{r})$ and $\mathbf{F}(\mathbf{r})$ are real, use of a special real-to-complex FFT algorithm and inverse gains a factor of roughly two in both memory space and FFT execution speed. Including a factor dt within the particle velocities and a factor $\frac{1}{2}dt^2$ within **F** saves multiplications in Eqs. (8) and (9). Grouping the particles into species of common mass and charge saves two words of memory space per particle. It also allows the options of including a factor of q_i/m_i in **F** to save multiplications and divisions in Eq. (9), and of excluding a factor of q_i from C and **D** to save multiplications in Eq. (2), although this requires multiplicative adjustment of $\mathbf{F}(\mathbf{r}_g)$, $C(g)$, and $\mathbf{D}(g)$ between species in a multispecies simulation. These options are typical of the class of techniques wherein operations are removed from the particle advance to the force computation when the number of particles greatly exceeds the number of gridpoints, but retained within the particle advance otherwise.

A. *Tailoring the Simulation to the Computer Structure*

With any computer system, the attainment of maximum execution efficiency for a large-scale simulation requires the programmer to consider the specific architectural and operational characteristics of the whole system. For the CHI system, the essential characteristics can be summarized as follows. Floating-point operations, which make up the major part of any scientific computation, are performed in the AP arithmetic unit, and initiated at a maximum rate of one addition and one multiplication per instruction, or six of each operation per microsecond, with results available two and three instruction cycles after initiation. The AP memory has a usable capacity at present of 60k floating-point words (1k = 1024), and a maximum access rate of 1 word every alternate instruction cycle, or 3 words per microsecond. The system presently has four disks, each with a capacity of 16,000k words, organized as 800 cylinders of 10 half-track records of 2k words. Disk rotational speed limits the peak transfer rate per disk to 1 word every 4.07 μsec. Because head repositioning at the cylinder boundary loses one record duration, the sustained transfer rate per disk drops to one word per 4.5 μsec or 4.9 μsec, depending upon whether one chooses maximum transfer rate or sequential record utilization. The disks have parallel and direct AP memory access for a peak combined rate of one word per microsecond. Although disk transfer and AP operations can proceed in parallel, the maximum memory access rate applies to all accesses combined, and can cause interference, in which case disk transfer has priority and AP instruction execution is delayed. The programmer controls the disk transfers and AP operations through programs written in the high-level Math System Language (MSL). These programs are interpreted at execution time by the MP, which also performs any explicit integer or index operations in the MSL programs. The interpretation proceeds in parallel with any initiated disk and AP operations, and requires about a half-millisecond for each MSL operator and each explicit MSL variable occurring as an operand or subscript.

For simulations with the desired 10^5 particles or more, it is clear that the 6-word particle descriptors (position and velocity) must reside on disk. To attain maximum transfer rate, all four disks are utilized in parallel, with particle descriptors being read from two disks into AP memory, updated there by the AP, and then written onto the remaining two disks. This flow is easily reversed at each successive timestep, with disk pairs merely exchanging read and write roles. Since each particle updated requires 12 words of disk transfer (one particle descriptor both read and written), or 3 words per disk, the maximum sustained transfer rate is one particle descriptor every $3 \times 4.9 = 14.7$ μsec if disk records are transferred sequentially. The maximum sustained transfer rate could be increased by 8% if records were transferred nonsequentially, but elaboration of either system software or user coding would be required, and has not yet been attempted. If partial records are used, the sustained transfer rate is reduced proportionally. Use of full records, and allocation of sufficient buffer space in AP memory to sustain the transfer rate, thus gives 14.7 μsec as the minimum particle-update time permitted by the disk transfer.

The attainable particle-update time is also limited by the AP computation time, which depends upon both the computations to be performed and the number of accesses to AP memory by the arithmetic unit. Since the minimum acceptable gridsize is usually 64×64, or 4k gridpoints, with larger grids often being desirable, one may be tempted to optimize the use of the 60k memory by reducing the number of needed arrays to a minimum. Approximating $\mathbf{G}(\mathbf{r}_g)$ of Eq. (22) by $\Delta \mathbf{F}(\mathbf{r}_g)$ allows a reduction to two force arrays, FX and FY, but each of these must be accessed five times per particle update: the nearest gridpoint and its four nearest neighbors. Similarly, the quantity $C(g) - \Delta \cdot \mathbf{D}(g)$ of Eq. (23) can occupy a single array, with ten accesses needed per particle in order to add the charge to the nearest gridpoint and add or subtract the dipole moment to its four nearest neighbors. Unfortunately, when this first version was implemented in AP code, the particle loop contained approximately 170 instructions, with an average of nine instruction cycles expected to be added during execution because of memory conflict. This gives a particle-update time of 30 μsec, or more than twice the minimum time permitted by disk transfer.

It was noticed that half the computation in the first version was associated with access of the four neighbors of a particle's nearest gridpoint, including the needed special treatment when the latter is a boundary gridpoint. For this reason a faster second version was coded, in which five arrays were added to represent $\mathbf{G}(\mathbf{r}_g)$ and $\mathbf{D}(\mathbf{r}_g)$ explicitly, namely GXX, GXY, GYY, DX, and DY. (GYX is not needed because $GYX = GXY$.) The particle loop of this version contained 90 instructions, with 45 floating additions, 33 floating multiplications, and 23 memory accesses. Since this code occupies the memory controller for $2 \times 23 = 46$ cycles of the loop, during which the disk transfer of a particle descriptor requires the memory controller for an average of $2 \times 12 = 24$ cycles, memory conflicts are expected to add $24 \times 46/90 = 12$ cycles, for a total of 102 cycles. This gives a particle-update time of 17 μsec, or 16% of greater than the minimum permitted by disk transfer.

This second version has nearly acceptable timing, but it was feared that modification of the mirroring algorithm or inclusion of a z-direction electric pump field would add an extra microsecond or two, and hence an attempt to tighten the code was made. The number of memory accesses and floating-point operations suggests that the code could be reduced by nearly a half by splitting the loop into two halves and merging these so that the last half of the operations on one particle were interspersed with the first half of the operations on the next particle. This third version was completed after considerable effort, and is the basic particle-update AP code now in use for fully periodic simulations. The particle loop contains 55 instructions, with 45 floating additions, 33 floating multiplications, and 24 memory accesses. Memory conflict is now greater, and is predicted to add an average of 21 cycles. The resulting particle-update time of 12.7 μsec is now well below the minimum permitted by disk transfer, and leaves the desired margin for minor future modifications.

All three versions of the code accumulate the charge and dipole contributions for the updated particle positions as an integral part of the particle-advance algorithm. They similarly compute the kinetic energy via Eq. (27) from the updated particle velocities. Simulation programs on large-memory general-purpose computers typically calculate these quantities in subroutines separate from that of the particle advance. To do so on the CHI computer would mean separate passes requiring a minimum of 14.7 μsec/particle each. Since the charge and dipole contributions are needed each timestep to compute the force arrays for the next particle advance, and the kinetic energy is desired each timestep for error detection and recovery, separate passes would double or triple the simulation time.

The existence of convenient floating-point masking operations in the AP leads to an additional advantage, provided that powers of 2 are chosen for the grid dimensions. Simulations on other computers often reduce each particle back into the basic grid when it passes the grid boundary, as allowed by periodicity. Since this destroys the unreduced displacement from the initial position as required for the diffusion quantities of Eq. (29), a small set of test particles is added whose unreduced motion is followed. The AP masking operation is so convenient that reduction to the basic grid is unnecessary. This obviates the need for a separate set of test particles, which not only saves disk space, but also improves the statistics by allowing the full set of particles to be used in the diffusion diagnostic.

These three versions of code represent the first experience on the AP for this programmer. The first two required under 1 man-week each. The final, tightly written version required 5–6 man-weeks, of which the first was spent in merging the floating-point operations from the two halves of the unfolded particle loop. The remaining 4 or 5 man-weeks were used in assigning parameters and intermediate results to the fast internal Data Pad registers. This may seem to have been a costly effort, but fortunately it constitutes the major part of the AP coding needed for the fully periodic electrostatic particle code. The two diagnostic codes for Eqs. (28) and (29) needed only $\frac{1}{2}$ man-day each, as did the two-dimensional real-to-complex FFT and inverse together, which utilized existing one-dimensional FFT codes.

B. *Programming in the Math System Language*

The remainder of the code is written in the high-level Math System Language (MSL). This language, with its operators and multidimensional arrays and subarrays, is quite close to the conceptual structure of applied mathematics. As such, it is a natural and concise language for scientific programming, and avoids much of the subscript and loop bookkeeping familiar in intermediate-level languages such as FORTRAN. As an example from the electrostatic particle simulation, the Poisson-solver code for obtaining the force arrays from the charge and dipole arrays contains more than 100 lines with triply nested loops in the FORTRAN version, but only six lines with a single-level one-line loop in MSL. This simplicity makes coding faster and easier, and greatly reduces the number of coding errors. Since programs and on-line execution from the terminals share the same language, code development and checkout is greatly simplified.

Most programs can be written entirely in MSL without the need for any AP coding by the user. Since AP operations are at the microsecond level and MSL interpretation is at the millisecond level, such a strategy can yield good execution efficiency, provided that vectors or arrays of kiloword size or larger dominate the problem. Vectorization of the particle-advance algorithm with its 78 floating-point operations is estimated to yield 60–80 msec/particle, which is within a factor of 5 of the time attainable with AP coding. Although this loss of speed, and the increased memory needed for storage of intermediate vector results, both dictate against vectorization for the final version of a particle code, it can be advantageous for an initial version or a smaller-scale problem.

The only remaining task in implementing the particle-advance algorithm is the MSL programming to integrate the disk transfers and execution of the AP code. For this, it is convenient to convert the previous per particle timings to the millisecond scale of MSL interpretation. The transfer time for a single record of 2k words is 8.33 msec, which is increased to a minimum average time of 10.0 msec per sequential record per disk on a sustained basis because of cylinder-boundary losses, or 5.0 msec per updated record with both pairs of disks. The final version of AP code requires 4.3 msec per record of 341 particle descriptors, including the allowance for memory conflicts. The use of partial records would decrease the single-record-transfer time and the AP execution time proportionally, but would not change the minimum average sustained-transfer time because this is governed by the 16.67-μsec rotation time of the disk. Therefore, if the overall simulation speed is to attain the maximum permitted by the disk transfer rate, the MSL program must have an interpretation time averaging no more than 5 msec per updated record. With this value the individual record transfers and the AP execution would gain on the MSL interpretation just fast enough to recover the loss at the preceding cylinder boundary, provided a sufficient number of AP-memory buffers exist.

The first version of the MSL particle-advance program used a seven-line central loop,

```
L   MOVE  READ1(H)→BUFFER(I)  INC I
    OP    UPDATE BUFFER(J), PARMS  INC J
    MOVE  BUFFER(K)→WRITE1(L)  INC K
    MOVE  READ2(H)→BUFFER(I)  INC I
    OP    UPDATE BUFFER(J), PARMS  INC J
    MOVE  BUFFER(K)→WRITE2(L)  INC K
    INC L  INC H  GO TO L
```

Here READ and WRITE are the input and output disk variables, BUFFER is the set of AP-memory buffers, OP UPDATE invokes the previously described AP-code to update the particles, and H, I, J, K, and L are index/subscript variables, which are incremented and tested by the INC operator. If incrementing (by unity in this case) of an index causes it to exceed its defined range, the index is reset to its initial value and program control passes to the next line. Thus the buffer indices repeatedly cycle through their range until the loop terminates after H has reached the end of its range. The use of three distinct memory indices and two distinct disk indices permits the necessary offsetting of the input, update, and output operations.

READ1 and WRITE1 use the even-numbered buffers, $I = 0, 2, \ldots$, and READ2 and WRITE2 use the odd-numbered buffers, $I = 1, 3, \ldots$, while UPDATE processes the odd and even-numbered sets alternately. A minimum of six buffers is needed if the four disk transfers and the AP execution are to be overlapped. Since the rotation of the four disks cannot be synchronized, an additional four buffers are needed in order to avoid interruptions in the disk transfers. Each such interruption adds a full rotation delay for the disk involved, and could increase the average transfer time to as much as $1\frac{1}{2}$ rotations, or 25 msec, per record per disk. Additional buffers beyond ten would give a somewhat smaller benefit, by mitigating losses at the cylinder boundaries. The optimal number of buffers involves a trade-off between execution time and memory space, and must be determined by extensive testing, including optimization of the input/update/output offsets.

Initial testing was performed with buffers of 2k words each. Little decrease in average time was found beyond ten buffers, and 9.3 msec/record was the minimum time obtained. Further investigation showed this to be governed by the 18.4-msec interpretation time for the cental loop. One obvious way of reducing the average time per record is to use two-record buffers of 4k words each, in order to process four records per pass through the central loop. Testing with ten such buffers gave an average time of 5.4 msec/record. Unfortunately, these large buffers occupy so much of the 60k-word memory that the eight force and charge arrays are thereby restricted to a maximum size of 2k gridpoints. Since most applications require 4k gridpoints per array, any increase in speed must be achieved with single-record buffers if it is to be generally useful, which thus dictates a reduction in the central-loop interpretation time.

Interpretation time depends primarily upon the number of operators and variables. Since the UPDATE operation is performed twice in the central loop, to update two adjacent buffers, these operations may be combined into a single UPDATE operation acting on a double buffer. This allows a reduction to the four-line central loop,

```
L   MOVE READ1(H)→BUFFER(I) READ2(H)→BUFFER(I+1) INC I
    OP WAIT BUFFER(J+1) OP UPDATE BUFFER(J), PARMS INC J
    MOVE BUFFER(K)→WRITE1(L) BUFFER(K+1)→WRITE2(L) INC K
    INC L  INC H  GO TO L
```

OP WAIT is a system operator which delays further interpretation until its operands are free of any outstanding operation, which is a disk input from READ2 in this case. It is necessary because UPDATE now operates on both BUFFER(J) and BUFFER(J + 1), but does not include the latter in its list of arguments. The operating system is designed to avoid misordering of the overlapping AP execution and disk transfers by delaying interpretation whenever an operand with outstanding operations is encountered. For this purpose, the operand is identified by its starting address in AP memory. Since BUFFER(J + 1) does not appear in the argument list of UPDATE, the latter operation could begin BUFFER(J) and reach BUFFER($J + 1$) before the input of the latter from disk is completed. The use of OP WAIT prevents this error. This second version of the central loop has eliminated one occurrence of the UPDATE operation, and one incrementing of each of the indices, I, J, and K, which are now incremented by two by the INC operator. The interpretation time for this version was found to be reduced to 15.9 msec, and the corresponding limit of 8.0 msec/record for average particle-advance time was attained for 12 or more buffers. This is still 60% greater than the minimum allowed by the disk transfer rate.

The interpretation time can be reduced further by eliminating the memory indices, I, J, and K, and introducing the set of aliases Bn for the subarrays, BUFFER(n). This requires enlarging the central loop by replicating the four lines of the second version, with one replica per pair of buffers, and with successive replicas corresponding to cyclically incremented values of I, J, and K. The contraction of the remaining variable and operator names to single characters gives a further small reduction in interpretation time. With eight buffers and optimal offsetting of the input, update, and output operations, the third version of the central loop has the form

```
L   OP U B2, B3, P  MOVE B0→W(L) B1→V(L)
                    R(H)→B6 S(H)→B7 INC L INC H
    OP U B4, B5, P  MOVE B2→W(L) B3→V(L)
                    R(H)→B0 S(H)→B1 INC L INC H
    OP U B6, B7, P  MOVE B4→W(L) B5→V(L)
                    R(H)→B2 S(H)→B3 INC L INC H
    OP U B0, B1, P  MOVE B6→W(L) B7→V(L)
                    R(H)→B4 S(H)→B5 INC L INC H GO TO L
```

All names have been reduced to initial characters, with S and V for READ2 and WRITE2. The addition of an operand to OP U replaces the use of OP WAIT. Every change in the offsets or the number of buffers now requires a rewriting of this central loop. Each line of this version was found to have an interpretation time of 10.4 msec, or 5.2 msec/record, which is only 4% greater than the disk-transfer limit. Average particle-advance times of 5.8 and 5.4 msec/record were obtained for 12 and 16 buffers, respectively, or 12% and 4% greater than the limit imposed by interpretation time. As many as 20 buffers might be needed to reach the interpretation-time limit, but available memory and the required 4k gridsize do not permit more than 12 buffers. Thus 5.8 msec/record, or 17.0 μsec/particle, represents the lowest particle-advance time obtained so far with 4k gridsize. With 78 floating-point additions and multiplications per particle, this corresponds to an execution speed of 4.6 megaflops (million floating-point operations per second).

The effect of overlapping the disk transfers, AP execution, and MP interpretation can be seen by estimating the particle-advance timing if these operations were not overlapped. The particle data would then be most efficiently handled using a single memory buffer of maximum size, which would be 12 records in this case. The appropriate MSL program would contain just half of the central loop from the first overlapped version, without memory indices, with an estimated interpretation time of 8 msec per pass. Disk input would require about 130 msec, including access time, and disk output time would be the same. These times add to give 268 msec for the buffer of 4k particles, or 65 μsec/particle. AP execution would now have no memory-conflict delays, and would require only 55 instruction cycles, or 9 μsec, per particle. The total nonoverlapped time of 74 μsec/particle is 4.3 times as great as with overlapped operations.

The particle-advance time of 17.0 μsec/particle dominates the simulation time. Diagnostics that require processing of the particle descriptors, such as the temperature and diffusion diagnostics of Eqs. (28) and (29), usually add the equivalent of a full timestep per diagnostic each time they are performed. Some time could be saved by processing only a portion of the particles, but the resultant increase in statistical fluctuations is seldom acceptable. These diagnostics have short AP operations and their timing is dominated by disk transfer and MP interpretation. They typically constitute 5–20% of the total simulation time, depending upon how frequently they are performed. The computation of the force and derivative arrays from the charge and dipole arrays requires 400 msec for a 4k gridsize in the fully periodic case. Thus the force calculation requires only 2% of the total time for a million-particle simulation, and this figure increases to only 20% for 10^5 particles. Because the diagnostics and force calculation constitute such a minor portion of the simulation time, little attempt to maximize their speed was made. Even a 25% improvement in their speed would save only 2–10% in total simulation time, the smaller percentage being the more typical one.

C. Miscellaneous Simulation Codes

For a system bounded in one direction, the force calculation becomes more complicated and requires 1.2 sec for a 4k gridsize. Although still insignificant for a million-particle simulation, this amounts to 40% of the total simulation time for 10^5 particles. Therefore, in a separate simulation code originally written for bounded systems of fewer than 10^5 particles, the force calculation is written entirely in AP code to achieve maximum execution efficiency. The particle-advance AP code is similar to the first of the three versions described earlier, wherein interpolation is used instead of extra force-derivative and dipole arrays, because gridsizes as large as 8k are required. The greater execution time of this version of AP code is less significant for this simulation code because the particle-advance calculation no longer constitutes the major part of the simulation time for systems of 8k gridpoints and fewer than 10^5 particles.

Fluid simulations exhibit some features not seen in the particle simulations. The advance of the grid quantities corresponds to the particle advance, but is more elaborate and requires a longer AP code. The three-dimensional version involved 358 floating-point additions and multiplications plus one division per gridpoint, and could not quite be fitted into the 476 words of AP program memory available for user programs. This necessitated the creation of an overlay structure to accommodate the small excess. An update time of 90 μsec/gridpoint was achieved, which corresponds to an execution speed of 4 megaflops.

A more significant feature of the large-scale fluid simulations is the topological complexity introduced by the difference operators that approximate the spatial derivatives in Eqs. (31)–(34). Whenever the grid quantities at one point are being updated by the AP, those at the adjacent gridpoints must also be in memory. For the $2\frac{1}{2}$-dimensional simulation, a disk record contains several adjacent rows of the two-dimensional grid. The MSL program merely has to ensure that the updating of one buffer not be started until after the succeeding buffer has been read from disk, and that the updating be completed before reusing the preceding buffer. The three-dimensional case is more complex because adjacent gridplanes must be accessed in addition to adjacent sections within the gridplane being updated. Disk transfers can be minimized by reading the grid quantities from disk one gridplane in advance of updating, and delaying the reuse of memory buffers for one gridplane beyond updating. This scheme limits the size of a gridplane because the memory must contain two gridplanes in addition to the buffer actually being updated at any instant. At least one additional buffer is needed for overlapped disk input, with three being desirable to prevent cylinder-boundary losses if the buffers contain a single disk record each. No additional buffers are needed for disk output, since updated buffers can be written to disk anytime between updating and reuse. If 56k words of the 60k memory are allocated for buffers, with 3 buffers for overlapped disk input, a gridplane is limited to 12 buffers, or 24k words, assuming single-record buffers. With eight grid quantities, ρ, **v**, **B**, and T, a gridplane is limited to 3k gridpoints, e.g., a 55×55 gridplane. A full record contains 256 gridpoints, only half of which are updated at any given timestep, and the update time is therefore 11.5 msec/record. Unlike the particle simulations, this requires only two disks, which are sufficient for a sustained transfer limit of 10.0 msec/record. The MSL program used a central loop similar to the second of the three particle versions, but with half as many MOVE operands, and an additional operand on the second line. The resulting interpretation time of 13 msec for the loop did not limit the average update time because multirecord buffers were used even though this restricted the gridplanes to slightly less than 3k gridpoints. A central loop analogous to the third particle version would be necessary for minimizing the time with single-record buffers, unless the analog of the second version for four disks were used.

For both particle and fluid simulations, the size of the simulated system is limited by disk capacity rather than memory size. Two half-disks are allocated for the operating system and user diagnostic output. This leaves two half-disks each for the particle or fluid data before and after updating, with two half-disks for a copy for checkpoint/restart purposes. Thus, a single copy of the data is limited to two half-disks, or 16,000k words, which is sufficient for $2.7 \cdot 10^6$ particles or $2.05 \cdot 10^6$ gridpoints. With the average update times of 17.0 μsec/particle and 90 μsec/gridpoint, these maximum systems require 13 hours and 26 hours for each thousand timesteps.

For the molecular-dynamics simulation, the direct use of two-body interactions causes the execution time to increase as the square of the number of particles. As a result, the requirements of reasonable execution time for a run was found to limit the system size to several thousand particles. Such small systems can be contained completely within the AP memory, with no need for disk transfer and its complications.

V. OBSERVATIONS AND SPECULATIONS

A. *System Performance*

The programming efforts described in the preceding section resulted in execution speeds of 4–5 megaflops on the CHI computer for a variety of particle and fluid simulations. This is three times the speed of corresponding simulations performed on the local IBM 360/91, and is comparable to the speed expected on a current general-purpose computer of the class of the CDC 7600. The cost of the CHI computer, however, is only a few percent of the cost of the latter. Much of this cost reduction is achieved through the substitution of disk memory for core memory. Such a design strategy is expected to be successful only when most of the data can be organized and processed in serial order, as is the case for the plasma simulations described here.

This design strategy has an additional benefit that one user cryptically summarized as follows: "The CHI does such large simulations because its memory is so small." A large general-purpose computer such as the NMFECC Cray at Livermore can accommodate simulations of up to 10^5 particles entirely within core memory, and this limit has been acceptable for many applications. The corresponding core-memory limit for the CHI is an unacceptable few thousand particles, and forces the coding of a disk-based simulation with overlapped disk transfers and computation. This version is limited by disk capacity to 3×10^6 particles, or 30 times the core-memory limit accepted by most users of the NMFECC Cray.

B. Programming Strategies

Computation efficiency has two distinct aspects: execution efficiency and programming efficiency. The former is essential for large-scale simulations, and has dominated the discussion so far. The latter is of predominant importance for the small-scale special-case calculations, whose cost arises mainly from the consumption of programmer man-hours rather than computer cycles. For this class of problems, the optimal programming strategy is to use the language that enables one to write and check out his code in the least time, and usually to ignore completely the question of execution efficiency, at least initially.

The high-level language available on the CHI, called the Math System Language (MSL), has proved itself to be well suited to both the small-scale users and the simulationists. It has a concise and natural syntax for combining mathematical operators with the familiar scalars, vectors, and multidimensional arrays and subarrays of applied mathematics. MSL is comparable to APL in level, but has a more natural syntax for scientific applications. By avoiding much of the subscript and loop bookkeeping found in scalar languages such as FORTRAN, it makes learning and programming both faster and easier. New users typically are able to use the machine as a sophisticated desk calculator within a week, and have completed a usable small-scale program for their own application within a month. This learning speed seems to be the same for users with no prior computer experience as for those with considerable experience in FORTRAN. Programming in MSL also yields shorter and simpler codes, sometimes by more than an order of magnitude relative to corresponding FORTRAN codes, as seen with the Poisson solver from the electrostatic-particle simulation. Since the number of coding errors seems to decrease at least linearly with code size, the availability of a high-level language such as MSL greatly simplifies code development and checkout for small-scale user and simulationist alike. We feel that the programming efficiency of MSL rivals the system's execution efficiency in contributing to overall computation efficiency.

Large-scale simulations have required two modifications of the small-scale special-case programming strategy in order to attain high execution efficiency. One modification is that the programmer must identify those sections of the computation which dominate the execution time, and code these critical sections in AP assembly language. In the more favorable cases, these critical sections may be quite short, and nearly all of the program can be written in MSL. This is true of the electrostatic-particle simulations and, to a lesser degree, of the fluid simulations. Fortunately, the users of the CHI have not yet had to resort to the more complex Macro Language and MP assembly language, although these have been used by system programmers to add utilities and minor MSL extensions to the system.

AP coding of the critical sections often can be avoided if these sections are predominantly of vector or array structure, or can be so reorganized. Since the fundamental MSL operations on entities of up to kiloword size are performed at the millisecond interpretation time scale, execution times as low as a microsecond per component are thereby attained. This is three orders of magnitude faster than a predominantly scalar code in MSL, and less than an order of magnitude slower than an optimized AP code. This vectorization approach has been found to yield acceptable efficiency on the small-scale problems as their scale or use expands. However, the gain in efficiency resulting from the increased parallelism and reduced memory access of an optimized AP code has proven advisable for large-scale simulations.

The other strategy modification for large-scale simulations is that the programmer can no longer ignore the actual physical structure of the computer system. Efficient integration of the AP and disk operations requires a detailed consideration of disk data structure and transfer rates, AP execution speed, including memory conflicts, and MSL interpretation time, followed by a careful synthesis of the simulation structure to fit this computer structure. The operating systems of large general-purpose computers are usually designed to maximize the utilization of system resources in an environment of many average users, few of whom tailor their programs to the system structure. This protection against the inefficiency of the many often prevents the few from gaining as much efficiency via such a tailoring process as can be gained on a dedicated system like the CHI. This tailoring strategy is therefore more important on the CHI than on most large general-purpose systems.

C. *User Behavior*

A widespread, and often deep, reluctance to optimize codes has been observed. This is expected and warranted among the small-scale users, where gains in execution efficiency are often not worth the programming effort.

Users in this group with little or no prior computing experience seem at least as amenable to general efficiency guidelines as are those with prior experience, provided these guidelines are easily followed. Surprisingly, many large-scale users are as reluctant as the small-scale users to optimize their codes, even though they have much more to gain for their efforts. Although all will write the critical sections of their simulations in AP assembly language, fewer than half of these users attempt to tailor their simulation structure to the computer or overlap the AP, MP, and disk operations. This tailoring and overlapping could be incorporated into an existing simulation code with less than a week's effort, for up to four times the execution speed, and could be expected to recover the user's investment of programming time within a few months at most. This same reluctance has been observed among users of large general-purpose computers, some of whom overcome their reluctance as the utilization of the computer approaches capacity. Since the potential gain in execution efficiency is greater on a dedicated computer like the CHI, we anticipate that the motivation to optimize will be correspondingly greater as the number of CHI simulationists increases.

This reluctance is particularly likely to manifest itself when it is decided to transform an existing simulation code on another computer into a version for the CHI. There seems to be a natural desire to perform a direct statement-by-statement conversion of the existing code, which is usually in FORTRAN, into an MSL code. This would result in a large, slow, scalar code, rather than the concise vector code to which MSL is naturally suited, and the simulation would be ill adapted to the structure of the CHI. The proper strategy is to reprogram the simulation, starting from the level of physics, and to tailor it to the computer structure. The simplicity of MSL makes this process easier and faster than direct code conversion.

D. *System Evolution*

The high degree of modularity of the CHI system design facilitates expansion of the system on several levels. At the lower level, the original configuration of Fig. 4a can be expanded toward that of Fig. 4b simply by enlarging the AP core memories and adding disk drives. The recent doubling of the floating-point data memory to 128 kilowords, and quadrupling of the program memory to 2048 instruction words is at this level, as is the recent addition of a 60-megaword disk drive that nearly doubles the system disk capacity.

The primary motivation for the data-memory expansion was to provide the greater disk-transfer buffering needed for particle simulations with extended grids of up to 10^6 gridpoints. This extension requires that the force and charge arrays be transferred from disk through memory to disk in segments in the same manner as the blocks of particle descriptors. Since only a few segments can be contained in memory at any one time, a topological complexity similar to that of the fluid simulations is thereby introduced. This requires that the streaming of particle descriptors through the system be closely coordinated with the streaming of array segments. This in turn dictates both a partial ordering of the particles initially and a resorting of particles at intervals throughout the simulation in order to maintain the coordination.

The one-dimensional version of the extended-grid electrostatic-particle simulation has been completed. It has been used for simulating systems of up to 0.5×10^6 gridpoints and 2×10^6 particles. This scale is sufficiently large to handle significant inhomogeneity effects, such as boundary effects and detailed density structures within the plasma.

The topological complications increase rapidly with the number of dimensions. In one dimension, it has not proved necessary to resort particles at

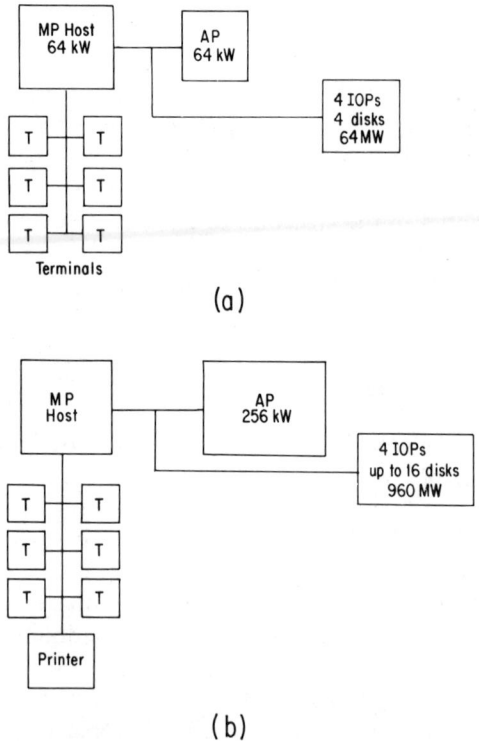

FIG. 4. Expansion of the CHI system on two levels: (a) original system; (b) first-level expansion with larger AP memory, more and larger disks, and a printer;

intervals of less than a thousand timesteps. In the two-dimensional version now being developed, such frequent resorting may be required that continual resorting as an integral part of the particle-update algorithm may be needed for maximum execution efficiency. This will almost certainly be necessary in the planned three-dimensional version.

At the higher level of system structure, the AP, its memories, the four IOPs, and the set of up to 16 disk drives can be considered as a basic module. This module can then be replicated a number of times to yield the expanded configuration of Fig. 4c. Additional user terminals, and peripherals such as tape drives and printer/plotters, could be attached to the MP, or a standard minicomputer could be inserted as an external interface to free the MP for more efficient control of the replicated AP/IOP/Disk modules. The advantage of modularity is that such an expansion could be accomplished in a relatively straightforward manner using the components already developed for the unexpanded system. This should result in a system capability approaching that of a Class VI computer, at perhaps an order-of-magnitude-lower cost.

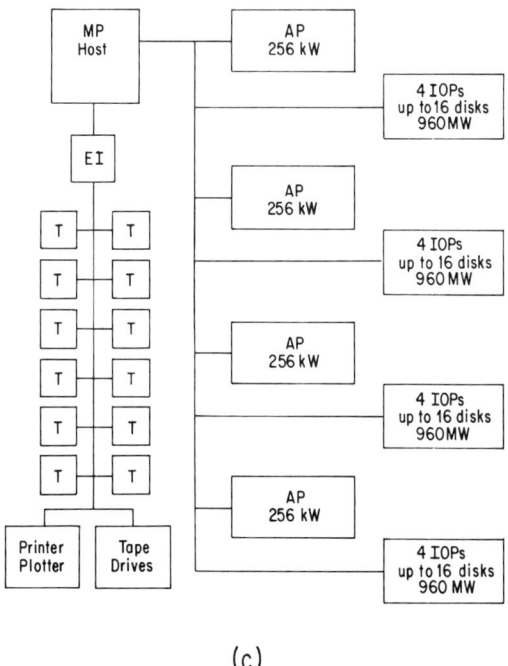

(c)

FIG. 4. (continued). (c) Second-level expansion by replication of the basic AP/IOP/Disk module, addition of terminals (T) and other peripherals, and insertion of an external interface (EI).

Acknowledgments

We are grateful for many useful discussions with Cheng-Chin Wu concerning the MHD fluid simulations, with V. K. Decyk regarding the bounded version of the particle code, and with Pieter Kokelaar concerning the overall administration and operation of the computer system.

The purchase of the system in its initial version was supported by the U.S. Department of Energy (then the U.S. Energy Research and Development Administration). The memory expansion was supported by the National Aeronautics and Space Administration, and the T300 disk drive was made available by TRW, Inc.

References

Dawson, John M., Huff, Robert, W., and Wu, Cheng-Chin (1978). *AFIPS Conf. Proc., 1978 Nat. Comput., Conf.* **47**, 395–407.

Kamimura, T., Dawson, J. M., Rosen, B., Culler, G. J., Levee, R. D., and Ball, G. (1975). Plasma simulation on the CHI microprocessor system. UCLA Plasma Physics Rep. PPG-248, University of California, Los Angeles.

Roberts, Keith V., and Potter, D. E. (1970). *In* "Plasma Physics" (Berni Adler, Sidney Fernbach, and Manuel Rotenberg, eds.), Methods in Computational Physics, Vol. 9, pp. 340–420. Academic Press, New York.

Index

A

Abarbanel, S., 252, 257
Adaptive grid
   Eulerian, 201
   flame propagation, 199
   Lagrangian, 201
   method, 162
Adaptive quadrature, 45
ADINC, 178
Aliases, 54
Anders, J. B., 251, 257
ARC3, 239
Array computer, 12
Asymptotic technique, stiff equation, 187
Asynchronous applications
   Monto Carlo, 44
   multiprocessor, 17
   quadrature, 44
Autocorrelation, 142

B

Backward Euler method, 102, 115
Backward substitution, 48
Bailey, H. E., 224
Baldwin, B. S., 224
Batcher algorithm, complexity, 21
Baudet, G. M., 14
Beam, R., 227
Beckwith, I. E., 251, 264

Berger, M., 127
Bergland, G. D., 71
Bidiagonal system, 18
Bin summing, 20
Biquadratic Lagrange, 144
Bisection method, cycle analysis, 137
BLAS, 8
Book, D., 178, 196
Boundary
   inflow, 259
   outflow, 259
   subsonic inflow, 258
Brigham, E. D., 70
Brown, D., 33
Buneman, O., 190
Burroughs
   BSP, 4
   ILLIAC IV, 4, 12, 16
   PEPE, 4
   "Wind Tunnel Proposal," 4

C

Cache loop, 134
CAL, 8
Calahan, D. A., 29, 30, 38
Castor, J. I., 162
Causality problem, 189
CDC (Control Data Corporation)
   7600, 8
   Cyber 203, 4

Index

CDC (cont.)
  Cyber 205, 4
  STAR-100, 4, 16
  "Wind Tunnel Proposal," 4
Chain slot time, 6
Chaining, 6, 133
CHEMEQ, 178, 192
Chime, 8, 27, 41, 132
Cholesky decomposition, 94
Chorin, A. J., 157, 158
Clauser approximation, 224
CM*, 4
Cochran, W. T., 55, 61, 64, 80
Colella, P., 156–158, 160, 164
Complexity, parallel, 15
Compress operation, 134, 167
Compressible flow
  Cartesian coordinates, 103
  Lagrangian coordinates, 103
Concus, P., 116
Conjugate gradient method
  computing time, 126
  generalized, 115, 117
  incomplete Cholesky, 39, 90, 120
  line-Jacobi, 117
  point-Jacobi, 117
  symmetric Gauss–Siedel, 118
  two-cyclic symmetric Gauss–Siedel, 119
Conservation equations, 176
Constant-time operations, 15
Cooley, J. W., 51
Cooley–Tukey algorithm, 57, 71, 82
  inverse, 68
Cornell AP
  memory, 281
  project, 280
Courant, R., 259, 263
Crank–Nicholson method, 102
Cray-1
  chain slot time, 7
  chime, 8
  constant-time operations, 15
  functional unit time, 7
  scalar registers, 8
  Super Vector (SV), 8
Crowley, W. P., 205
Cyclic reduction, 30, 33, 89, 190, 191, 207,
  incomplete, 96
  multiprocessors, 98
  permutation, 92

D

Data base
  pencil, 237, 240, 242
  planar, 237
DeBar, R., 158
Decompress operation, 167
Deiwert, G. S., 243, 247
Delta form, 227
  factored, 235
Denelcor HEP, 4
DFLUX, 178
Difference scheme, time-split, 252
Diffusion
  central difference approximation, 107
  curvilinear coordinates, 105
  equation, 101
  velocity, 177, 203
Dissipation, numerical, 228
Dollimore, J., 72
Dollimore's algorithm, 74
  inverse, 75
Dongarra, J., 31
Dubois, P. F., 122
Dwyer, H. A., 162

E

Edson's algorithm, 71
Eggleton, P., 162
Eisenstadt, S., 43
Elliptic partial differential equation, 67
Emery, M. H., 206
Equation of state, 17, 25, 131
  algorithm timings, 148, 150
  compress–sort method, 146
  computer timings, 26
  enthalpy, 204
  TIL-based method, 145
ETBFCT, 196
Euler method, 225
Expand operation, 134

F

Fan-in algorithm, 18
Farouki, R., 282
Fast Fourier transform, 51

Index

FCT
  algorithm, 196
  vectorization, 198
FFT, *see* Fast Fourier transform
FFTPACK, 78
Fiebig, M., 252
Fields, statistical behavior, 280
Finite Element Machine, 4
Flame propagation, 194, 199, 200
Floating Point Systems AP-120, 4
Flow
  compressible, 170
  convective, 196
  high Reynolds number, 222
  ILLIAC IV, 217
  implicit, 217
  reactive, 173–175, 184, 196
  supersonic, laminar, 244
  three dimensional, 271
  transonic, turbulent, 247
  wing–elevon junction, 252
Fluid dynamics
  stiff phenomenon, 193
  time scales, 133
Flux
  antidiffusive, 197
  diffusive, 197
  limiting, 197
  molecular diffusion, 202
Flux-corrected transport, 178, 192, 196
Flux Jacobian, 234
  inviscid, 226
  matrix, 225
  viscid, 226
Fong, K., 131, 38
Fornberg, B., 67
Forward substitution, 47
FPS AP
  hardware, 280
  software, 281
Friedrichs, K. O., 259, 263
Fritts, M. J., 178, 205, 206
Fry, M. A., 199
Functional unit time, 6

G

Gather operation, 134
Gaussian Elimination, 30, 33
  parallel complexity, 21
  partial pivoting, 20, 21
Gear, C. W., 115, 188
Gear method, 188
Gedanken flame problem, 178, 179, 195, 200
Gelinas, R. J., 155, 162
Gentleman–Sande algorithm, 59, 62, 68, 82
  inverse, 71
Geophysical flow
  condensation, 274
  convection, 274
  horizontal processes, 272
  parallelism, 277
  recursive processes, 273, 274
Geophysical vertical processes, 272
Glassman, J. A., 64
Glimm, J., 157
Godunov, S. K., 156
Godunov method, 164, 165, 169
Goodyear MPP, 4
Gordon, S., 204
Gottlieb, D., 252, 257
Greenbaum, A., 103, 40
Grid ordering
  natural, 110
  odd–even line, 112
Gropp, W. D., 162
Gustavson, I., 43

H

Heller, D., 96
Hestenes, M. R., 39
Hill, J. G., 160
Hockney, R. W., 67, 190
Holley, B. B., 264
Hybrid algorithms, 21
Hydrodynamics
  artificial viscous pressure, 155
  basic equations, 144
  heat conduction, 155
  two-dimensional, 43
  viscosity, 155

I

ICCG bandwidth minimization, 95
ICL DAP, 4

ILLIAC IV
  architecture, 230
  data flow, 232
  input/output, 233
  latency time, 241
  memory hierarchies, 232
  vector length, 233
Implicit stiff equation, 187
Interpolation, bilinear, 24
Iterative methods, 39

J

Jacobi iteration, 14
Jacobian, 185
JANAF table, 204
Jenkins, G. M., 55
Jones, W. W., 178, 193
Jordan, T. L., 3, 131

K

Karush, J., 117, 27
Kernighan, B. W., 209
Kim, J., 237
Knight, J. C., 260
Knuth, D. E., 145
Kogut, J. 281
Kohn, J., 145, 149
Korn, D. G., 67, 82
Kuck, D., 38
Kunz, P., 280
Kutler, P., 224

L

Lagrangian
  mesh, 205
  triangular grid, 206
Lambiotte, J., 67, 82
Lapidus, A., 219
Large-time operations, 20
LASNEX, 85, 90
Lawson, C., 8
Lax, P. D., 155

Leap-frog
  time integration scheme, 272
Ledgard, H. F., 209
Linear recursion, 18
Linear system
  banded, 38
  bidiagonal, 34
  5-diagonal, 40
  full, 131
  general, 38
  tridiagonal, 21, 85, 86, 112
Linear systems
  bidiagonal, 30
  tridiagonal, 30
LINPACK, 31
Log_2-time operations
  bidiagonal system, 17
  BLAS, 17
  divide and conquer, 18
  linear recursion, 17
  polynomial evaluations, 17
  tridiagonal system, 17
LRLTRAN, 126
LU factorization, 48
LU decomposition
  scalar algorithm, 86
  vector algorithm, 87

M

M chime, 138
M-section, 136, 138
  cycle analysis, 140
  Fortran program, 139
MacCormack, R. W., 252
Mach number, 264
Madsen, N. K., 30
Majda, A., 162
Marsh, M. C., 160
Marshal plan, 241, 242
Martin, J. T., 126
Matrix
  full zone, 111, 117
  line form, 112
  negative definite, 112
  positive definite, 112
  rational polynomial, 115

symmetric, 112
two-cyclic, 113
Matrix multiplication
　block tridiagonal, 27
　complexity, 27
　general, 28
　sparse-banded, 27, 47
　SV performance, 27, 29
McBride, B. J., 204
McDonald, B. E., 190
McMahon, F. H., 132, 160
Memory, scratch pad, 6
Merriam, M. L., 241
Method of lines, 114, 121
MFLOPS, 12
MIMD, 3
　data flow, 4
MIPS, 10
Mitra, N. K., 252
Moin, P., 237
Monte Carlo, 17
　asynchronous computing, 46
　neutron transport, 46
Moving finite element method, 155
Multiprocessor, 13
　speedup, 14
MUSCL, 159

N

Navier–Stokes equation
　compressible, 251
　geophysical fluid model, 272
　high Reynolds number, 252
　implicit, 240
　implicit solutions, 234
　linearized, 257
　Reynolds averaged, 234
　space differencing, 224
　thin layer, 223
　three dimensional, 219
　time-split differencing, 252
　transformed, 219
　unsteady, Reynolds-averaged, 219, 239
Noh, W. F., 158, 169
Nozzle
　Laval, 251
　nonaxisymmetric, 252
　slotted, 251, 257
　suction slots, 251
　two-dimensional slotted, 263

O

Oliger, J., 258
Operations
　constant time, 15
　large time, 15
　$\log_2$ time, 15
Oran, E. S., 176, 178, 197
Orbits, D. A., 39
Osher, S., 162

P

Paracomputer, 15
Parallel algorithm
　accuracy, 22
　consistency, 22
　redundancy, 22
Particle-in-cell (PIC), 17
Paul, G., 67
Pease, M. C., 60
Pease algorithm, 60, 71, 81, 82
Perfect shuffle, 12
　connection, 13
Peterson, W. P., 78
Peyret, R., 219
PFORT, 80
Piecewise-parabolic method, 167, 168, 169
Pipeline computer, 10
　efficiency, 10
Plauger, P. J., 209
Plemmons, R. J., 97
Poisson equation
　geophysical flow, 274
　parallel solution, 275
Polynomial evaluation, 19
　speedup, 20
Pomraning, G. C., 101
Prandtl
　mixing length, 224
　number, 220, 254
Prophet, H., 204

Q

Quick sort, complexity, 21

R

Radix sort, 145
Random choice method, 158
r-cyclic reduction, 21, 34
Reactive flow
  Eulerian, 196
  Lagrangian, 196
Recursive doubling algorithm, 131
Relaxation, chaotic, 14
Renormalization group, 281
Reynolds number, 220
Richter, R., 282
Rodrigue, G. H., 30, 40, 131
Rusanov, V. V., 155

S

S-1, 4
Sameh, A. H., 38
SCATTER
  complexity, 20
  Monte Carlo, 20
  PIC, 20
  problem, 20
Scatter operation, 134
Sedgewick, R., 21
Shampine, L., 115
Shapiro, S. L., 282
Shock
  smearing methods, 156
  thickness, 155
  tube problem, 156
Shrewsbury, G. D., 247
SIMD, 4
  array, 3
  vector, 3
Singleton, R. C., 68
SL/1, 260
SLIC, 162, 169
  algorithm, 158
Slot suction, 258
Slow-flow algorithm, 178, 194

Sod, G. A., 157
SPLISH, 178
Startup time, 182
Steger, J. L., 226
Stiefel, E., 39
Stiff equation
  asymptotic method, 191
  chemical kinetics, 191
Stockham's algorithm, 64, 71, 78, 81, 82
Stokes's hypothesis, 220
Stone, H. S., 21, 73
Strang, G., 252
Streamfunction, 264
Stull, D. R., 204
Sundstrom, A., 258
Sutcliffe, W. G., 158
Sutherland's law, 254
Synchronous computer, 10

T

Table lookup, 17, 21, 129
  bank conflict, 23
  cache-loop strategy, 143
  complexity, 24
  compress-sort strategy, 143
  enthalpy, 204
  one-dimensional, 130
  speedup, 121
  TIL strategy, 143
  vector arguments, 130
  vectorization, 205
Tarter, C. B., 101
Tempterton, C., 62, 67, 78
Text, 157
Theta method, 234
Thin-layer approximation, 223, 240
Thomas, P. D., 252
Thompson, J. F., 265
TI ASC
  estimating parallelism, 275
  FORTRAN compiler, 270
  four pipe, 270
  processing times, 181
  speedup, 277
  startup time, 276
  two pipe, 180
  vector operations, 185

Timestep splitting, 188
Tobochnik, J., 282
Transform
  aliased discrete complex, 53
  computing time, 80
  discrete complex, 54
  discrete complex periodic Fourier, 54
  even number of sequences, 70
  fast Fourier, 51
  Fourier even quarter-wave, 77
  Fourier odd quarter-wave, 75
  Fourier of even sequence, 74
  inverse, 62
  inverse discrete complex periodic Fourier, 54
  inverse Fourier even quarter-wave, 77
  inverse Fourier odd quarter-wave, 76
  inverse Fourier of even sequence, 74
  inverse odd Fourier, 72
  inverse real periodic, 69
  multipole, 67
  n-point, 53, 55, 57
  $n/2$-point, 53, 55
  $n/3$-point, 57
  odd Fourier, 72
  quarter-wave cosine, 77
  quarter-wave sine, 75
  real periodic, 68, 69
  sequence, even number of points, 70
  sine, 72
  software, 78
  symmetric, 71
  symmetric sequence, 68
  vector algorithms, 70
Transmit indexed list operation, 134
Transposed Pease algorithm, 64
Transposed Stockham algorithm, 61
Tree, adaptive quadrature, 45
Tscharnuter, W. M., 162

Turbulence
  eddy-viscosity model, 240
  modeling, 224

U

University of Texas TRAC, 4
Upwind differencing, 154

V

Van Driest damping, 224
van Leer, B., 158
Varga, R., 103
Vector bit, 160
Vector scan, 136
Velocity, contravariant, 221
Vinokur, M., 219
Viscous approximations, 222
Viviand, H., 219

W

Wang, H. H., 68
Warming, R. F., 227
Warshaw, S., 149
Watts, D. G., 55
Weinberg, A. M., 101
Wigner, E. P., 101
Williams, F. A., 176
Winkler, K. H., 162
Wood, D. M., 282

Y

Young, D. M., 117
Young, T. R., 178, 191, 205